FUNGI OF EUROPE:

INVESTIGATION, RECORDING AND CONSERVATION

FUNGI OF EUROPE:

INVESTIGATION, RECORDING AND CONSERVATION

General Editor

D.N. Pegler

Editors

L. Boddy B. Ing P.M. Kirk

Production Editor

S. Dickerson

Published by

The Royal Botanic Gardens, Kew

on behalf of the co-sponsors

The conference at which these papers were originally presented was co-sponsored by the Royal Botanic Gardens, Kew, the British Mycological Society and the CAB International Mycological Institute.

Typeset at the Royal Botanic Gardens, Kew, by
Pam Arnold, Christine Beard, Dominica Costello,
Margaret Newman, Pam Rosen and Helen Ward

Cover design by Media Resources, Royal Botanic Gardens, Kew

ISBN 0 947643 54 0

Printed and bound in Great Britain by
Whitstable Litho Ltd., Whitstable, Kent.

PREFACE

I am most pleased that the Royal Botanic Gardens, Kew was able to host the Congress of European Mycologists when it met for the first time in England. Kew is an institution that has not forgotten the importance and the significance of the mycological kingdom and so it was a great encouragement to our mycologists to receive their European colleagues into our institution. This Congress was particularly significant because it brought together into a single forum the mycologists of both eastern and western Europe after a lengthy period of political separation. The 1992 Congress will be remembered as the one that rebuilt links, much to the benefit of mycology throughout Europe.

This volume is significant in that it contains a great many approaches to the study of fungi in addition to straightforward taxonomy. In particular it has sections on fungi in European ecosystems and on the conservation of European fungi. It is obvious from the selection of papers presented here that the fungi play an important role in a number of very different ecosystems from coastal sand dunes to the alpine zones of high mountains. As mycological research progresses we are beginning to realise the extraordinary size of the kingdom mycota. When areas are studied in any detail they are found to have many more species of fungi than of plants. The recent studies and reports on biodiversity often forget the importance of the fungi. I am, therefore, glad to see a section of seven papers on aspects of conservation of fungi. Fungi species are already in decline in certain habitats for a large number of reasons varying from acid rain to the overharvesting of the most popular edible species. The fungi play such a vital role in the dynamics of many ecosystems and so it is essential to see that they are conserved just as strictly as plants and animals. Two papers focus on mycorrhizae, a good example of the importance of fungi for the integrity of ecosystems. Quite apart from their ecological role those of us who enjoy supplementing their diet with the many delicious species of fungi that grow in Europe must surely be the most interested in their preservation.

Finally I would like to thank D N Pegler and his co-editors for their good work in editing this important new contribution to mycological knowledge.

Ghillean T Prance
Director
Royal Botanic Gardens, Kew

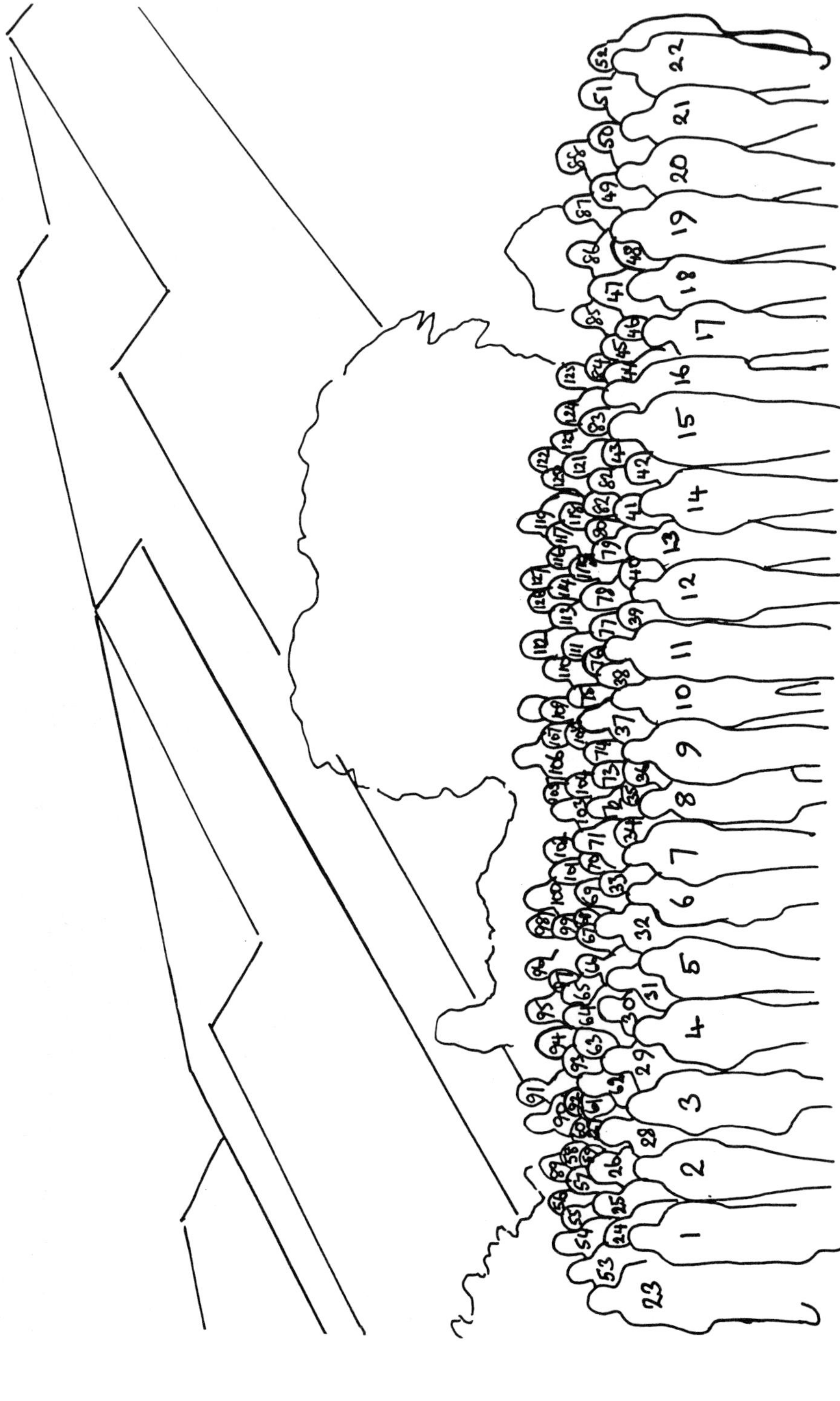

PARTICIPANTS AT XI CONGRESS OF EUROPEAN MYCOLOGISTS, KEW 1992

1, M. E. Harkonen; 2, T. M. Saarimäki; 3, V. Salo; 4, R. Cook; 5, S. Gay; 6, M. Lawrynowicz; 7, F.-E. Eckblad; 8, A. Skirgiello; 9, F. D. Calonge (Vice President); 10, E. Parmasto (President); 11, L. Ryvarden; 12, S. Dickerson; 15, E. B. Gareth Jones (BMS President); 16, S. Wasser; 17, I. Melo; 18, A. C. Leonard; 19, D. N. Pegler (Secretary General); 20, J. Marcinkowska; 21, A. M. Bujakiewicz; 22, D. Benkert; 23, J. V. R. Marriott; 24, J. C. Frankland; 26, M. A. J. Thomas; 27, B. Ing; 28, G. Gulden; 30 E. Vellinga; 31, G. Pacioni; 32, M. Nauta; 33, J. Baar; 34, M. Barluzzi; 35, C. Barluzzi Martini; 37, C. Lado; 39, M. Martin; 41, M. T. Telleria; 43, E. E. Emmett; 44, A. Piltaver; 45, G. Dickson; 47, J. Webster; 48, M. Gregory; 49, M. M. Noordeloos; 50, T. Laessøe; 51, P. J. Fisher; 52, S. Helfer; 53, M. Matavulj; 54, J. A Schmitt; 55, G. Arnold; 56, E. I. Kortekangas; 57, P. H. Molitoris; 58, B. Senn-Irlet; 59, K. Van der Gucht; 61, C. Grgurinovic; 62, G. Kinsey; 63, M. Rotheroe; 64, S. Svensson; 66, N. Hallenberg; 67, J. Stalpers; 69, A.-E. Torkelsen; 70, S. Petkovsek; 71, J. Ginns; 72, T. Kuyper; 73, M. Bondarceva; 74, S. Ujang; 75, A. Henrici; 76, H. Wollweber; 78, E. Wollweber; 79, R. Courtecuisse; 80, B. B. Buyck; 81, R. Watling; 82, J. van Brummelen; 83, C. Perini; 84, B. Gahne; 86, F. Buscot; 87, K.-H. Rexer; 88, C. Lange; 89, A. J. S. Whalley; 90, B. M. Spooner; 91, Y.-J. Yao; 92, O. Foitzik; 94, E. Quanten; 95, G. Fourre; 96, P. Fourre; 97, K. Hammon; 98, A. Runge; 99, A. Leuchtmann; 100, F. Runge; 101, B. Declercq; 102, K. Høiland; 103 E. Bendiksen; 104, M. Karadelev; 105, E. Ohenoja; 106, A. Dahlberg; 107, J. Stenlid; 108, K. Veghelyi; 109, H. Knudsen; 110, F. Bellu; 111, B. Ivancevic; 112 B. C. Sutton; 113, M. Candusso; 114, P. Renvall; 115, V. Demoulin; 116, V. Muhkin; 117 T. Niemela; 118, P. J. Keizer; 119, A. Fraiture; 120, A. Kovalenko; 121, E. Arnolds; 122, S. Vanev; 123, R. Fellner; 124, J. Lennie; 125, M. Svrcek; 126, S. Watkinson.

CONTENTS

1. INTRODUCTION

This volume records the proceedings of the XI Congress of European Mycologists, which was co-sponsored by the Royal Botanic Gardens, Kew, the British Mycological Society and the CAB International Mycological Institute, and held at the Royal Botanic Gardens, Kew on 7–11 September 1992.

The Congress of European Mycologists is recognized by the International Mycological Association (International Union of Biological Sciences, Section for General Mycology). The Standing Committee of the Congress serves as the Committee for Fungi in Europe within the IMA. The objectives of the Congress are (i) to promote all forms of mycological investigation within Europe, (ii) to enhance contact between mycologists within Europe, (iii) to exchange results, information and opinions by means of paper-readings, poster sessions, field meetings, workshops and committees, and (iv) to prepare a European input into other international bodies and initiatives. Congresses meet approximately every three years, each within a different European country, and alternating with the International Mycological Congresses. This is the first occasion that the Congress has been hosted by England.

The XI Congress departed from previous programmes by introducing a specific theme for the conference, and structuring the week to provide a symposial format. The chosen theme was 'Fungi of Europe: Investigation, Recording and Conservation', bringing together the scientific expertise from throughout Europe. In this way an attempt was made to bring about a unified approach to the many problems of threatened species and the urgent need for their conservation during this time of environmental threat. Some two hundred scientists participated, representing twenty-eight countries.

Three major symposia were included in the programme, namely 'Fungal Recording and Mapping', 'Fungi in European Ecosystems' and 'Conservation of European Fungi'. The results of the sessions are published in this volume. In addition, some seventy posters were presented. An important part of the Congress was the open, general meeting of the European Council for the Conservation of Fungi, which concentrated on the need for a European Red Data List.

The symposium on 'Fungal Recording and Mapping' (papers 2–6) examines the problems involved in the recording of the mycota of Europe, a task still far from complete. Contributions from Belgium, France, Germany, The Netherlands and the region of Fennoscandia are included. Régis Courtecuisse, from France, and André Fraiture, from Belgium, consider the practicalities of preparing computer-based programmes, involving many individuals spread over wide geographical areas. The need for internationally acceptable systems is emphasized, as well as the requirements for the initial compilation of complete and accurate checklists. It is only when such recording and mapping is well advanced that the observations considering conservation requirements can be stated. Maryke Naute & Elsa Vellinga present a set of elegant results based on a national project from The Netherlands, which included half a million records. They emphasize the need for the introduction of corrective factors in their analyses, in order to cope with problems arising from the diverse origin of records and irregular recording. A series of examples show how such results can be applied to the needs of both nature management and conservation. Similarly, Johannes Schmitt shows how regular observation by one individual over many years can equally contribute to the overall picture of decline or otherwise within the mycota of Europe. Leif Ryvarden presents the unusual distribution patterns found within the very specialised mycota of Fennoscandia, results which might usefully be compared with those by Victor Mukhin, who looks at the wood-decaying fungi of the West Siberian Plain.

The main symposium of the meeting deals with 'Fungi in European Ecosystems' (papers 7–16), which emphasizes the extraordinary range of habitats encompassed within the continent of Europe. Mediterranean habitats were covered by Claudia Perini and her co-workers, who have examined the woodlands of Tuscany over many years and show how restrictive in their distribution many of the macrofungi can be. Carlos Lado looks at the Myxomycetes of Mediterranean woodlands and finds that seventy per cent of the world's myxomycota occur in the area. By contrast, Anna Bujakiewicz investigates the alpine mycota and the associated floristics of the Babia Gora Massif, north-west of the Tatra Mountains of central Europe. Maurice Rotheroe looks at the fungi of sand dunes, revealing the range of systems which exist even within this specialized habitat. Dieter Benkert finds a whole range of genera within the Pezizales which are parasitic on mosses, even though parasitism was thought to be rare within the order. Discomycetes are again covered by Mirko Svrcek, whose fifty year-old study in former Czechoslovakia emphasises the conservation needs debated in the final symposium. It is suggested that the discomycete decline in the area could be as much as 50–80 per cent during the period of study. Jan Stenlid investigates the processes influencing the succession of wood-decay fungi within boreal forests. Margarita Bondarceva places great emphasis on the adaptation by macromycetes to their environment and proposes an ecological classification which can provide a useful tool for comparative purposes between contrasting geographical areas. Finally, Solomon Wasser gives the reader a valuable detailed account of radionuclide uptake by fleshy fungi, following the Chernobyl Atomic Power Station accident in 1986. With such macromycetes exhibiting a measurable uptake of over 500 times that shown by the surrounding soil, these fungi provide excellent indicators of radioactive contamination. The final symposium, 'Conservation of European Fungi' (papers 17–23) considers the needs resulting from the observations presented within the earlier symposia. Eef Arnolds, hitherto the chairman of the European Council for the Conservation of Fungi (ECCF), provides an expert introduction to fungal conservation within Europe and presents an overview based on an analysis of Red Lists so far published from eleven European countries. The survey emphasizes the current limitations of our knowledge, with scarcely any data available from throughout southern Europe. In order to get the European conservation needs onto a sounder basis, Bruce Ing, the newly elected chairman of ECCF, proposes a first attempt at a list of those macrofungi which appear to be presently endangered throughout Europe, rather than from within individual countries. This is followed by a series of papers on topics which outline potential or real threats to the European mycota. Rostislav Fellner emphasises a problem of particular importance to eastern European countries, but by no means restricted to that area, namely the effect of air pollution on ectotrophic mycorrhizas in forests. He shows how ectomycorrhizal fungi can be good indicators of the disturbance of forest stability. Peter-Jan Keizer considers how practices such as mowing and grazing of road verges can have a marked affect on macromycetes production, by the lowering of nutrient content within the soil. Similarly, Jacqueline Baar & Thomas Kuyper look at the effect of the removal of litter upon the survival of mycorrhizal fungi. Giovanni Pacioni outlines the special problem of mushroom and truffle picking and the attempts towards effective legislative measures within Italy. Finally, Stephan Helfer brings to our attention a peculiar dilemma which can face the fungal conservationist. How to save those parasitic fungi which themselves place their host plants under threat !

Clearly, headway is being made in describing the distribution of, and ascribing roles to, fungi in different ecosystems throughout Europe. However, most of this information derives from observations of fruitbodies, but these are, of course, only the 'tip of the ice-berg' — it is the mycelium which obtains nutrients and achieves vegetative spread. Thus, ideally, we need to know about the spatial and temporal distribution of mycelia (which has been alluded to by Jan Stenlid). While, obviously, fruitbodies do reflect mycelial distribution to some extent, there are caveats. Firstly, the absence of fruitbodies does not necessarily imply the absence of a fungus. Secondly, the presence of a fruitbody on a particular substratum is not always indicative of the presence of the mycelium of that fungus within the substratum, either currently or in the past. Thirdly, since mycelial and reproductive development are often affected differently by exogenous and endogenous factors, fruitbody production is often sporadic or seasonal, the latter being well recognized. The foregoing must be borne in mind when interpreting data on sporocarp distribution.

As with previous Congresses, much emphasis was placed on the field study of fungi and, linked to the paper-reading sessions, two pre-Congress residential field meetings, two post-Congress residential field meetings and two half-day field excursions were arranged. This gave the participants ample opportunity to investigate the extensive, and at times unique, mycobiota of England.

The pre-Congress meetings were both held in northern England, and based on the shores of Lake Windermere, at Ambleside, Cumbria. These meetings covered both the lowland and upland regions of the Lake District, offering the opportunity to explore some of the finest Atlantic oakwoods of Europe, especially around Derwentwater, in addition to the high level grassland and moorland.

After the Congress, two excursions were arranged in southern England. One meeting was held in the New Forest, at Lyndhurst, Hampshire. This represents one of the largest tracts of broad-leaved forest left in England, containing stands of beech and oak of great age and botanical richness. Its southern, warmer position shows affinities with the Mediterranean flora, and is particularly rich in Basidiomycetes. The other meeting was based at Stroud in Gloucestershire, and this particularly concentrated on an investigation of the hypogeous fungi ('truffles'), being centred in the historically important areas studied by C. E. Broome and the late Professor Lilian Hawker. The localities covered the Cotswolds Hills, the famous Leigh Woods above the Avon Gorge, near Bristol, and sites on chalk in Wiltshire.

During the week of the Congress, excursions were arranged. The first excursion visited the ancient woodlands of the Ruislip Woods complex, the largest wooded area of Greater London. The following day the particularly rich mycota of Windsor Great Park was examined, offering a locality which boasts the finest collection of mature trees in Europe.

The Organizing Committee is most grateful for the Congress to be supported by the Commission of the European Communities — DGXII. Gratitude is also expressed to the sponsors and also to the Royal Society, the Society of General Microbiology, the New Phytologist Trust, the British Council, and Comp Plus Ltd for additional financial support. The Committee also thanks the Director of the Royal Botanic Gardens, Kew, Professor G. T. Prance, for putting the facilities for the conference at our disposal, and for all those individuals who contributed to its success. Special thanks must be given for the tireless efforts of both Suzy Dickerson, Secretary of the Organizing Committee, and to Tim Nodder of the Friends of Kew, for his overall management of the Congress. Thanks to Dr. Gordon Dickson, Ted Green, Prof. David Hawksworth, Dr. Bruce Ing for all their many efforts in the organization of the field meetings.

Finally I should personally wish to thank my co-editors, Lynne Boddy, Bruce Ing and Paul Kirk for all their effort and support during the production of this volume. A special thanks is also given to Suzy Dickerson who, as Production Editor, has guided each production process in such an efficient, speedy and apparent effortless manner, to Ann McNeil and Dominic Hilton for the checking of textual details, and to Gill Butterfill for the production of the index.

D. N. Pegler & L. Boddy
March 1993

Courtecuisse, R. (1993). Recording and mapping the French mycota (The French recording and mapping programme for mycota in the European context. Principles and first results). In D.N. Pegler, L. Boddy, B. Ing & P.M. Kirk (Editors). Fungi of Europe: Investigation, Recording and Conservation, pp. 5–12 Royal Botanic Gardens, Kew.

2. RECORDING AND MAPPING THE FRENCH MYCOTA (THE FRENCH RECORDING AND MAPPING PROGRAMME FOR MYCOTA IN THE EUROPEAN CONTEXT. PRINCIPLES AND FIRST RESULTS)

R. COURTECUISSE

Department of Botany, Faculty of Biological and Pharmaceutical Sciences, B.P. 83, F. 59006, Lille Cedex, France

Abstract

A mapping programme is proposed to record the mycota of France, based upon a network established at the Departmental and Regional levels, with central co-ordination and controls. Proformas and computerized mapping for recording data are discussed, together with the results obtained after two years.

Introduction

The overall aims of this programme are to increase the knowledge of the French mycota on the taxonomical, ecological, chorological, bio-indicative points of view, and to take part in current European programmes with similar aims. A conservative finality is also very important.

The need to realize such a programme has become necessary because France has no formal check-list for its mycoflora, and because maps, such as those published by Lange (1974), showed a mycological desert in France, even for species as common as *Lactarius necator* (Gmel.: Fr.) Pers. and many others. Lange's work was the result of the first collective attempt to map European macrofungi.

Two years ago, the opportunity was taken to initiate a national inventory and mapping programme. I made contact with many French colleagues and friends, and I received a very good response. I also received sponsorship from the Société Mycologique de France.

Some papers that I published (Courtecuisse, 1991a–f) to announce the programme were echoed by several French colleagues (Fourré, 1991; Mabon, 1991; Malaval, 1991, Mornand, 1991, 1992; Pouzaud, 1992) showing the interest to this programme in various areas.

Development of a geographical structure

A geographical structure has been established to deal with the processing of the data.

In France there are 96 Departments or administrative subdivisions. A co-ordinator, in each department, collects the data from mycologists and societies, and compiles the available data, and sends the results to the regional level. We have 23 regions, each gathering several departments, each provided with a co-ordinator for this programme. The processing is carried out at this level, and the revised documents are sent to the national co-ordinator. Fig. 1 shows this geographical structure and the hierarchy from departments (thin lines) to regions (thick lines). Many regions are perfectly organized and effective at present. A few others, still lacking one departmental co-ordinator or meeting other problems will be effectively organized, hopefully by the end of 1992.

Development of a computer structure

Contributions are sent in paper form or as computer disks, employing computer software developed by a computer specialist at Lille University.

First inventory contributions

Fig. 1 shows the current state of already received check-lists (1: exhaustive lists, 2: important lists, 3: fragmentary lists). Some departments have not sent data yet, but documents exist and are under current processing. Nowhere are data completely lacking, and I feel that it is quite a satisfactory statement as this formal programme was initiated only two years ago. On the other hand, the national check-list is already under current preparation.

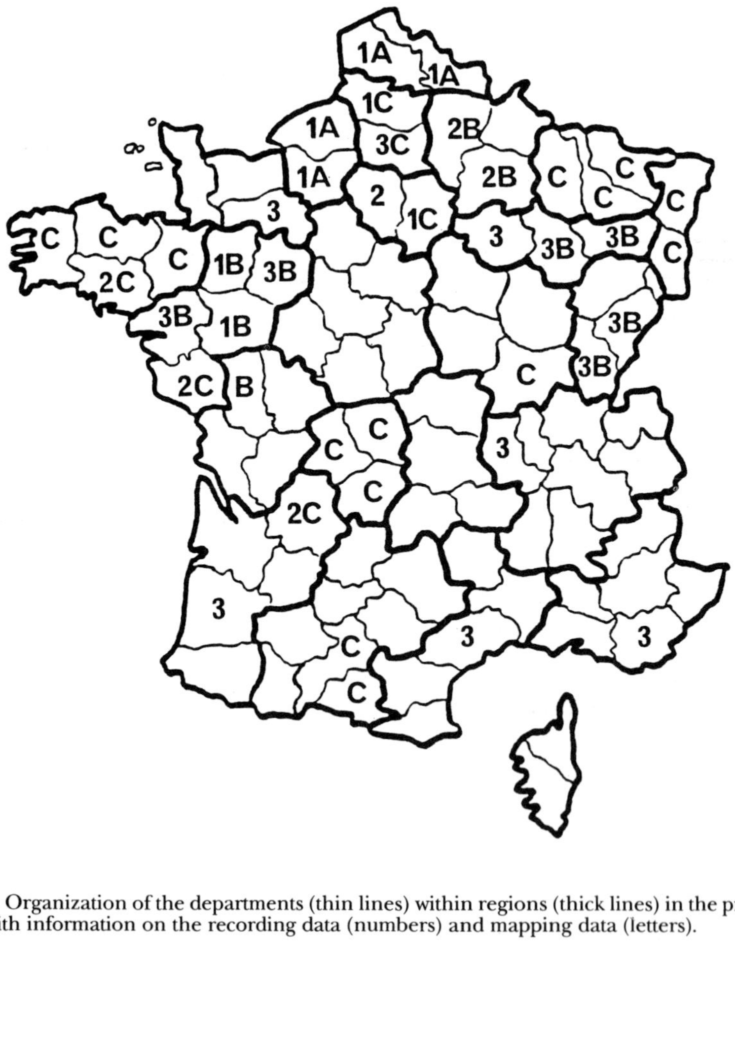

FIG. 1. Organization of the departments (thin lines) within regions (thick lines) in the programme with information on the recording data (numbers) and mapping data (letters).

```
UTILISATEUR :Courtecuisse_Regis___________   CODE_REGION:14 CODE_SECTION: 0
------------------------- Nom de l'espèce ---------------------------
Genre :Pluteus_______________
Epithète :aurantiorugosus______________________
Autorités:(Trog.)Saccardo_________________________________
-------------------------------- Synonymes -------------------------------
Genre1 :Pluteus_________   Epithète 1:coccineus______________________
Genre2 :________________   Epithète 2:___________________________________
Genre3 :________________   Epithète 3:___________________________________
--------------------------- Illustrations ---------------------------
Auteur1:Courtecuisse________   Année1 :1984
Publication1:Bull.Soc.Mycol.Nord_37:15______________________

Auteur2:Orton__________________   Année2 :1986
Publication2:Brit.Fung.Fl._4:50______________________________

Auteur3:______________________   Année3 :____
Publication3:____________________________________________
----------------------- Répartition Départementale -----------------------
Pour le 3ème caractère des champs "Département"
 * => Espèce non encore signalée.

Département N°1:59_     1_Découvreur :G.Vanhelle___________________
Département N°2:___     2_Découvreur :_______________________________
Département N°3:___     3_Découvreur :_______________________________
Département N°4:___     4_Découvreur :_______________________________
Département N°5:___     5_Découvreur :_______________________________
Département N°6:___     6_Découvreur :_______________________________
Département N°7:___     7_Découvreur :_______________________________
Département N°8:___     8_Découvreur :_______________________________

------------------------------ Ecologie ------------------------------
Ecologie :___________________________________________________________
_____________________________________________________________________
_____________________________________________________________________

Altitude_Min :____  Altitude_Max :____
Année de récolte N°1:1983                Année de récolte N°2:____
Année de récolte N°3:____                Dernière année de récolte:____

---------------------------- Liste Rouge ----------------------------
0 Pas sur liste rouge / 1 Menacée d'extinction / 2 Fortement menacée
3 Menacée / 4 Potentiellement menacée / 5 Eteint
   Statut_liste_rouge :2
-------------------------- Numéro d'herbier --------------------------
N°1 :RC/F8309291
N°2 :___________
---------------------- Position Systèmatique ----------------------
Famille :Pluteaceae______________
----------------------------- Confusions -----------------------------
Espèce N°1 :______________________________
         Caractère écologique N°1     :______________________________
         Caractère macroscopique N°1 :______________________________
         Caractère microscopique N°1 :______________________________

Espèce N°2 :______________________________
         Caractère écologique N°2     :______________________________
         Caractère macroscopique N°2 :______________________________
         Caractère microscopique N°2 :______________________________
```

FIG. 2. Structure of the proposed form for the recording data.

The inventory contributions are as shown on Fig. 2, here at the regional scale: binomial and authorities for each taxon, some synonyms, some bibliographical references to ascertain the interpretation referred to. The number of the departments where the fungus is cited, with eventually the name of the mycologist who first collected the taxon in the geographical unit, especially if it is a new record (marked by the asterisk). Some ecological data are added, along with years of occurrence, especially if the species is seldom or sporadically seen, a tentative quotation within the red data list scale, an herbarium number to allow future confirmation of the determination and finally the systematic position and data about possible confusions (all the items are not yet completed).

The mapping programme

Generally speaking, it is the author's opinion that when you wish to start with a mapping programme, the most important parameter is to give motivation to as many contributors as possible. So, you evidently have to choose a system as easy to use as possible, and as cheap as possible for users. Since the great majority of contributors will be amateur mycologists, you have to take into account that any serious difficulty will dramatically reduce their number. No mapping programme can meet success if only a few people are working on it.

So a grid system is proposed which has been accepted by the French mycological community. The grid system is based on the cutting of the territory in grades after the Paris meridian. This grid is clearly printed on all the maps published by the IGN (Institut Géographique National), these maps being already used by field workers, and are available in any bookstore. Fig. 3 shows the national matrix with the elementary squares each measuring around 10 × 14 km. Some maps published by the IGN are exactly coinciding with two or four of these elementary squares. The nomenclature of this system is based on the serial numbers of the IGN maps themselves. So it is extremely easy to use this grid system.

By contrast, almost no maps in France show the UTM grid pattern. The UTM co-ordinates are given in the margin of some maps, but as a rule, using a nomenclature for squares within a grid system is much simpler than finding the exact co-ordinates for each place where you collect fungi. Since the nomenclature of the UTM grid is only used by the French army, it is almost impossible to obtain it.

In order to take part in any future European collective mapping programme for fungi, I strongly recommend that users and contributors keep, for each species, the exact co-ordinates of each place where it is found. This may be done in any system (degrees, grades, UTM, Lambert) since computer programs convert easily co-ordinates from one system to any other.

Some contributions have been received already (Fig. 1; A: important contributions, B: fragmentary contributions, C: works under current processing). This figure shows a fragmentary statement, but again, I must emphasize the fact that the programme was initiated only two years ago.

Although I recommend to the contributors to map as many species as they can at the same time (I mean not to work with limited groups before dealing with others), a list of 130 taxa to be mapped as a priority has been proposed, in order to publish some maps within a few years, and therefore to stimulate more people to take part in the project.

Of course I am reluctant to publish maps at this early stage of the work, but just an example with the Fig. 3, showing the presently known distribution of *Cystolepiota bucknallii* (Berk. & Br.) Singer & Clém. This map has been made with the specific contributions received since the beginning of the programme and shows a wide motivation. More data are expected from other parts of the country.

The working structure for this programme also efficiently contributed to the mapping of the pathogenic *Armillaria* species following a request from J.J. Guillaumin (INRA

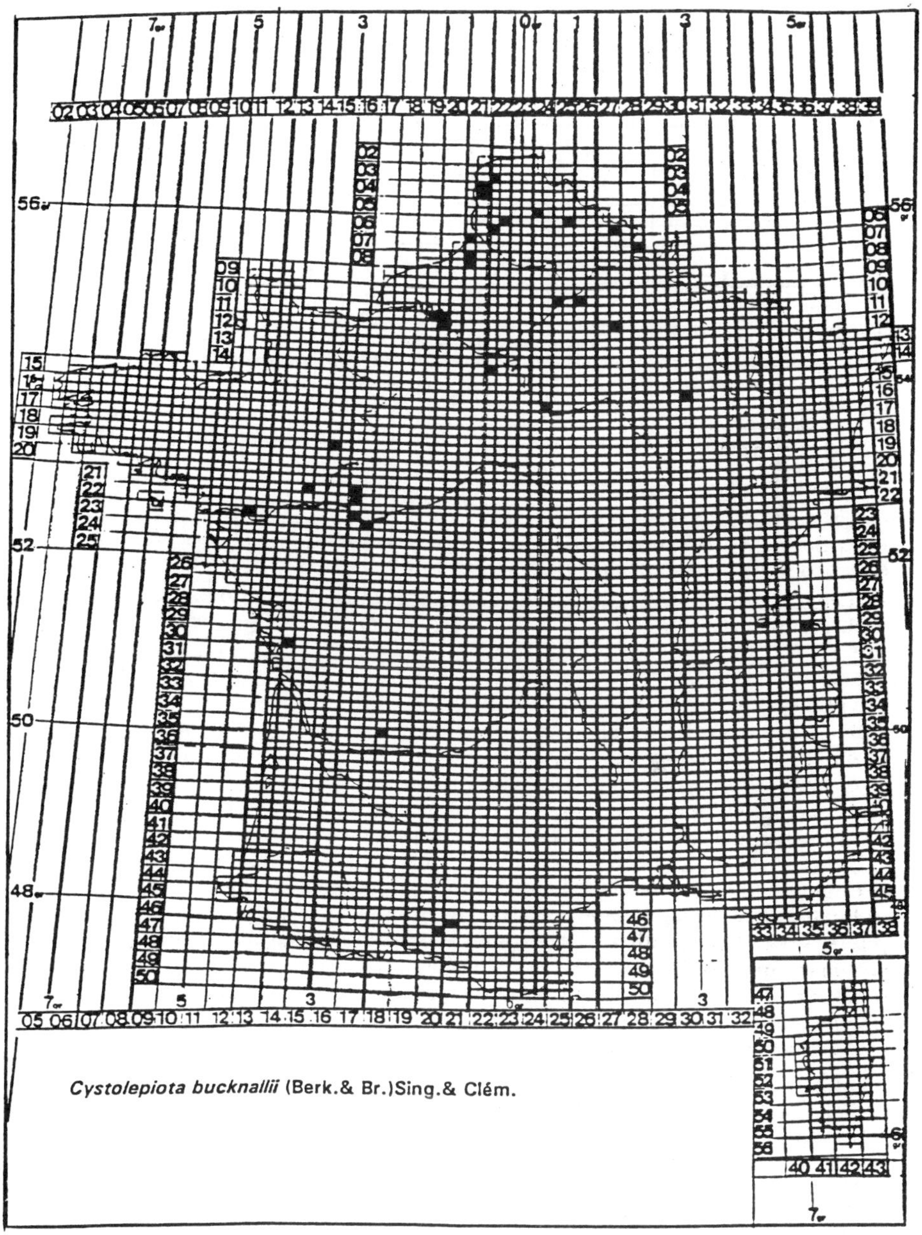

FIG. 3. National mapping matrix showing a provisional map of *Cystolepiota bucknallii*.

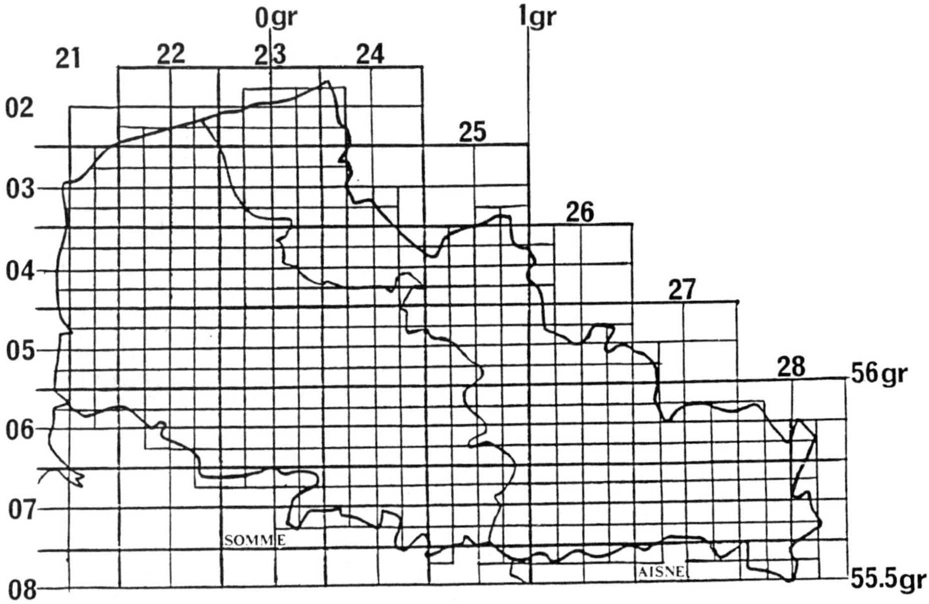

FIG. 4. Regional grid for Nord-Pas-de-Calais region, with smaller mapping units.

Clermont-Ferrand). It may also be said that the grid system may be used with smaller units at the regional or departmental scales. For example, Fig. 4 shows the regional grid for Nord-Pas-de-Calais, with 5 × 7 km squares. Contributions may be sent as hand-written forms or on computer tables (Fig. 5). This may also be sent as lists of square numbers for each species, such data being enclosed in the national data base which will be automatically printed on the computer maps in the near future.

The specialists team

In order to assess the reliability of data submitted by amateur contributors, a specialist group has been established. These specialized mycologists, working within their own group, receive rare, problematical, doubtful or critical specimens for control and determination, and will check the final results, to trace any chorological, ecological, taxonomical aberrations.

Contribution to the red data lists

The final aim of the programme concerns the conservation of fungi. The preparation of a formal check-list of French fungi and of a chorological atlas of the known species will partly serve as a basic step in dealing with the preparation of red data lists, at

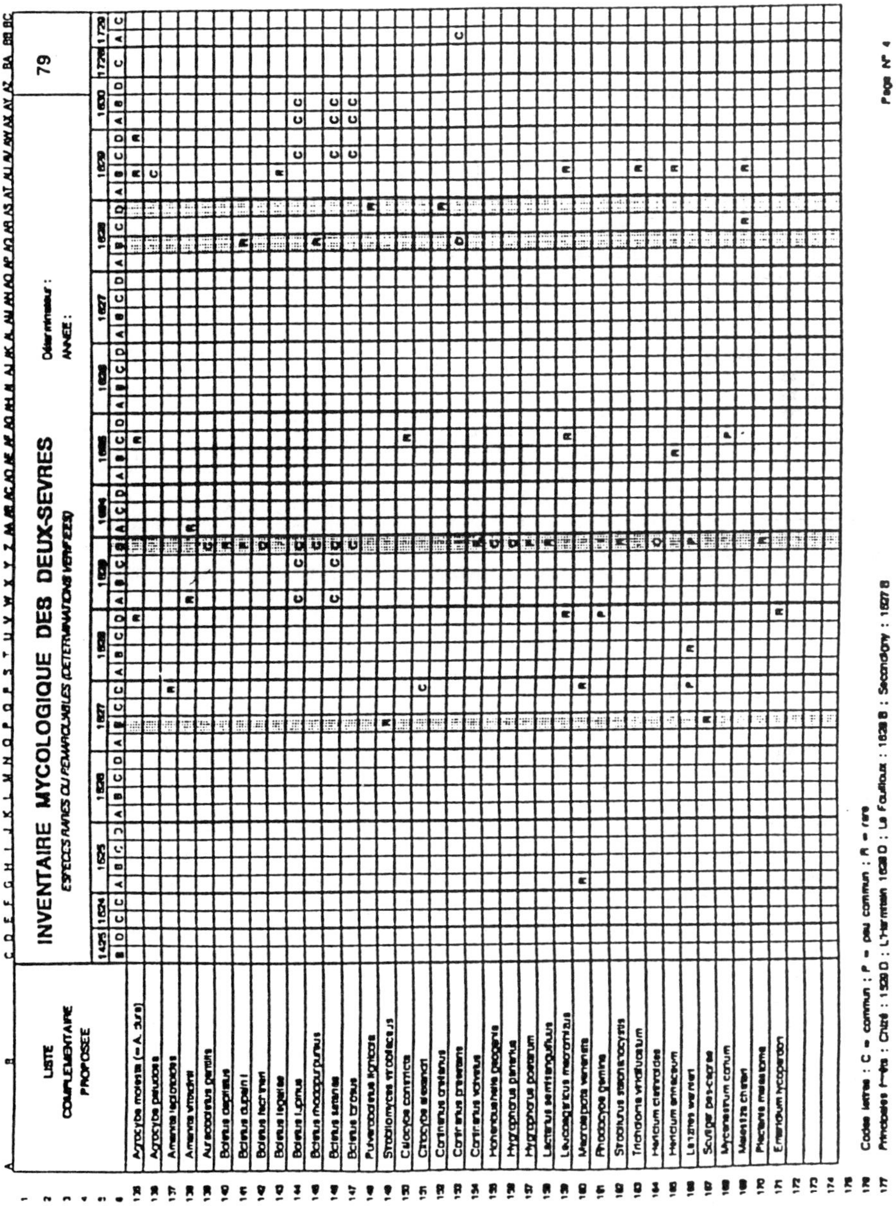

FIG. 5. Example of computer table for compiling mapping data.

departmental, regional and national levels. Some elements already exist for France under this respect.

An example concerning red data lists: *Hypocreopsis lichenoides* (Tode) Seaver is a species which, following work carried out by myself, is now considered to be threatened with extinction in Nord-Pas-de-Calais region. Other examples in the same area and in the same category are *Entoloma hirtum* (Velen.) Noordel., *Psathyrella sphagnicola* (Maire) Favre, *Russula sericatula* Romagn., *Cortinarius lividoviolaceus* Henry. Strongly threatened species include *Agaricus lanipes* (Moeller & J. Schaeff.) Singer, *Lactarius flavidus* Boud., *Russula cremeoavellanea* Singer.

Conclusion

As a conclusion, I must repeat that the French programme for inventory and mapping the fungi began only two years ago. The organization and efficiency still needs to be improved. Much progress is expected in the near future and further extensions will be considered. For example, conservation of threatened herbaria and libraries with processing of the enclosed data, development of the computer processing and especially of the computer mapping, actual conservative actions and others.

References

Courtecuisse, R. (1991a). Présentation succincte d'un projet d'inventaire mycologique national et de cartographie des champignons français. *Bull. Féd. Mycol. Dauphiné Savoie* (Montmélian) **121**: 4–8.

Courtecuisse, R. (1991b). Notule sur le projet d'inventaire mycologique national et de cartographie des champignons français. *Miscel. Mycol.* (Mons, Belgium) **30**: 28–30.

Courtecuisse, R. (1991c). Présentation sommaire d'un programme d'inventaire mycologique national et de cartographie des Mycota français. *Bull. Soc. Mycol. Nord.* (Lille) **48**: 3–15.

Courtecuisse, R. (1991d). Mapping and inventory of the French Mycota. *Mycologist* **5**(4): 187.

Courtecuisse, R. (1991e). Present state of the inventorying, mapping and red data list programs concerning French Mycota. *Eur. Council Conservation Fungi Newslett.* **5**: 8 (Vilm meeting).

Courtecuisse, R. (1991f). Le programme d'inventaire mycologique national et de cartographie des Mycota français. 1ère note: Présentation générale. *Bull. Trimestriel Soc. Mycol. France* **107**(4): 161–203.

Fourré, G. (1990). Le recensement des cryptogames. *Le Courrier de l'Ouest.* 8 Novembre 1990.

Lange, L. (1974). The distribution of Macromycetes in Europe. *Dansk Bot. Ark.* **30**(1): 7–105.

Mabon, G. (1991). Inventaire et cartographie des champignons en France. *Cah. Mycol. Nantais* **3**: 4–7.

Malaval, J.C. (1991). Présentation d'un programme d'inventaire mycologique national et de cartographie des Mycota français et données complémentaires pour les observations de l'année 1990. *Actes Mus. Rouen* **1991**: 31–70.

Mornand, J. (1991). Inventaire et cartographie des champignons en France. *Bull. Trimestriel Soc. Etudes Sci. Anjou* **81**: 12–15.

Mornand, J. (1992). Etat d'avancement de l'inventaire des champignons en Maine-et-Loire. *Bull. Trimestriel Soc. Etudes Sci. Anjou* **83**: 3–10.

Pouzaud, F. (1992). Champignons sur liste rouge. *Monit. Pharm. Lab.* **1996**: 35–36.

Fraiture, A. (1993). Mapping of fungi in Belgium and the G.-D. of Luxembourg – recent developments. In D.N. Pegler, L. Boddy, B. Ing & P.M. Kirk (Editors). Fungi of Europe: Investigation, Recording and Conservation, pp. 13–20 Royal Botanic Gardens, Kew.

3. MAPPING OF FUNGI IN BELGIUM AND THE G.-D. OF LUXEMBOURG — RECENT DEVELOPMENTS

A. FRAITURE

Jardin Botanique National de Belgique, B-1860 Meise, Belgium

Abstract

The first part of the paper describes the new computer programs used to draw the maps. Besides the usual distribution maps of species, these programs automatically create maps by reassembling the data of several taxa (map of the prospected squares, of the floristic richness, ...) and produce statistics. The second part of the text presents different distribution patterns among the treated species. The third part comments on the evolution of the mycoflora. At the present state of knowledge, it is not possible to give unambiguous proof that any of the treated species is actually declining. On the other hand, it is obvious that the frequency of several species (usually lignicolous) is increasing. Some of them are even neophytes.

The Computer Programs Used to Draw the Maps

The first fascicle of the *Distributiones Fungorum Belgii et Luxemburgi* (Heinemann & Thoen, 1981) contains 80 distribution maps. The second fascicule (Fraiture et al., 1993) will contain 52 distribution maps, 12 of them consisting of an updating of maps published in 1981.

The maps for the second part of the "Distributiones Fungorum" will be the first to be computer-drawn. The programs used have been written (in FORTH) by Alain Empain, bryologist in the Jardin Botanique National de Belgique. Each map consists of an assembly of about 50 files, all of them being registered in the same coordinate system, like in a superposition of several overlays. These files concern:

(1) The topographical background: The files were created with a digitizer. The coordinates of each element of the topographical background (river, border, coast) were registered in a separate file. The programs give the opportunity to add or to remove easily each of these elements, as well as to represent them, if necessary, in a different way (for example, the coasts by a large grey line, the borders by a black dotted line, the big rivers by a fat black line and the small rivers by a fine black line). These programs are very flexible and can be used for other countries or regions.

(2) The phytogeographical districts: They are represented on the maps by white or grey zones. The squares situated on the limit between two districts appear in a different grey.

(3) The coordinate grid: The grid-system used is that of the Institut Floristique Belgo-Luxembourgeois (I.F.B.L.). It is based on the 1/50000th maps of the Institut Géographique National (I.G.N.) of Belgium. In order to obtain clearer maps, only the limits of those I.G.N. maps were used. However, a file containing the complete grid (4 × 4 km squares) is available. The programs are also able to use other coordinate grids (e.g. U.T.M.).

(4) The coordinates of the distribution of the different species: They were coded on the base of the "handwritten-dossiers", in order to constitute a very simple ASCII file, the first line of which containing the title to be printed on the map, and the other lines the coordinates of the observations. The coordinates of a same 4 × 4 km square may be introduced several times for the same species. This file has the following appearance:

<title> Amanita phalloides
D423
L743
B532
...

The symbol used by default by the computer is a black square. In some cases, it is however useful or necessary to use two or several kinds of symbols to distinguish, for example, ancient from recent data, doubtful from checked data, etc. In the file, each observation coordinate must then be completed by a number indicating the group to which it pertains, and the head of the file must define, for each number, the kind of symbol to use and the legend to print on the map. This kind of file has the following appearance:

<title> Gloeophyllum odoratum
<symb> 4 05A9 > 1980
<symb> 3 150T 1967–1980
<symb> 5 13A9 < 1967
3 G814
4 E533
5 G843
...

The kind of symbol to be used is here defined by a group of 4 digits and/or letters. The first determines the shape of the symbol (0 = square, 1 = circle, ...), the second concerns the size of the symbol, the third the "colour" (white, more or less pale or dark grey, black) and the last the width of the black line which makes the border of the symbol.

In the files concerning the second fascicule of the "*Distributiones Fungorum*", only the coordinates of the observations were coded. The programs presented here were however conceived in order to be connected with a data bank containing, for example, for each observation, the place, the date, the biotope or substrate, the number of fruitbodies observed. This information can be indicated on the maps by the use of different symbols. The programs for the operating of the data bank (encoding and sorting of the data, connection with the programs for the drawing of the maps) are in progress.

Another advantage of the computer is that it allows the creation of synthesis-maps, reassembling the data of several species (for example, all the treated species, or different species of a same genus). Three types of synthesis-maps can be created by the programs.

(1) The map of the prospected squares from the beginning of the "*Distributiones Fungorum*" shows all the squares in which at least one of the 120 treated species was found. The files containing the data of the different species are merged into one file, which is cleaned (removal of headings and coefficients, as well as of duplicate coordinates). The map is then drawn, on the base of this large file, with the same programs that are used for the single species maps.

(2) The map of the intensity of the exploration (Fig. 1): The same method is used, but without removal of duplicate coordinates. The observations are then numbered for each 4 × 4 km square and the squares are distributed among several frequency categories, each of them being represented on the map by a different symbol. All these operations are carried out automatically by the computer after it has received the list of the concerned species.

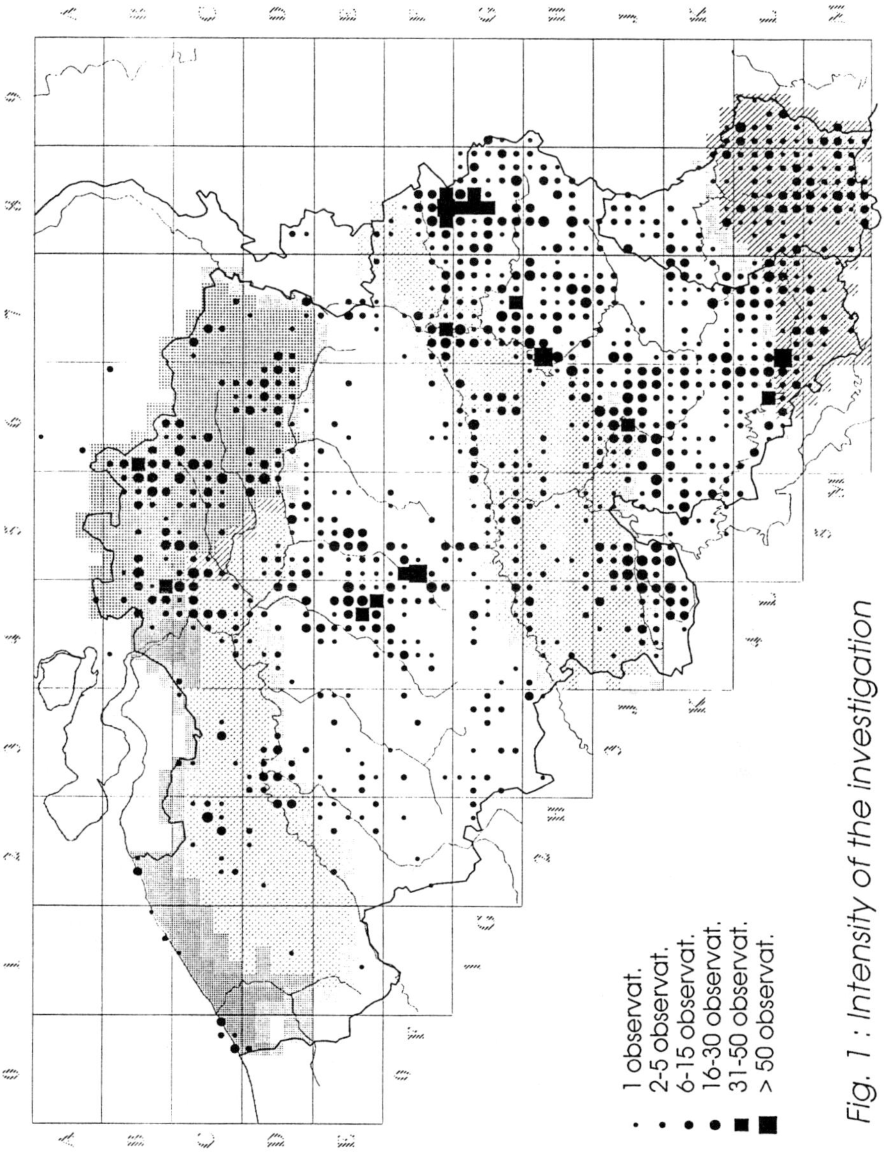

Fig. 1 : Intensity of the investigation

(3) The map of "floristic richness" present the number of species observed in each square. It is drawn with the same program used for the previous map, except that the removal of the duplicate coordinates is made in each file before they are merged.

Statistical analysis can also be performed on the data. The computer can give, for each phytogeographical district, the percentage of squares in which a species was observed. The squares situated on the limit between two districts are considered as a separate territory. Calculations are made either on the number of squares in which the species was observed or on the number of records concerning the species in the district. As the exploration of the territory is not yet complete, an attempt was also made to calculate the percentages on the number of explored squares in each district (i.e. the number of squares in which at least one of the 120 treated species was recorded).

The main program for the map drawing described above generates, for each of these maps, a detailed description in PostScript language. This allows production of proofs on a 300 dpi (dots per inch) laser printer and also final copy on a 2500 dpi professional photocomposer. This method is very adaptable and provides for maximal quality. It is also possible to display the maps on the computer screen.

Some Comments about the Distribution of the Mapped Species

Several different distribution patterns are observable. Numbers between brackets refer to map numbers: 1–80 = first fascicule of the "*Distributiones Fungorum*", 81–132 = second fascicule.

— *Phallus hadrianii* (79) and *Psathyrella ammophila* (66) are restricted to the sand dunes of the maritime district.

— *Gyroporus castaneus* (26) is found only growing on acid sandy soils.

— Many species are restricted to calcareous zones, i.e. the "mosan" district and some parts of the "lorrain" and "brabançon" districts. The southern border of the "mosan" district (Calestienne) is composed of calcareous rocks on which calcicolous fungi are often particularly abundant. *Micromphale foetidum* (43), *Cortinarius bulliardii* (Fig. 2) (110), *Cystolepiota bucknallii* (103), *Tremiscus helvelloides* (7 and 122), *Boletus satanas* (90), *Sarcosphaera coronaria* (84) and *Verpa conica* (3) are calcicolous species. The latter was also found on the calcariferous coastal sands.

— Conversely, a certain number of taxa were only found in the "ardennais" district, which is situated at a higher altitude than the other districts and mainly composed of acid oligotrophic clayey soils. These fungi are e.g. *Stropharia hornemannii* (104), *Cortinarius brunneus* (Fig. 3) (109), *Hygrophorus olivaceoalbus* (95), *Lactarius repraesentaneus* (118) and *Pholiota scamba* (106).

— *Astraeus hygrometricus* (73 and 131) and *Pisolithus arhizus* (74 and 132) were almost exclusively found on mine refuse disposal sites.

— *Tephrocybe palustris* (34) and *Galerina paludosa* (62) develop in *Sphagnum* bogs ("Kempen" and "ardennais" districts).

— Other species present a distribution which mainly corresponds with that of their host-tree. *Fomes fomentarius* (16 and 124, also associated with *Populus*), *Mycena pelianthina* (45, mainly on neutral and calcareous soils) and *Mycena crocata* (46, id.) are associated with *Fagus*. *Lactarius controversus* grows only under *Populus* and is almost exclusively found in artifical stands of these trees. *Cortinarius armillatus* (107), *C. pholideus* (112), *Piptoporus betulinus* (15) and *Russula claroflava* (71 and 129, on peat soils) are associated with *Betula*. *Gyrodon lividus* (92) is associated with *Alnus* and *Auricularia auricula-judae* (6) grows mainly on *Sambucus*.

— In contrast, some species have a very large ecology and distribution: *Laccaria amethystina* (35), *Lactarius necator* (130, mainly associated with *Betula* and *Picea*), *Ganoderma lucidum* (86), *Chalciporus piperatus* (91) and *Xerocomus badius* (94).

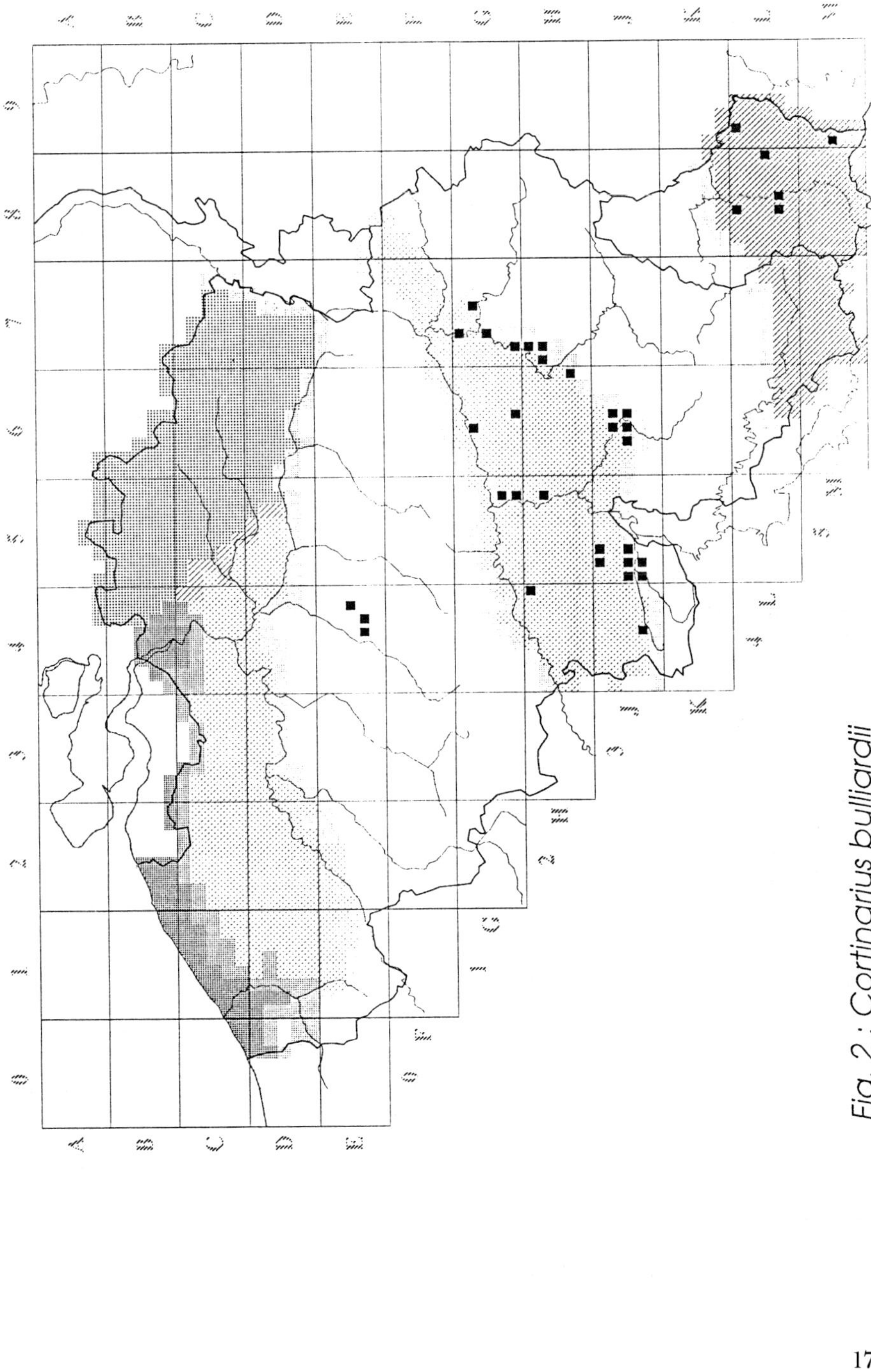

Fig. 2 : *Cortinarius bulliardii*

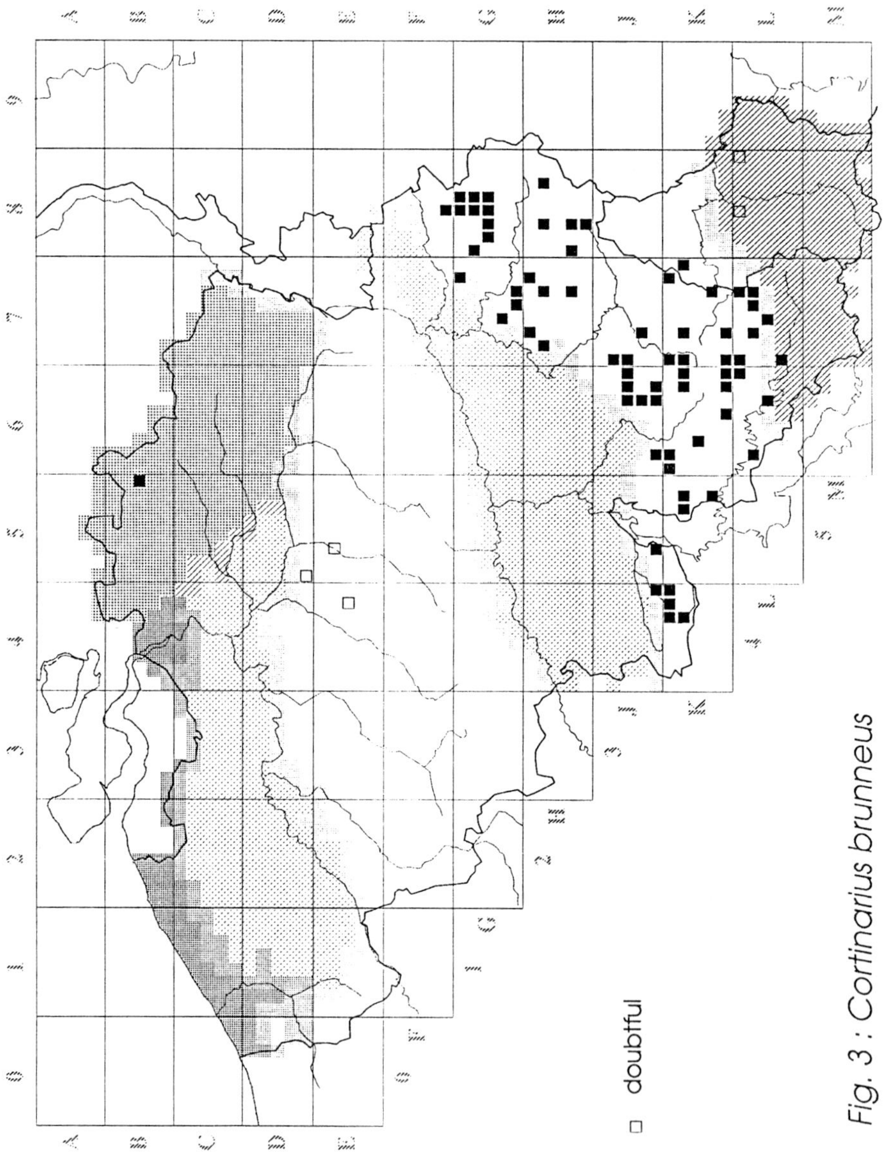

Fig. 3 : Cortinarius brunneus

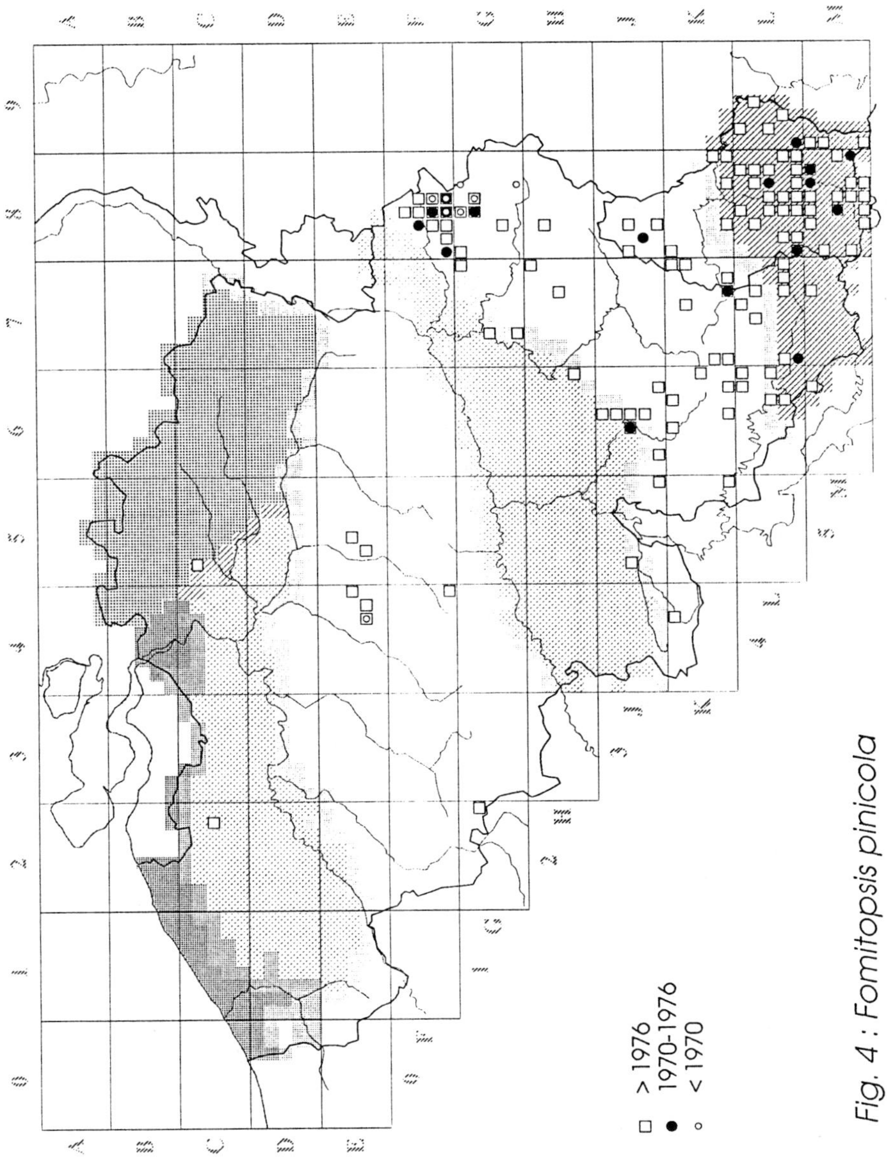

Fig. 4 : Fomitopsis pinicola

Some Comments about the Evolution of the Mycoflora

At the present state of knowledge, we are not able to give any significant proof that certain species are actually declining, even if many mycologists, both amateur and professional, have the idea that the general tendency is a progressive impoverishment.

If *Tricholoma equestre* s.str. and *Tricholoma auratum* are not synonyms, it is probable that the latter, after having been previously common on acid sandy soils in the northern part of Belgium, is now declining markedly (map 96).

Conversely, a few species are evidently becoming more common. Some of them are even neophytes and have been recorded for the first time rather recently:

— *Clathrus archeri* (120) is typically a neophyte. It originates in the southern hemisphere and was first found in Europe in 1920 (Vosges, France). The first record in Belgium in 1968, first record in the G.-D. of Luxembourg in 1980. The species is rather constant in its known locations but remains rare in Belgium, somewhat less rare in Luxembourg.
— *Pycnoporus cinnabarinus* (19) was first discovered in Belgium in 1966. Twenty-five years later, it is known from more than one hundred 4 × 4 km squares.
— *Neobulgaria pura* (83) was discovered in the G.-D. of Luxembourg in 1964 and in Belgium in 1965. Now, it is almost common on *Fagus* in certain parts of the two countries.
— *Stropharia hornemannii* (104), *Fomitopsis pinicola* (Fig. 4) (85) and *Gloeophyllum odoratum* (87) are also in progression in the two countries. None of these species was observed before 1936 and they must probably be considered as neophytes too.

References

Fraiture, A., Heinemann, P., Monnens, J. and Thoen, D. [Dutch translation by Rammeloo, J.] (1993). Distributiones Fungorum Belgii et Luxemburgi, II. 64 p. [maps 81–132; to be published]. Jardin Botanique National de Belgique, Meise.

Heinemann, P. and Thoen, D. [dutch translation by Rammeloo, J.] (1981). Distributiones Fungorum Belgii et Luxemburgi, I. 96 p. [maps 1–80]. Jardin Botanique National de Belgique, Meise.

Nauta, M. & Vellinga, E.C. (1993). Distribution and decline of macrofungi in the Netherlands. In D.N. Pegler, L. Boddy, B. Ing & P.M. Kirk (Editors). Fungi of Europe: Investigation, Recording and Conservation, pp. 21–46 Royal Botanic Gardens, Kew.

4. DISTRIBUTION & DECLINE OF MACROFUNGI IN THE NETHERLANDS*

M. NAUTA & E.C. VELLINGA

c/o Institute for Forestry & Nature Research, P.O. Box 23, NL-6700 AA Wageningen, The Netherlands

Abstract

Since the start of the mapping project of macrofungi of the Dutch Mycological Society over 500,000 records have been collected, originating from the literature and herbaria (mostly older data) and from reports of forays and single observations (mostly recent data). Selective study of the country, the diverse origin of the data, the increase in numbers of recorded taxa and, most of all, the increasing numbers of records hamper the interpretation of absolute numbers of records of a species in terms of decline or increase of that species. The use of percentages of numbers of records of a species per period is a good method to indicate changes in frequency. Other correction methods are derived from this one and imply a restriction to regions or biotopes or to a set of grid units with continuous observations. This last method allows statistical testing. Decline and increase occur in all functional groups, but ectomycorrhizal fungi and saprophytes of non-fertilized grasslands are most threatened.

Introduction

Observed changes in the distribution of macrofungi since the 1960s motivated the Dutch Mycological Society to organize a national mapping project (Arnolds, 1985). Since 1981 all kinds of records of macrofungi have been collected and stored in a computerized database, amounting now to half a million records. This allows insight into changes of the mycoflora during this century. Within the scope of preparing the first volume of a distribution atlas of macrofungi in the Netherlands 450 species are selected to study the changes and compare earlier distribution with present distribution (Nauta & Vellinga, 1992). The selected species occur in most of the biotopes in the Netherlands; very common saprophytes are selected as well as species from endangered habitats. The difficulties in interpreting the data and possible correction methods are discussed here.

Mapping Project

In the Netherlands the mapping project of macrofungi, started in 1981, is rather successful (Jansen, Nauta & Vellinga, 1992). Data are recorded from recent times as well as from the past, and from forays as well as the literature. The records are based on as small a scale as possible, at least on 5 × 5 square km grid units, but preferably on 1 × 1 square km. A national grid system is used. The data are thoroughly controlled and checked before entering the database. The total number of records now exceeds 500,000,

*Communication no. 4 of the project 'Distribution atlas of macrofungi in the Netherlands'.

and the number of grid units (5 × 5 sq. km) from which one or more species are recorded amounts to 1548, i.e. 92% of the total number of grid units. Table 1 shows the frequency of grid units per category of number of species.

TABLE 1. Frequency of grid units per category of number of records

frequency classes (number of species observed)	number of grid units
0	129
1–10	271
10–50	426
50–100	256
100–200	302
>200	293

Possibilities, Methodological Problems and Correction Methods

With such a database there are more possibilities than just producing distribution maps of species. Increase or decline of macrofungi and changes in distribution patterns and their possible causes can be analyzed. However, the composition of the database gives rise to numerical problems of several kinds.

The first problem is the selective study of the country: traditionally mostly wooded areas of the Netherlands were investigated, such as the woods on sandy Pleistocene soils in the south, centre and east, and also the forests and estates of the old (inner) dunes (see Fig. 1). Nowadays the areas on marine clay in agricultural use and peat grasslands are also inventoried, although not systematically. The north-eastern part (Drenthe) was not or hardly investigated until 1960. The new polders in the former Zuider Zee have been inventoried since the mid-seventies. Interpretation of data from species which mainly occur in the formerly neglected parts of the Netherlands will be difficult, especially when questions concerning increase or decline have to be answered.

The second difficulty arises from the diverse origin of records: a greater part of the records of former times comes from herbaria or the literature, and the few excursions which are recorded are not necessarily complete. Nowadays records consist mostly of field data and the forays are more completely recorded than formerly. This is, in general, not a major inconvenience, but for those species with a disproportionate number of records originating from herbarium collections in certain periods it can become a factor of dominant importance.

The third problem is the increase in number of recorded species: from c. 1450 taxa recorded before 1950 to 3850 taxa recorded now. This increase is probably due to taxonomical changes, and also to the fact that in earlier times some species were not well known and not or rarely recorded (e.g. *Xylaria carpophila* (Pers.) Fr.). However, the newly recorded taxa from only a minor part of the records of the database, c. 1.5%. For the majority of species this is not a problem.

The main problem of the composition of the data is the disproportionate number of records throughout the years which is partly caused by the other factors mentioned (Fig. 2). Abcut three-quarters of the data comprise observations since 1980. This also accounts, though to a lesser extent, for the number of grid units in which fungi are recorded (Fig. 3). The number of records is an indication for the degree of investigation of the country. For the analysis of a time series of a species the data have to be corrected, since in general the

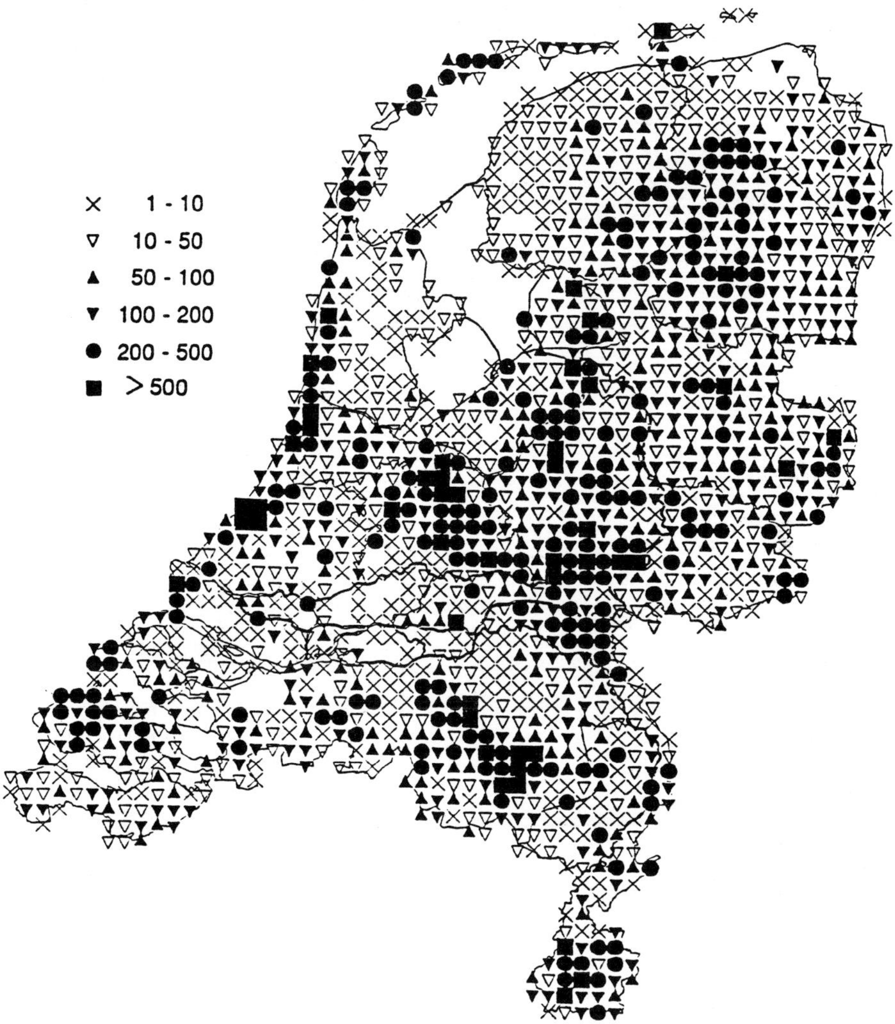

FIG. 1. Observed number of species per grid unit.

number of records of a species and the number of grid units in which a species was observed will increase (Fig. 4). Only strongly declining species will exhibit another pattern.

Correction of the data for the increasing number of records throughout the years can be carried out by calculating the percentage frequency of a species per period. The number of records of a species in a certain period is calculated as a percentage of the total number of records of all species in that period. This is legitimate as the composition of species remained more or less the same during the years.

This can be illustrated by some examples. A standard pattern is shown by *Amanita rubescens* (Fr.) S.F. Gray, the most common species of *Amanita* in the Netherlands. The species forms ectomycorrhiza with several deciduous and coniferous tree species and occurs mostly on sandy soils (Fig. 5). The pattern of the bar-diagrams (Fig. 4) shows an

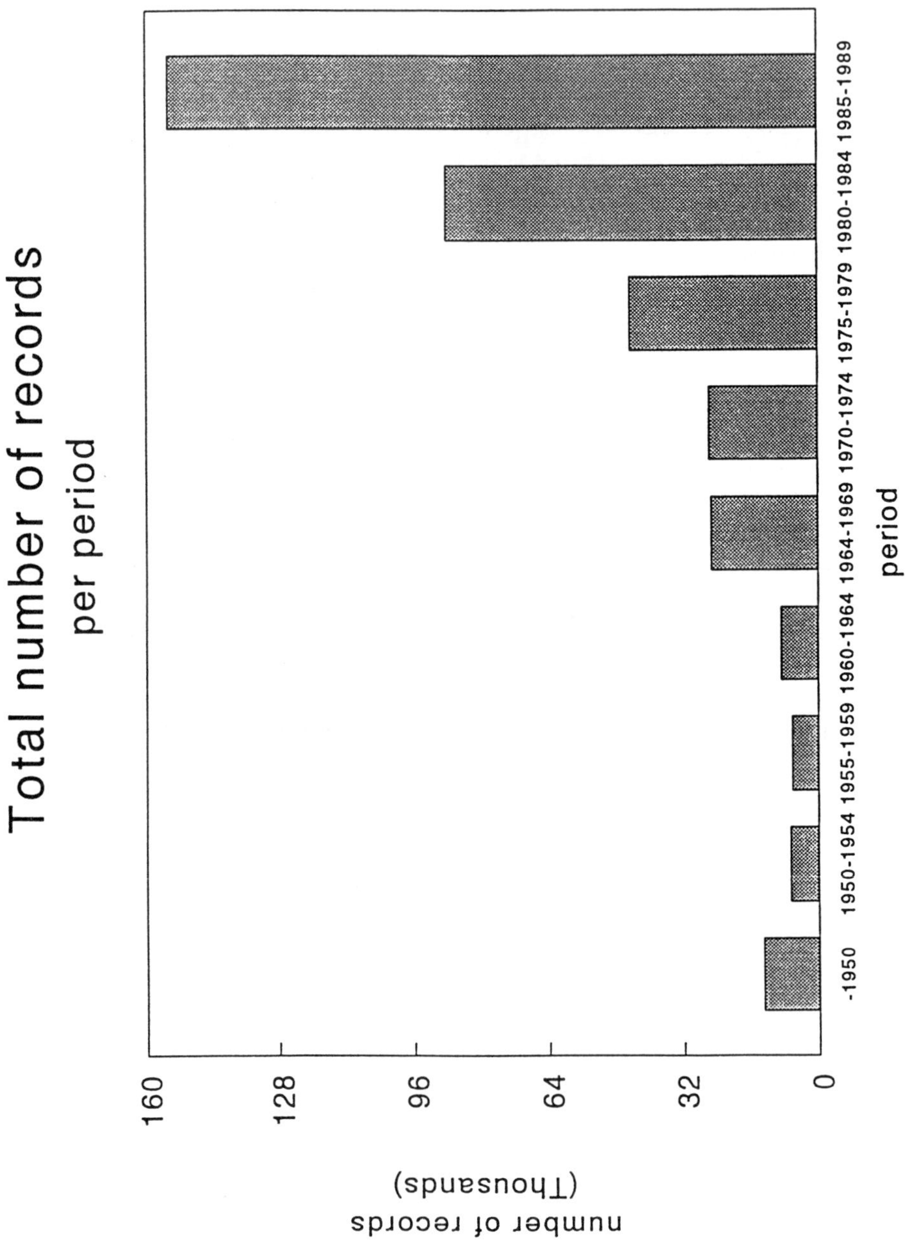

FIG. 2. Total number of records per period.

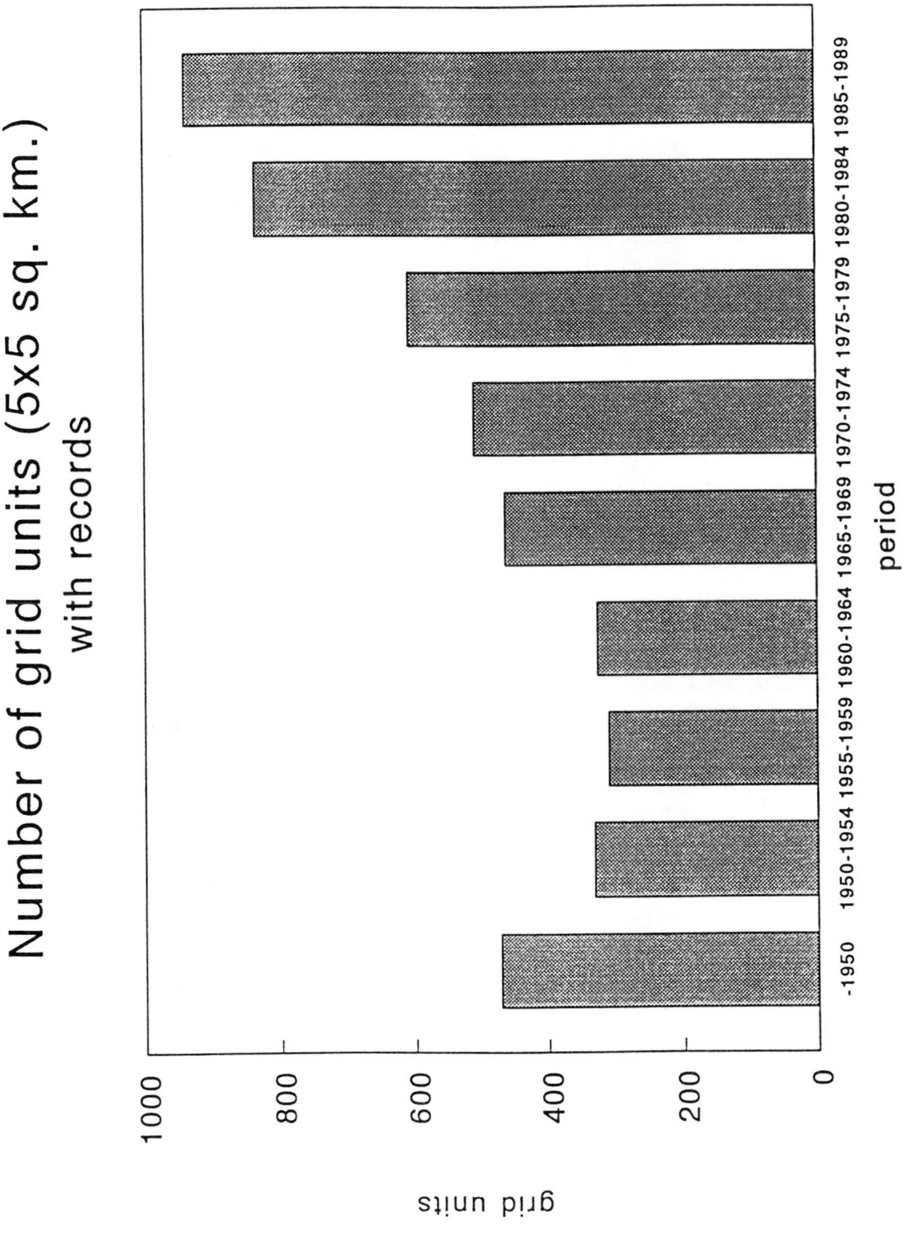

FIG. 3. Total number of grid units with one or more recorded species per period.

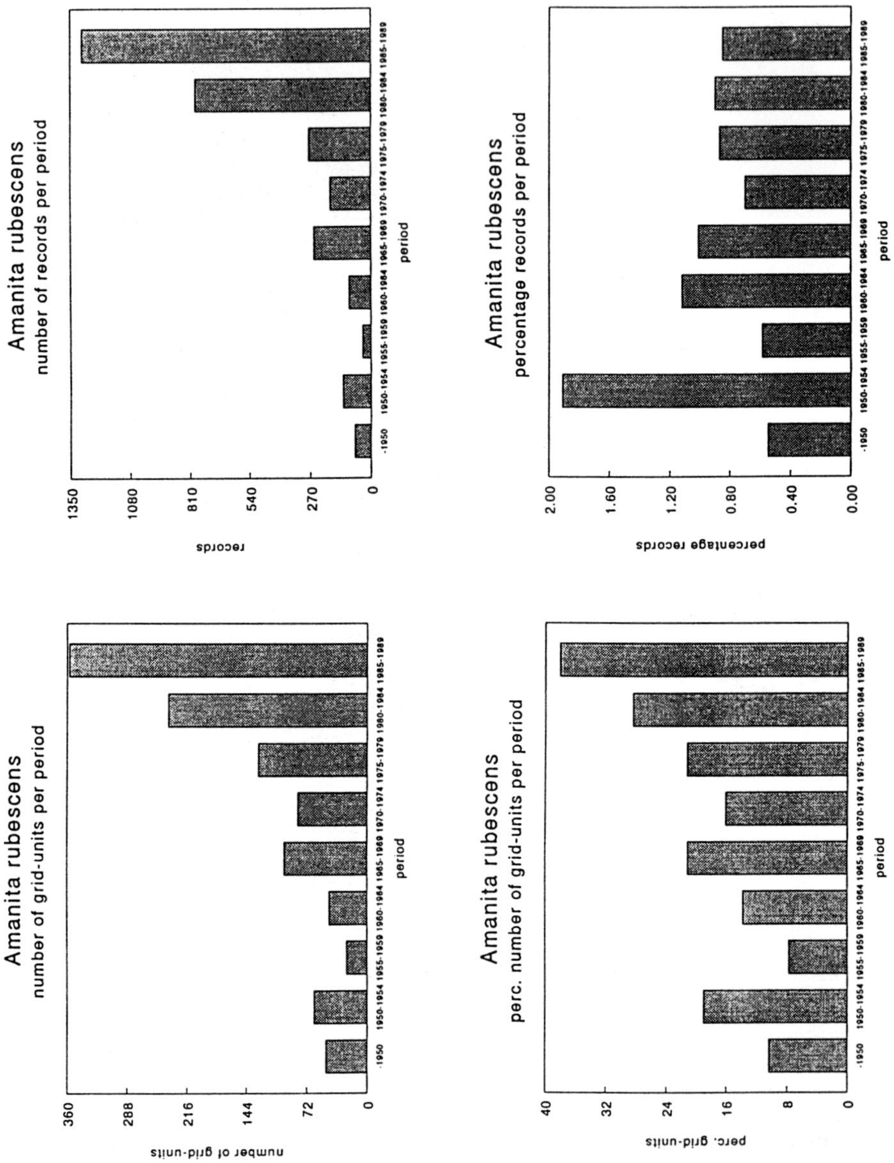

FIG. 4. *Amanita rubescens* – the absolute and percentual numbers of records and grid units per period.

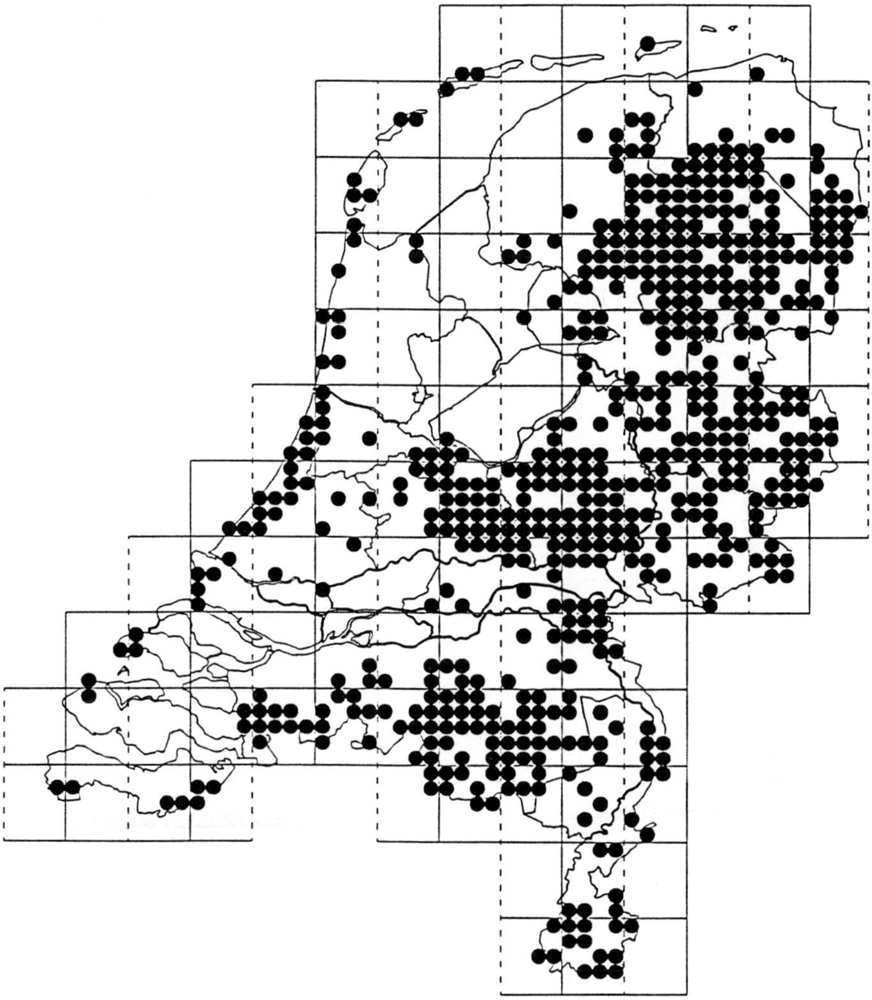

FIG. 5. Distribution of *Amanita rubescens* in the Netherlands.

increasing number of records, and an increasing number of grid units in which the species was observed. The percentual frequency of records per period remains more or less equal. The high peak in the bar-diagram in the period 1955–1959 can be an artefact due to the disproportionate number of records from the Rijksherbarium, Leiden.

The percentage number of grid units shows a rising curve for most species. This curve is much more difficult to use for interpretation since in general it is increasing, and the total number of grid units is limited. If a species is declining in numbers of carpophores, the number of grid units in which the species is observed will react more slowly, if it decreases at all. In conclusion, the percentual frequency of records of a species per period provides the best insight into the decline or increase of a species.

An example of a declining species is *Chroogomphus rutilus* (Schaeffer: Fr.) O.K. Miller. Figure 6 shows the distribution in the Netherlands where it occurs mostly on sandy soils,

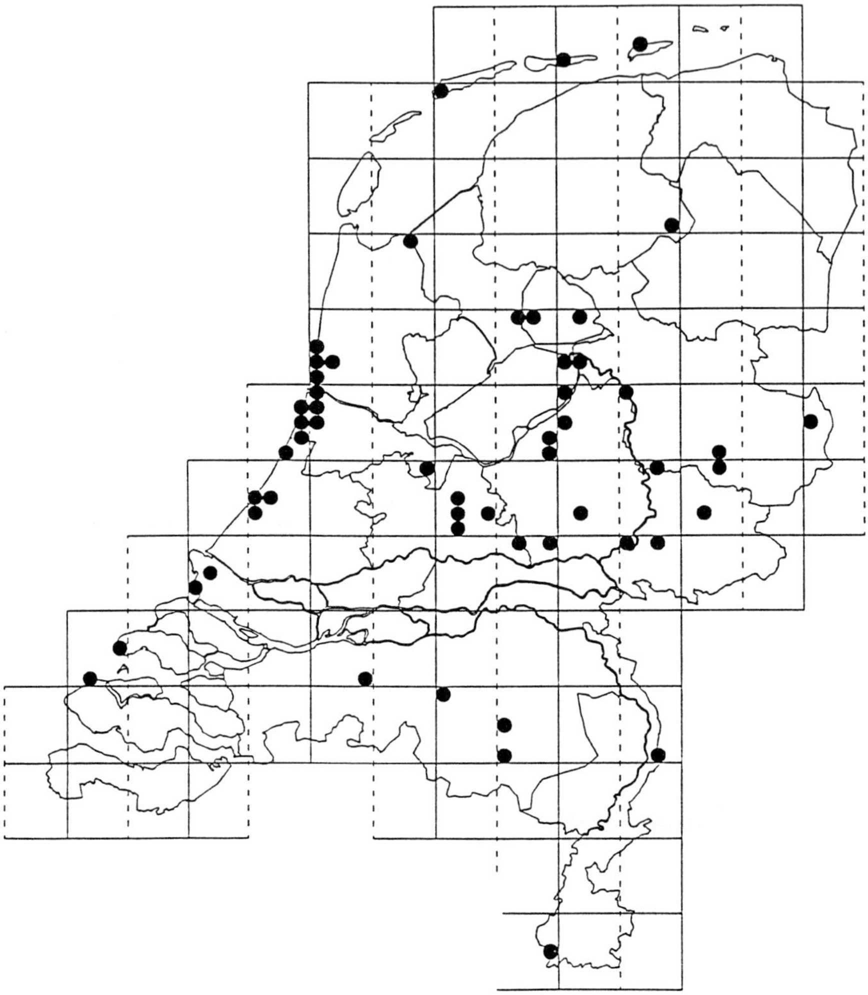

FIG. 6. Distribution of *Chroogomphus rutilis* in the Netherlands.

especially in the more or less calcareous dune-areas. Its habitat is pine-forests where it forms ectomycorrhiza with *Pinus*. It is regarded as a threatened species (Arnolds, 1989) and it is nowadays rather rare. The absolute number of records per period (Fig. 7) shows an undulating curve, while the decline is very obvious considering the percentage of number of records. The number of grid units in which the species was observed during the years shows fluctuations. It is known from fewer grid units now than before 1955. The percentage of grid units shows also an obvious decline. The conclusion must be drawn from the figures that this species is strongly declining.

Detailed Correction Methods

Some results give rise to an additional and more detailed method of study. This can include for instance a restriction of the studied areas, that is a study by region. In the case

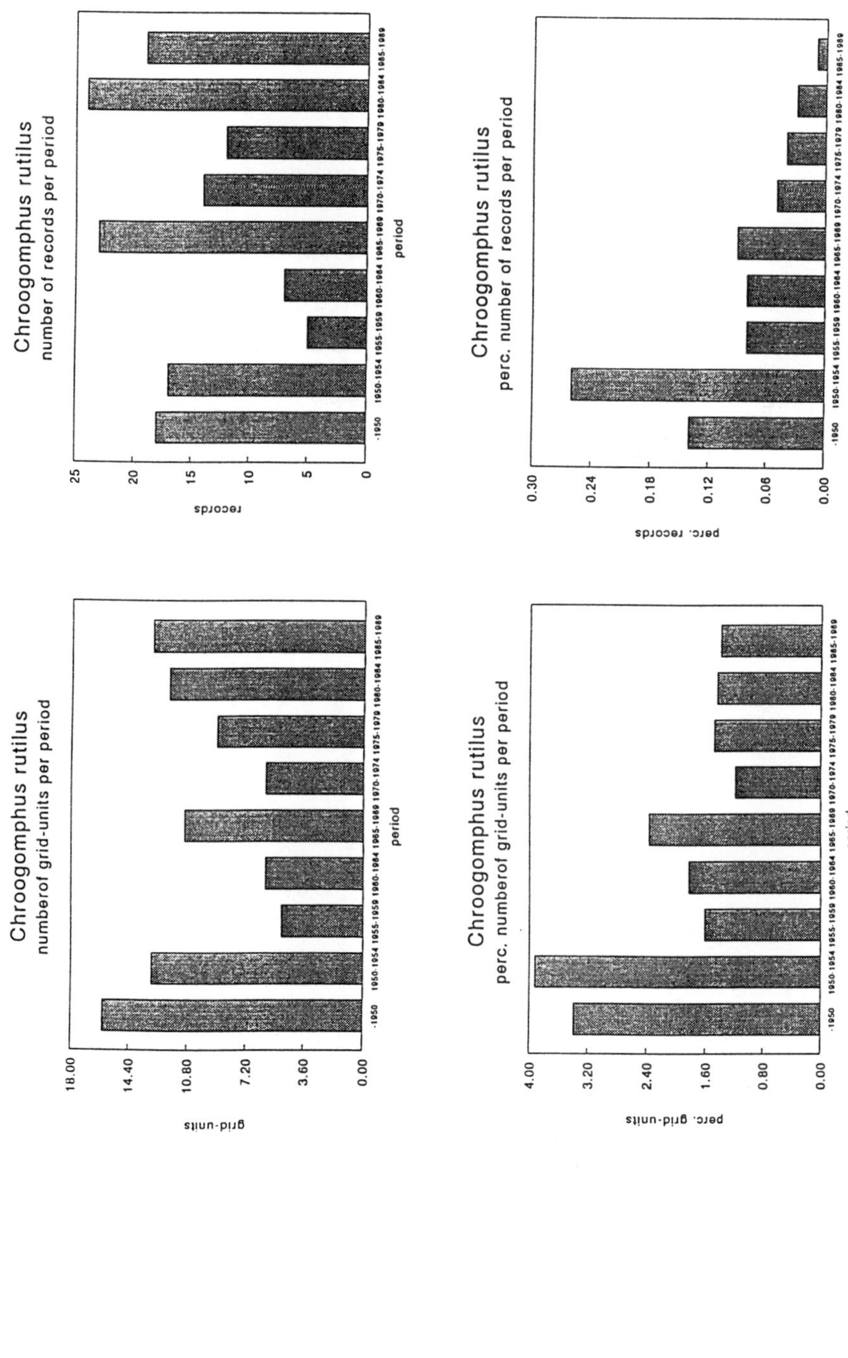

FIG. 7. *Chroogomphus rutilus* – the absolute and percentual numbers of records and grid units per period.

of *Chroogomphus rutilus* this could comprise a study of its occurrence in the dune-area apart from the Pleistocene area in order to establish the amount of decline. The method encloses a comparison per region per period of the percentage of number of records of a species. *Chroogomphus rutilus* is declining less in the dune-area than in the Pleistocene area (Fig. 8a & 8b), but a new biotope has also been found in the new polders.

Another additional method can also be a restriction of biotopes. The method is the same as described above.

Disadvantages of Correction Methods

Every correction method has its disadvantages, and so have the ones above. The influence of selective study of the country can not be excluded; the geographic as well as the biotope enlargement plays an important role. Their role is in fact so important that statistical testing of the data is more or less useless, even after correcting as described. Therefore a more reliable method is developed, taking into account these disadvantages. A selection is made from grid units which are continuously visited throughout the years, and a second criterion is a number of at least 50 observed species per decade. With these

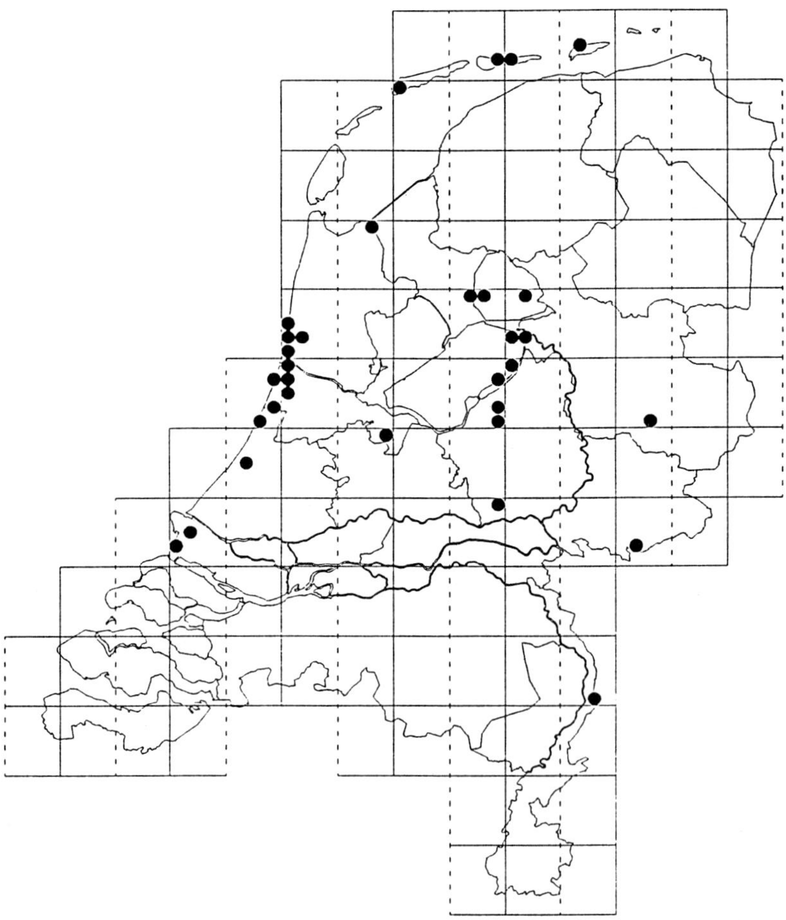

FIG. 8a. Distribution of *Chroogomphus rutilus* in the Netherlands before 1970.

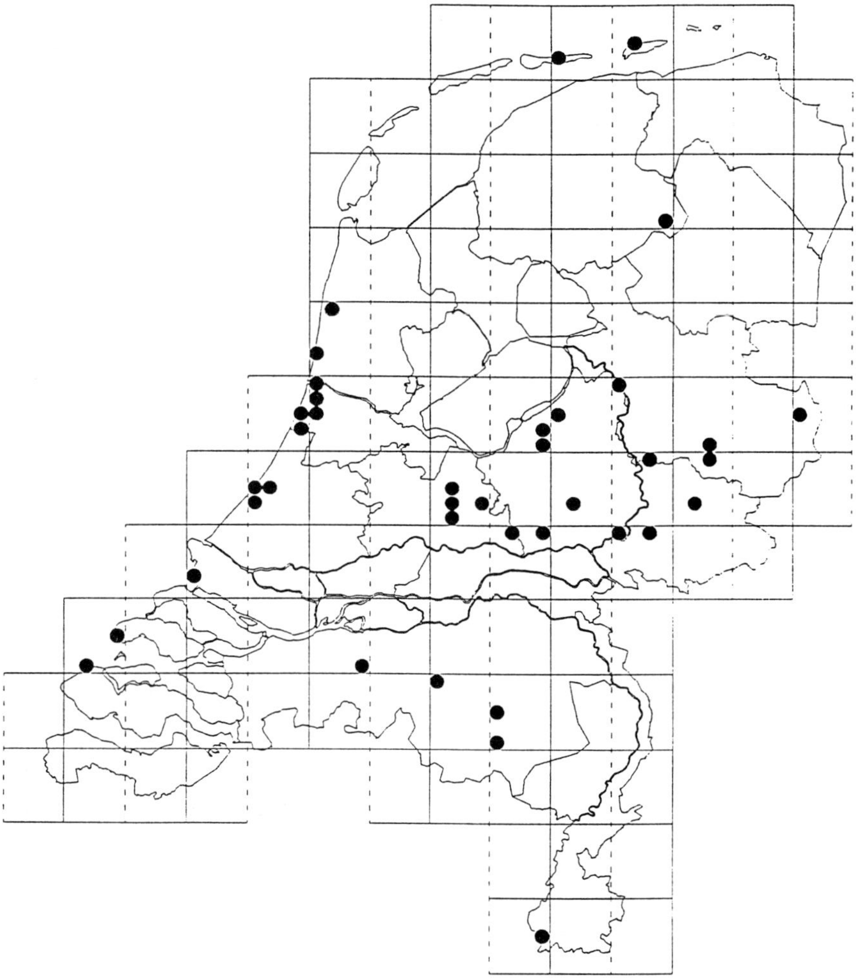

FIG. 8b. Distribution of *Chroogomphus rutilus* in the Netherlands since 1970.

criteria 40 grid units were selected with which statistical testing is possible. These grid units are distributed in the centre and west of the Netherlands (Fig. 9), being the most visited areas. The periods are not necessarily of equal length, to allow the number of records to be more or less equal. Of course, the number of degrees of freedom has to be sufficient for statistical testing. The fitting of a line with multiple and linear regression becomes possible and gives a good impression of decline or increase of a species throughout the years.

A disadvantage of this method is the exclusion of the north-eastern and the other parts which were not well investigated in former times. Furthermore, this method is only useful for well-known species with a distribution throughout the country without emphasis on the excluded areas, so in fact only for the species of forests on sandy soils. For many species only a tendency can be given, but comparing all the patterns of the species several general patterns emerge.

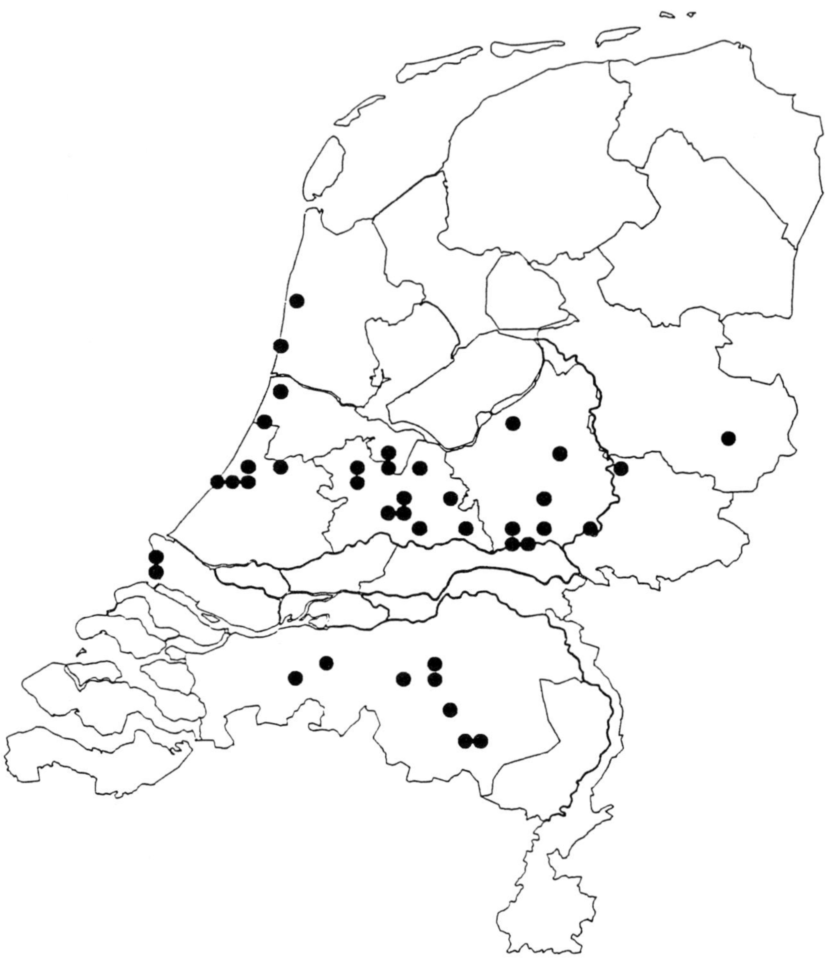

FIG. 9. Distribution of 40 grid units with continuous observations.

Examples of Patterns

Some examples of common patterns of decline or increase of species are given below. In general, species are declining in the Netherlands due to environmental pollution and habitat destruction, and partly also from changed forest management. Increase and decline both appear in all ecological groups. The exact causes and correlation with environmental parameters are not yet known for each species, and a more appropriate method seems to be to recognize and cluster patterns of species instead of considering each species on its own. After this the possible causes for these patterns can perhaps be found.

About 70 species of the selected 450 are suitable for statistical testing with the above described method.

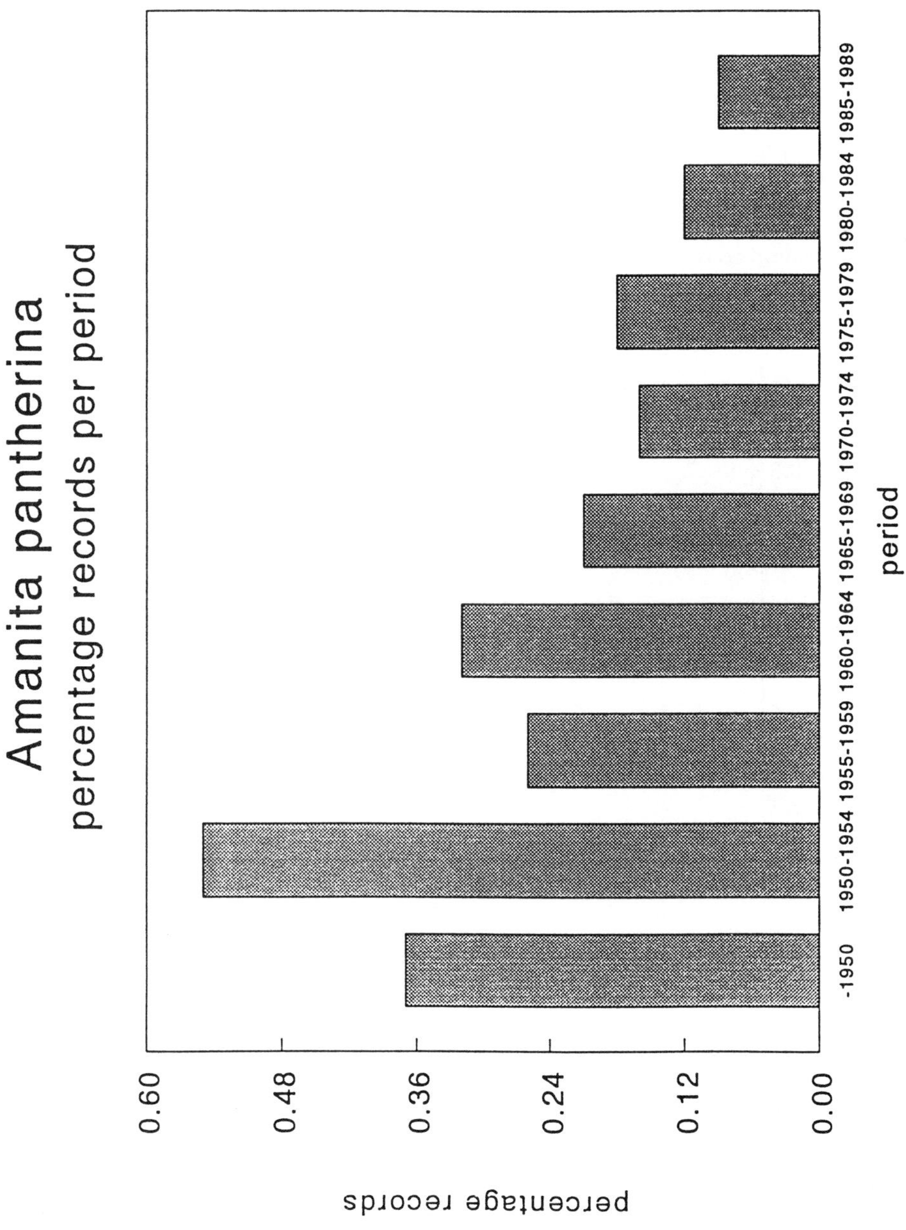

FIG. 10. *Amanita pantherina* – the percentual numbers of records per period from all grid units.

A. Ectomycorrhizal species

Of the selected 185 ectomycorrhizal species, 58% are declining and 26% have not been recorded since 1980, thus more than 80% are declining. Examples of the latter are as follows.

Amanita pantherina (DC: Fr.) Krombh. (Fig. 10) is not mentioned on the Red List (Arnolds, 1989), and is distributed all over the country. It is a rather common species, forming ectomycorrhiza with several trees (probably with a preference for *Quercus*). However, the figures show a tendency to decline (Fig. 10). The high peak in the first half of the 1950s is probably an artefact. For this species the 40 grid unit method is also suitable, and proves a significant decline (Fig. 11).

Another pattern is shown by *Boletus erythropus* (Fr.: Fr.) Krombh. This species is considered threatened (cat. 3) on the Red List (Arnolds, 1989) and is rather commonly distributed all over the Netherlands. It has a preference for Pleistocene soils and inner dune forests, and occurs in the south-eastern part of the Netherlands as well. The decline after the 1960's is rather obvious (Fig. 12). Many ectomycorrhizal species, such as species

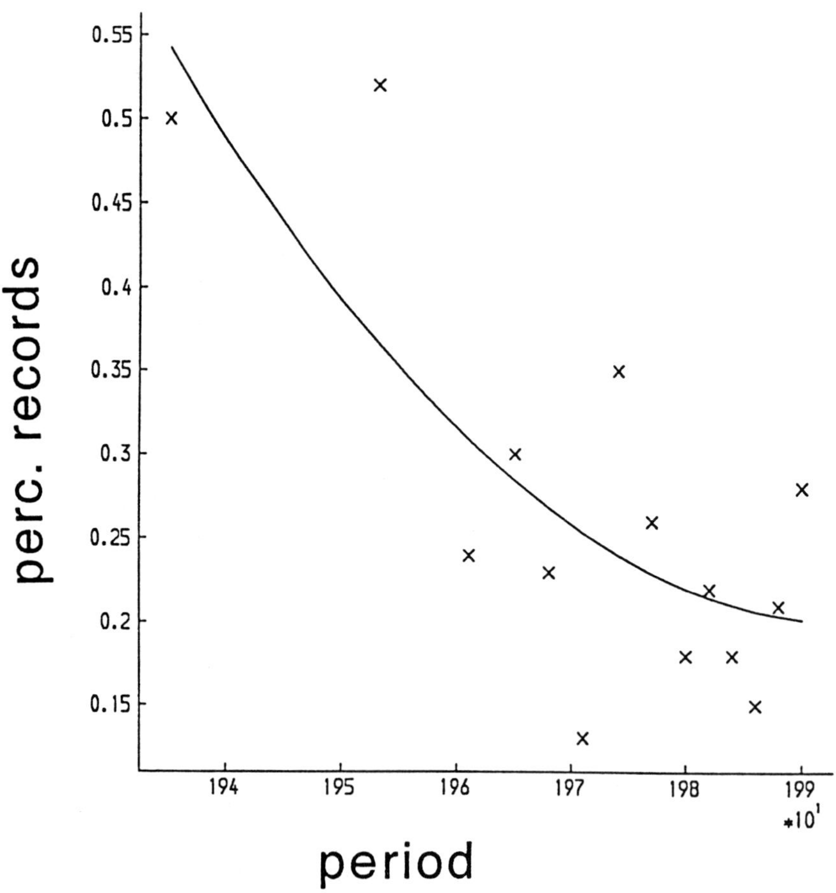

FIG. 11. *Amanita pantherina* – the percentual numbers of records per period in 40 selected grid units, with a fitted line.

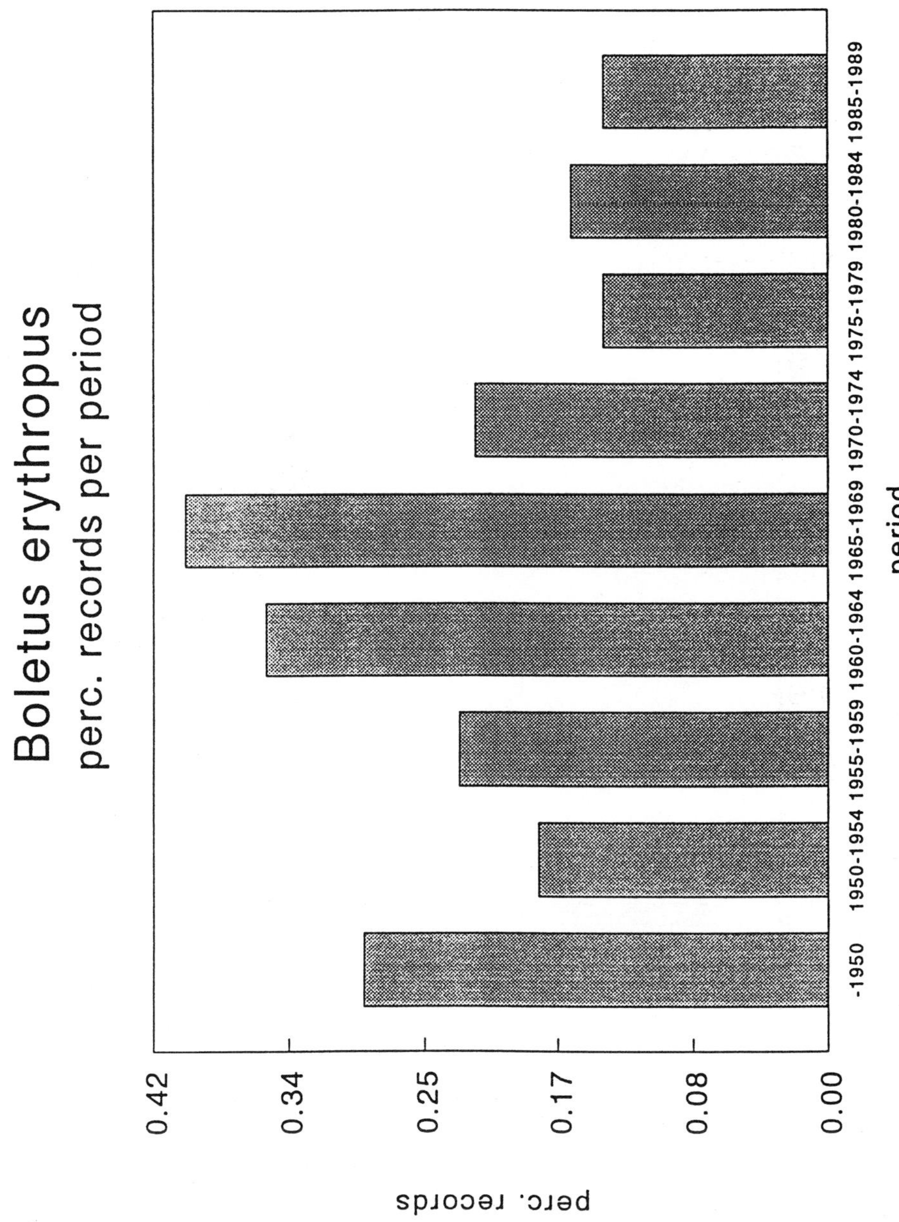

FIG. 12. *Boletus erythropus* – the percentual numbers of records per period from all grid units.

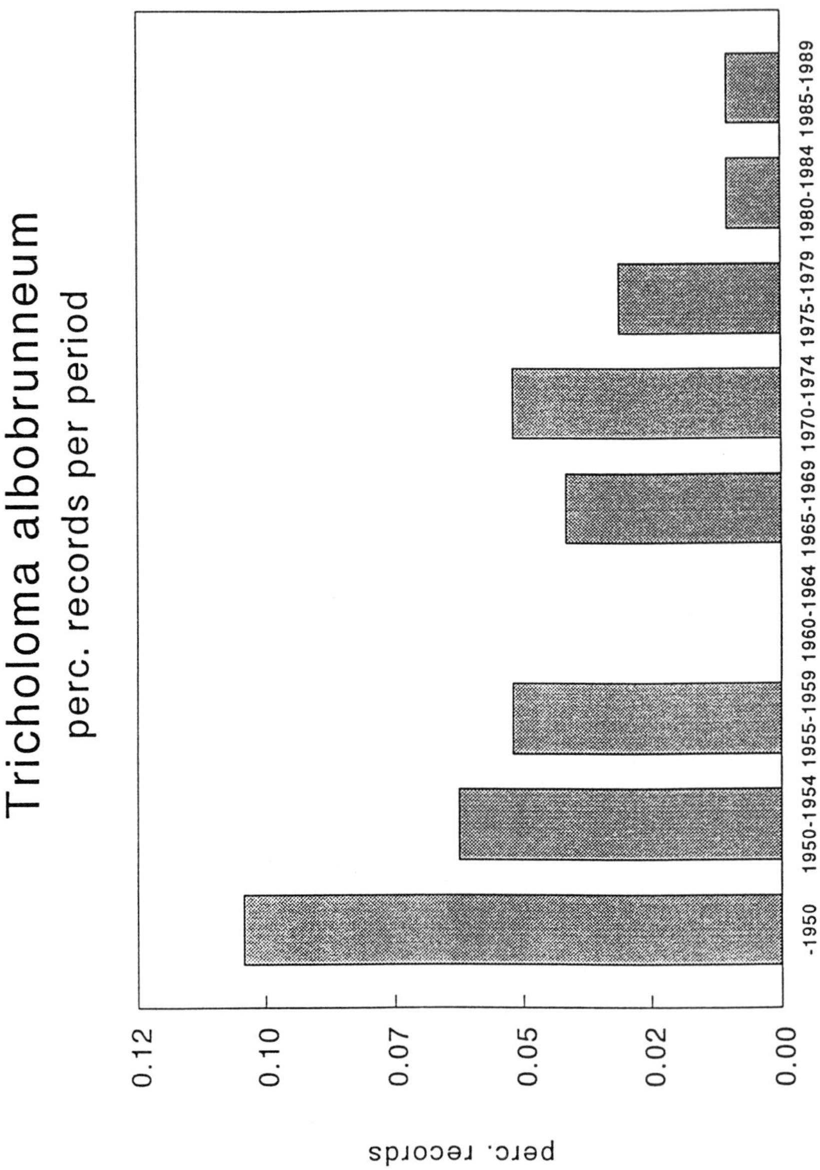

FIG. 13. *Tricholoma albobrunneum* – the percentual numbers of records per period from all grid units.

of *Lactarius* DC ex S.F. Gray and *Russula* (Pers.: Fr.) S.F. Gray, show this same kind of pattern of decline since 1970 (see also Nauta & Vellinga, 1992).

Tricholoma albobrunneum (Pers.: Fr.) Kummer is a rather rare species, occurring only on nutrient-poor sandy soils (Veluwe, Noord-Brabant) in open pine-forests where it forms ectomycorrhiza with *Pinus*. It is a strongly declining species (Fig. 13) and is regarded as a strongly threatened species on the Red List (Arnolds, 1989). The distribution maps show that the species has not been recorded from Noord-Brabant after 1975 (Fig. 14a & 14b). The probable cause for this is air-pollution and the consequent changes in undergrowth of the forests. However, it has recently appeared in the new polders.

An example of an increasing ectomycorrhizal species is the very common *Paxillus involutus* (Batsch: Fr.) Fr. It grows in various kinds of woods, where it can also live as a saprophyte. The pattern shows a general increase (Fig. 15).

B. Non-fertilized grassland

Seven per cent of the selected species of non-fertilized grassland have not been recorded after 1980, 58% are declining, and only 7% are increasing.

Anellaria semiovata (Sow.: Fr.) Pearson & Dennis (*Panaeolus fimiputris* (Bull.: Fr.) Quél.) is also considered threatened on the Red List (cat. 3, Arnolds, 1989). The species shows an

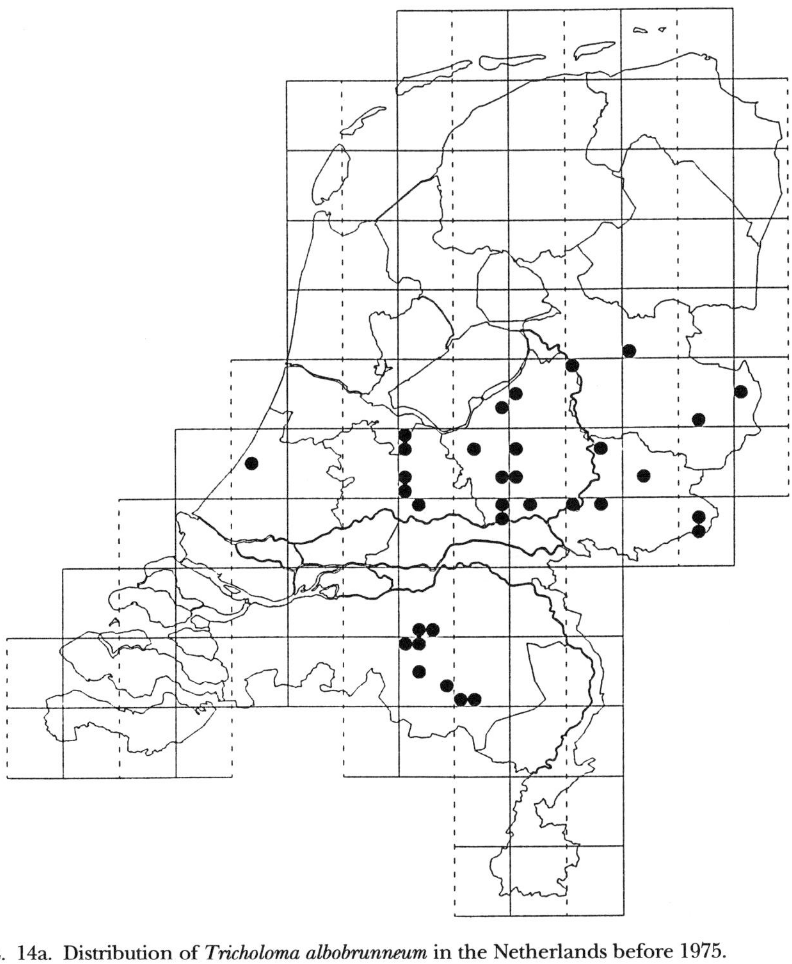

FIG. 14a. Distribution of *Tricholoma albobrunneum* in the Netherlands before 1975.

undulating declining curve, but has a tendency to recover since 1970 (Fig. 16). In earlier times this species was rather widespread and common in the Netherlands, but declined seriously due to intensified cattle-breeding. It is distributed now again all over the Netherlands, on rather poor soils. It occurs on cow and horse dung, and the recent increase in number of records as well as number of grid units is due to grazing by cows and horses in nature reserves.

In particular species of *Hygrocybe* (Fr.) Kummer, *Camarophyllus* (Fr.) Kummer and *Geoglossum* Pers. are declining. An example is *Camarophyllus pratensis* (Pers.: Fr.) Kummer, which is rather rare, occurring in poor grasslands on moist to dry, acid to basic loam and sand, and occasionally in forests. It shows a general decrease (Fig. 17). On the Red List it is threatened (cat. 3, Arnolds, 1989), but it is probably more common abroad (Arnolds, 1990).

C. Litter and wood saprophytes

Of the litter and wood saprophytes, 4% are not recorded after 1980, 40% are declining, and 33%, mainly wood saprophytes, are increasing.

An example of a litter saprophyte is *Phallus impudicus* L.: Pers. The pattern shows a stable to rising curve (Fig. 18). This species is very common in several forests with a thick

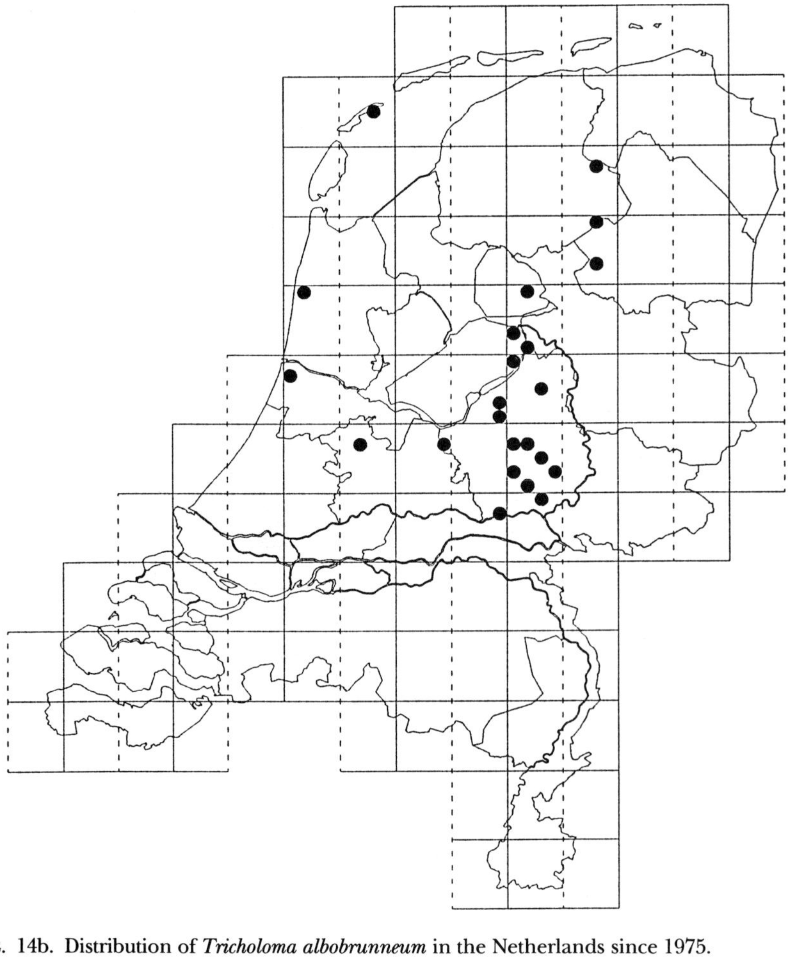

FIG. 14b. Distribution of *Tricholoma albobrunneum* in the Netherlands since 1975.

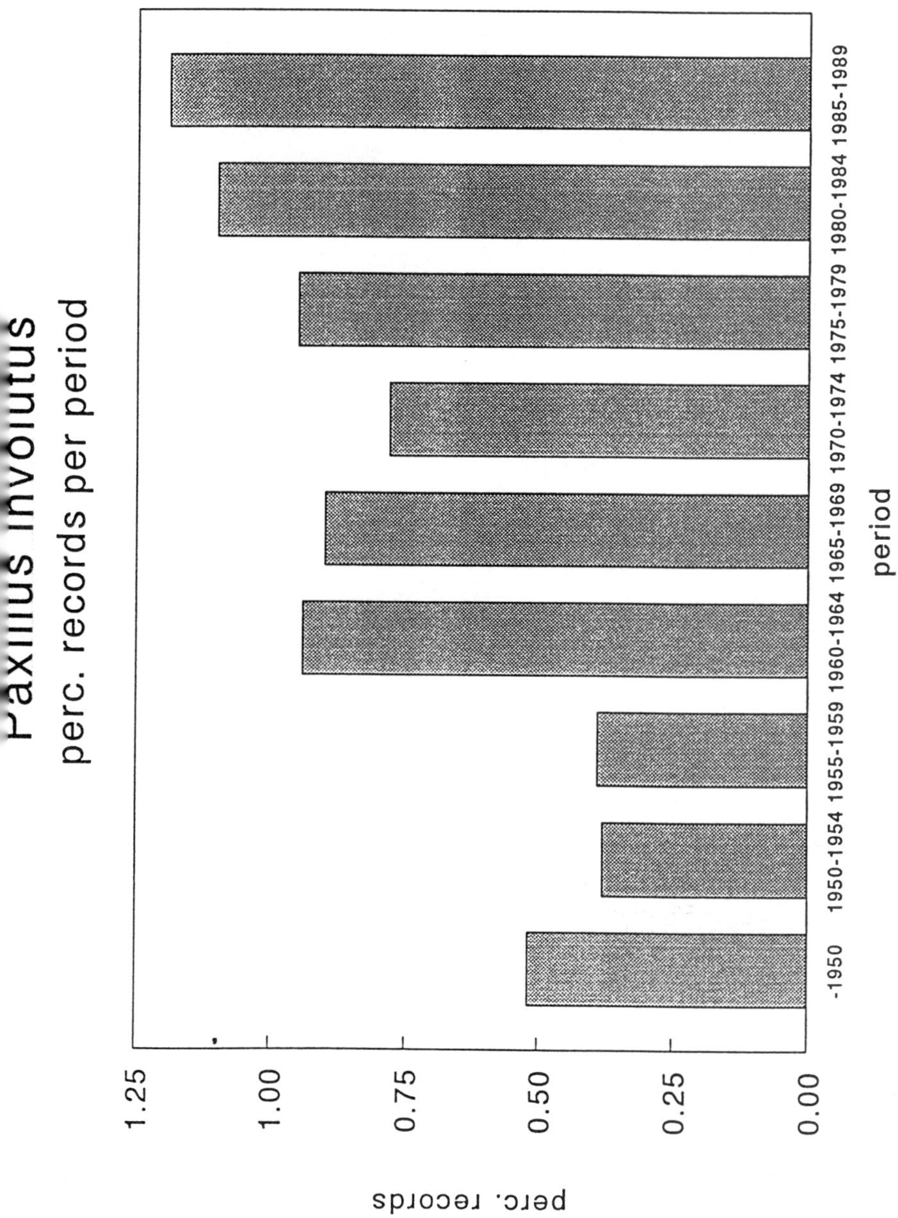

FIG. 15. *Paxillus involutus* – the percentual numbers of records per period from all grid units.

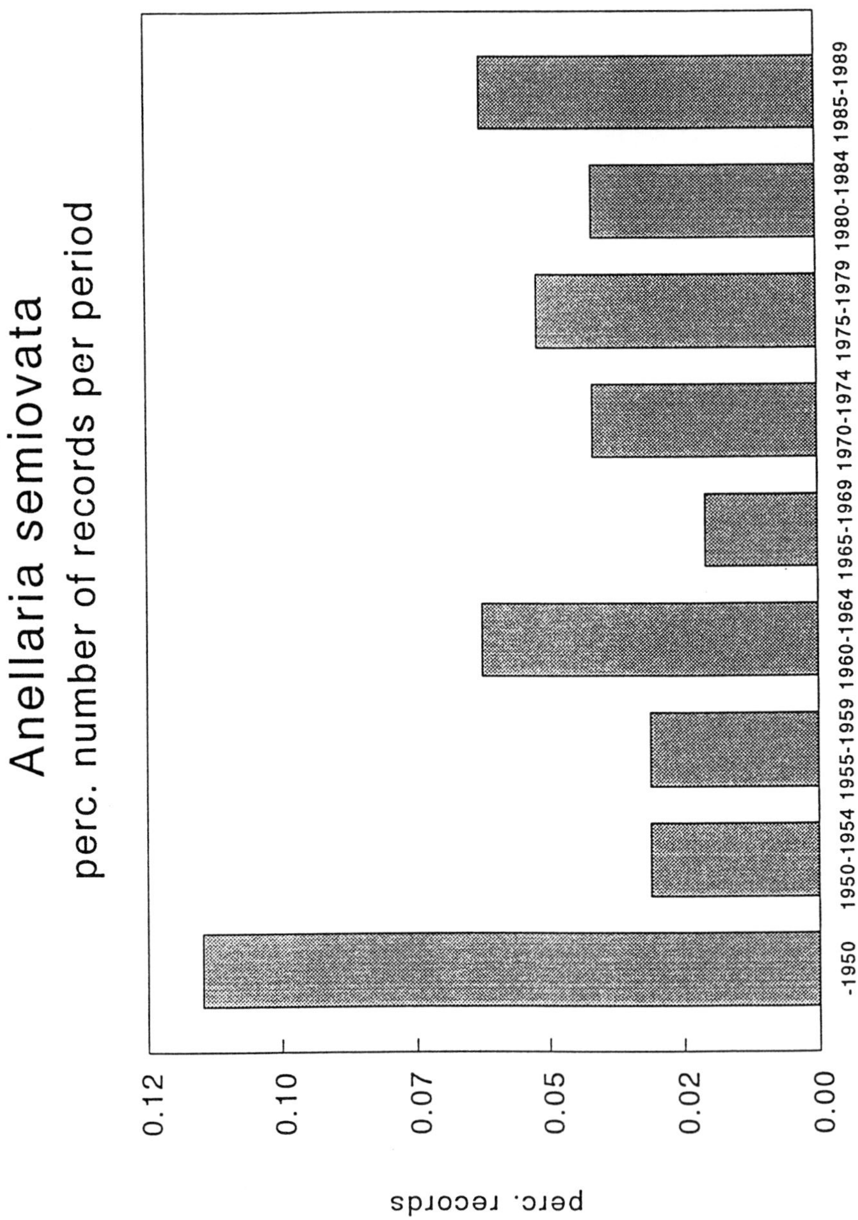

FIG. 16. *Anellaria semiovata* – the percentual numbers of records per period from all grid units.

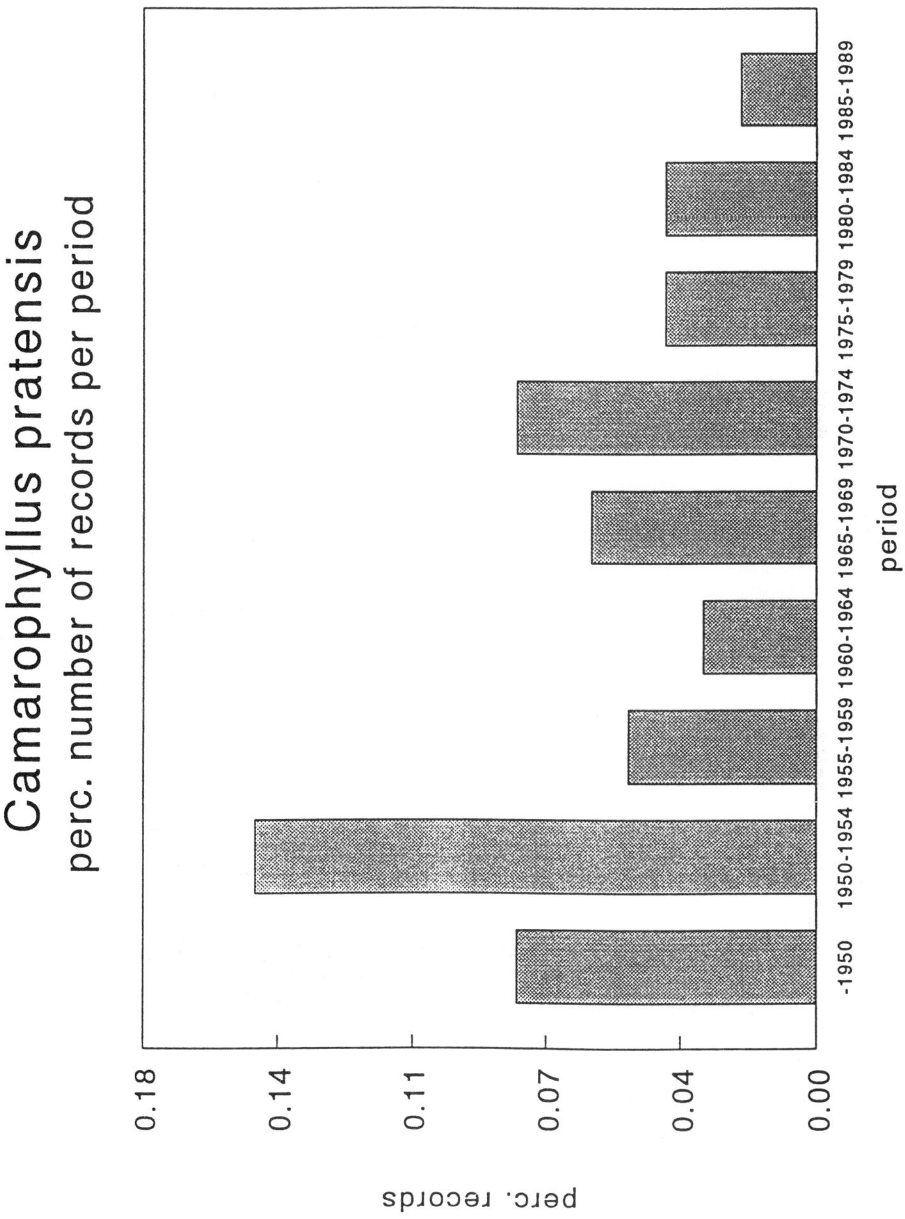

FIG. 17. *Camarophyllus pratensis* – the percentual numbers of records per period from all grid units.

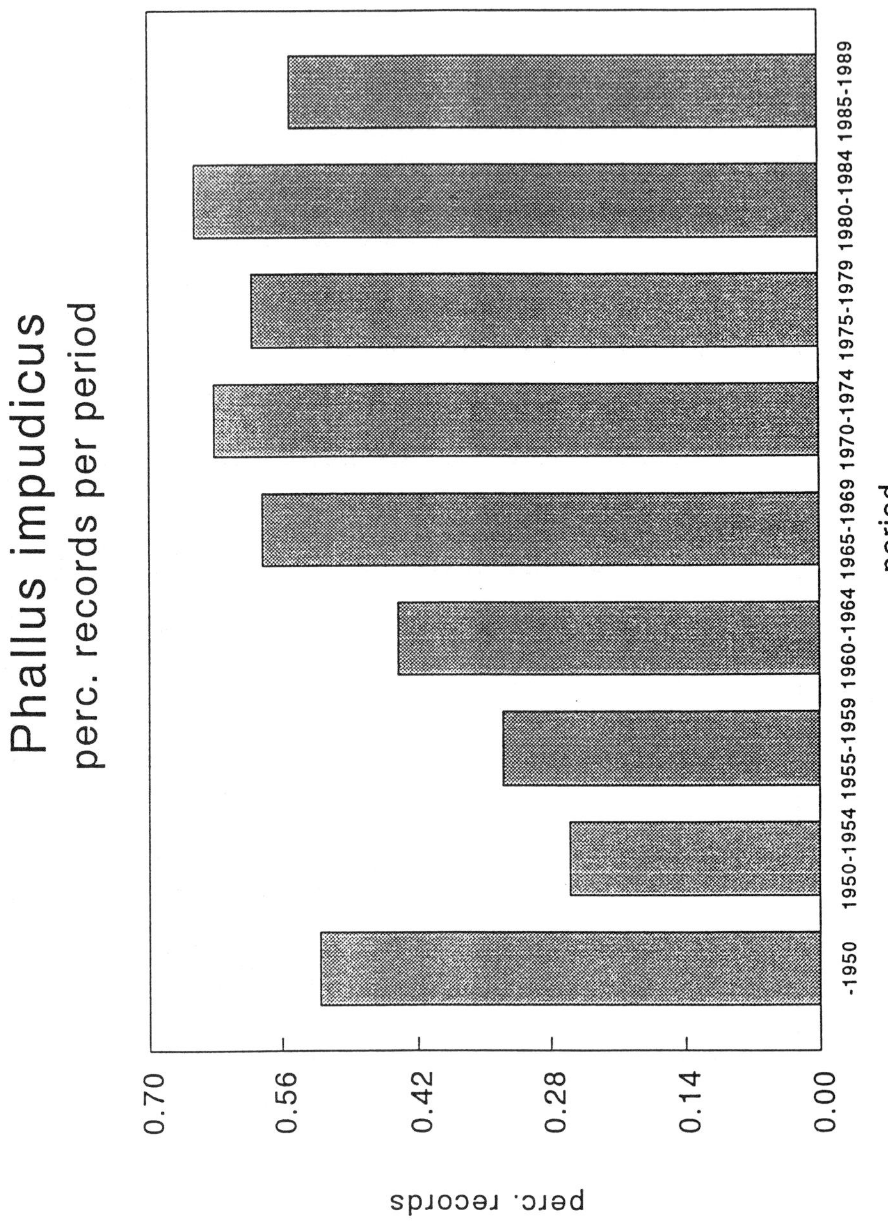

FIG. 18. *Phallus impudicus* – the percentual numbers of records per period from all grid units.

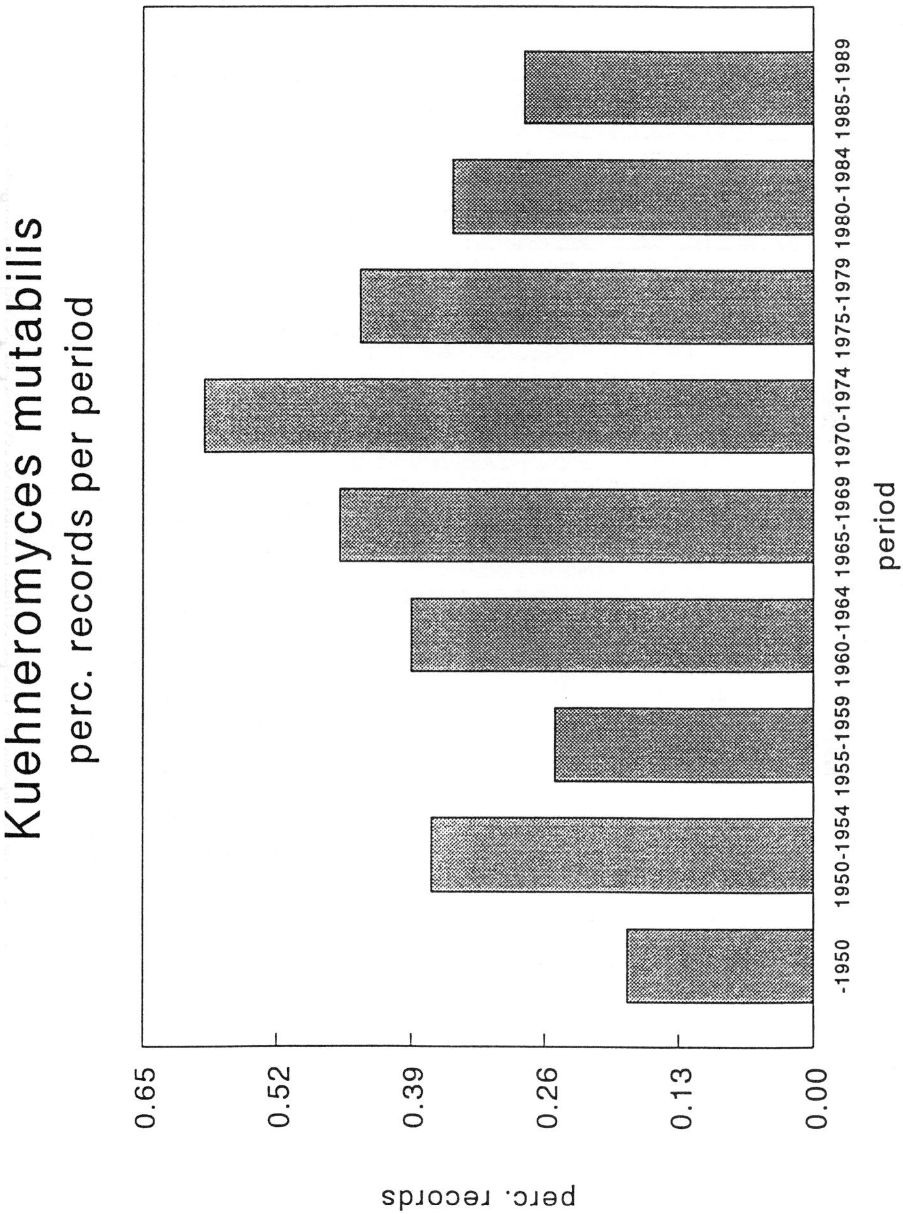

FIG. 19. *Kuehneromyces mutabilis* – the percentual numbers of records per period from all grid units.

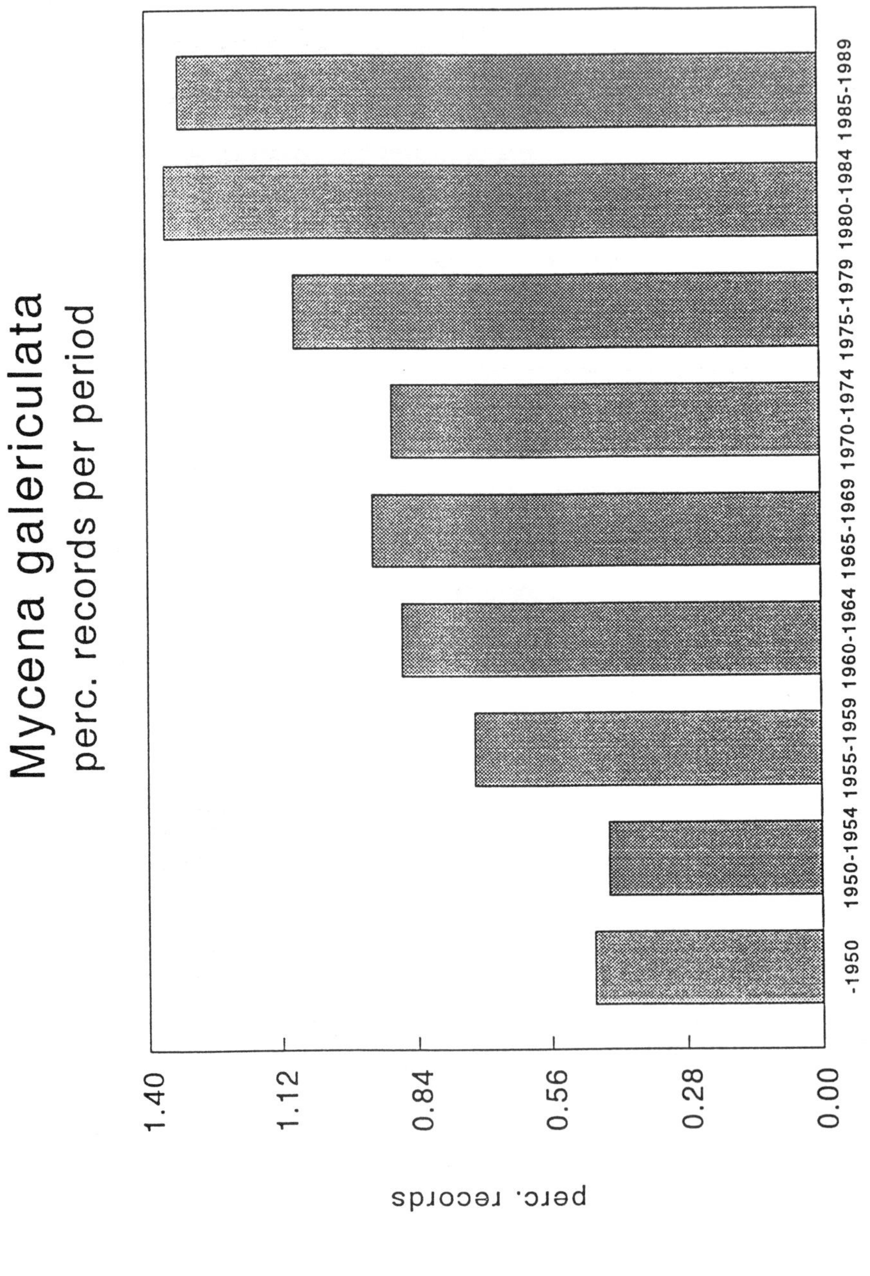

FIG. 20. *Mycena galericulata* – the percentual numbers of records per period from all grid units.

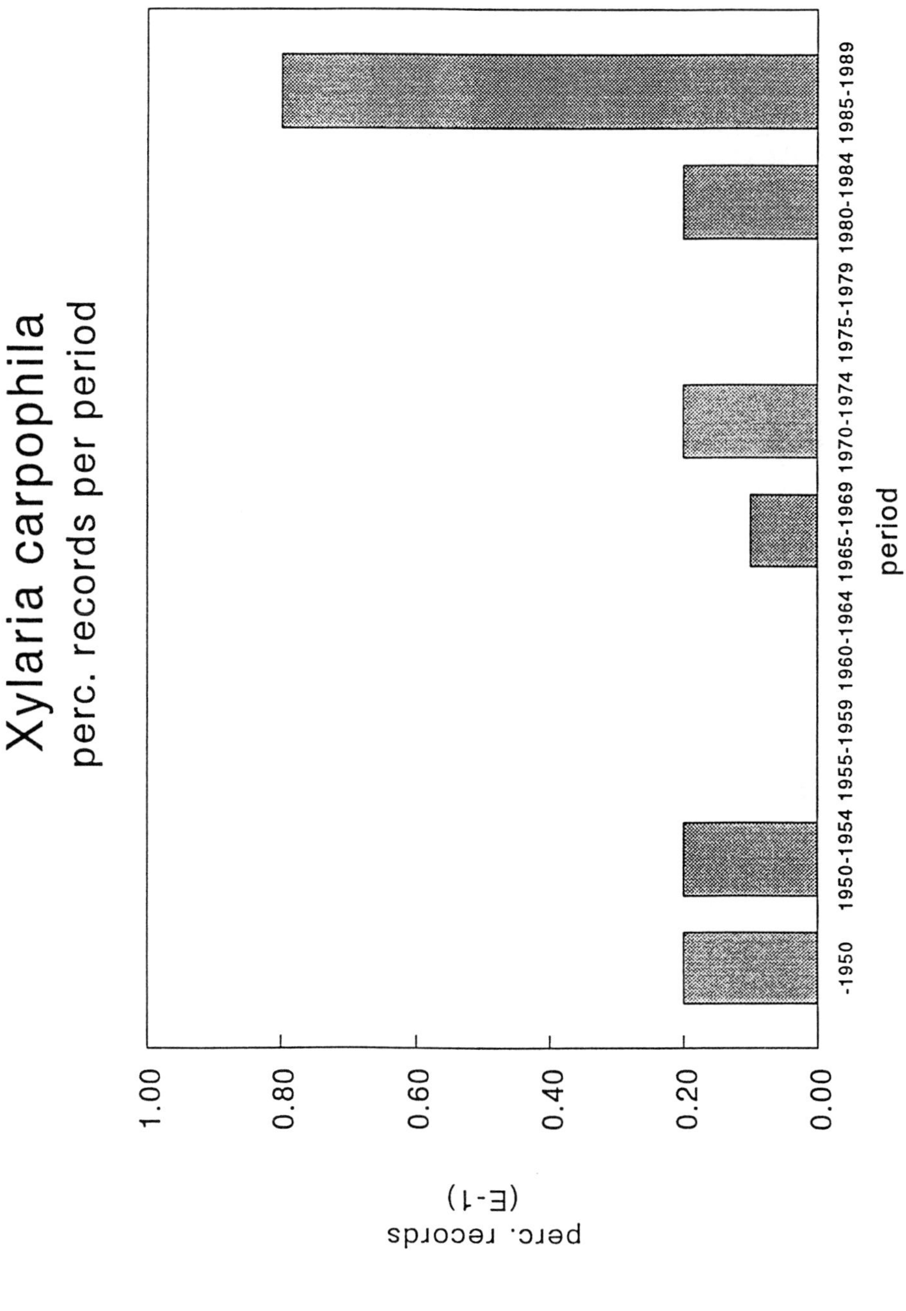

FIG. 21. *Xylaria carpophila* – the percentual numbers of records per period from all grid units.

layer of litter.

Of the species occurring on wood the common *Kuehneromyces mutabilis* (Schaeffer: Fr.) Singer & A.H. Smith shows a typical pattern (Fig. 19). It reveals a decline after 1975 of which the cause is not known. Perhaps this is due to changing forestry. The species occurs on stumps of deciduous trees. Another declining species is *Lenzites betulina* (L.) Fr.

Mycena galericulata (Scop.: Fr.) S.F. Gray shows an obvious increase in the country as a whole (Fig. 20), and also in the selected 40 grid units. It is widespread (see for a distribution map Nauta & Vellinga, 1992), and nowadays one of the most common macrofungi in the Netherlands, occurring on stumps, twigs, and branches in all kinds of habitats.

Xylaria carpophila, a common species on cupules of *Fagus* also shows a rising curve (Fig. 21). This species was, however, scarcely known in the past, which may well be the reason for the noted increase.

Conclusion

In general these correction methods are an attempt to use the data concerning macrofungi for nature management, nature conservancy, and to give an opinion concerning the state of nature. Using percentages of numbers of records is a quick, easy and reliable method to get insight in increase or decline of a species. The use of percentages of numbers of records from selected, well investigated, grid units also enables statistical testing. It is not the only available method, however, to calculate changes in the mycoflora. Another method, for example, is a comparison of foray reports of a limited number of areas in two periods (Arnolds, 1985: 12–24), which also enables statistical testing. Using all foray reports in the Netherlands the relative foray frequency can be calculated (Arnolds & Jansen, 1992). Statistical testing is, however, not possible with this method since the effect of the different geographical distribution of data in various periods is too large. Nevertheless, the general results of these methods are largely similar.

Possible causes driving the observed patterns of changes of the mycoflora will be investigated in future research.

Acknowledgements

The data are based on the records stored in the database of the Biogeographical Information Centre. The records are collected by the contributors of the mapping project who are hereby warmly thanked; the data are owned by the Dutch Mycological Society. The data were available through the co-operation of the BIC and the Institute for Forestry and Nature Research. The atlas project is subsidized by the Ministry of Agriculture, Nature Management and Fisheries.

References

Arnolds, E. (editor) (1985). Veranderingen in de paddestoelenflora (mycoflora). *Wetensch. Meded. Kon. Ned. Natuurhist. Ver.* **167**: 1–101.

Arnolds, E. (1989). A preliminary red data list of macrofungi in the Netherlands. *Persoonia* **14**: 77–125.

Arnolds, E. (1990). Hygrocybeae. In C. Bas, Th.W. Kuyper, M.E. Noordeloos and E.C. Vellinga (editors). Flora agaricina neerlandica. Vol. 2, pp. 70–133. A.A. Balkema, Rotterdam/Brookfield.

Arnolds, E. and Jansen, E. (1992). New evidence for changes in the macromycete flora of the Netherlands. *Nova Hedwigia* **55**: 325–351.

Jansen, E., Nauta, M. and Vellinga, E. (1992). Paddestoelen karteren. *Natura* **89**: 202–203.

Nauta, M. and Vellinga, E.C. (1992). Towards a distribution atlas of macrofungi in the Netherlands. *Mycologist* **6**: 6–10.

Schmitt, J.A. (1993). Fruiting period changes of selected agarics in the last 20 years in Saarland. In D.N. Pegler, L. Boddy, B. Ing & P.M. Kirk (Editors). Fungi of Europe: Investigation, Recording and Conservation, pp. 47–69 Royal Botanic Gardens, Kew.

5. FRUITING PERIOD CHANGES OF SELECTED AGARICS IN THE LAST 20 YEARS IN SAARLAND

J.A. SCHMITT

Fachrichtung Biochemie, Universität des Saarlandes, D-6600 Saarbrücken, Germany

Abstract

One hundred of the most widespread species of Agaricales (s.l.) were investigated between 1970 and 1989 with respect to their monthly fruiting periods, evaluating about 10,000 excursions. Linear regression of the beginning and the end of fructification in each of the twenty years documents a trend of change, if the regression lines show slopes different from zero. Only 18% of the species showed no significant changes in their Yearly Fruiting Periods (FP). In 82% of species, the FP shifted and/or shortened (41%) or enlarged (34%). A seasonal shift of the FP from autumn to winter can be explained, for example, by a lower availability of photosynthetic products from woody plants for mycorrhizal fungi in the growing season. The shortened FP document in general a deterioration of life conditions. The enlarging of the FP in 34% of fungal species is in no case combined with an increase of populations or fructifications, that means, with no improvement of life conditions. Parallel to this investigation, for each of the 100 species the Monthly Fructification Incidence (MFI) in the 20 years-period is given. Only 60% of them show a MFI of 100%, that means they were found in each year in one or more definite months.

Introduction

Under deteriorating environmental conditions for a fungus species, its populations, reproduction rates and also the growing and fruiting periods during a year are influenced, that means shortened and/or shifted as a special aspect of fungus decline. The decline of fructifications of many fungus species is well known in many European countries according to their published Red Lists and by the work of the "European Council for the Conservation of Fungi". A summarizing overview was given by Arnolds (1991), but little is known on changes in yearly fruiting periods.

In the Saarland region of South-western Germany, the fungus flora has been investigated intensively over the last 40 years or so; a summary of this work is given in the two tomes of our "Atlas of the Fungi of Saarland", see Derbsch & Schmitt (1984, 1987) and in the six parts of a project, supported by our Ministry of Environment, entitled: "Supply and Dynamics of the Fungus Flora in Selected Areas of the Saarland in Longtime Observation" (Schmitt 1990/91). On the basis of these data, this paper is the summary of part four of this project entitled: "Fruiting Period Changes of Selected Agarics in the last 20 Years in Saarland".

Results and Discussion

In the years between 1970 and 1989, about 11,000 field excursions in different areas of the Saarland, distributed over all months of the year, have been carried out to document the fungus flora. These studies could only have been undertaken with the assistance of my friends and colleagues and I wish to thank them herein for their kind help.

Fig. 1a: Yearly Fruiting Periods FP of
Amanita vaginata

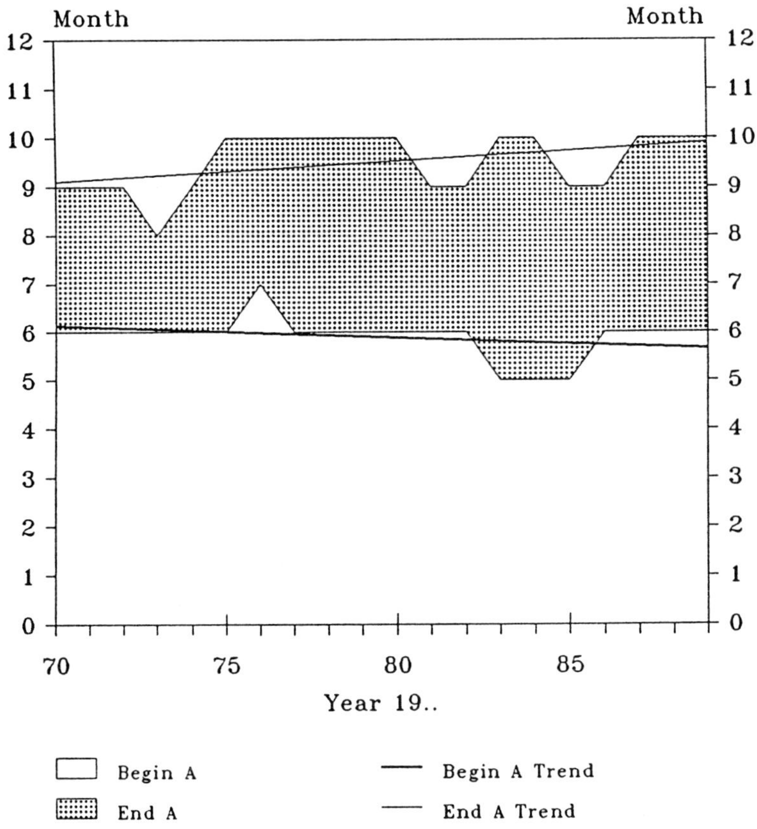

Datei: Kew3b

For the investigation of changes in the fruiting periods, the most widespread agarics were first selected, because the needed data on them are numerous and of statistical relevance. One hundred of the most widespread species of Agaricales s.l. (Agaricales, Boletales, Russulales and Polyporales), which were found, i.e. noted as found species in an excursion list, in each of the twenty years and summaries for more than about 200 in the 20 years. The 100 species belong to different ecological groups: half are mycorrhiza-forming symbionts, the others are saprophytes. For each species in each year, the months of detected fructification are noted and the data of the Yearly Fruiting Periods (FP) documented graphically as the example of *Amanita vaginata* (Bull.: Fr.) Vittad. in Fig. 1a on the basis of monthly fructification. A more detailed documentation, based on weekly

Fig. 1b: Monthly Fructification Incidence MFI of Amanita vaginata

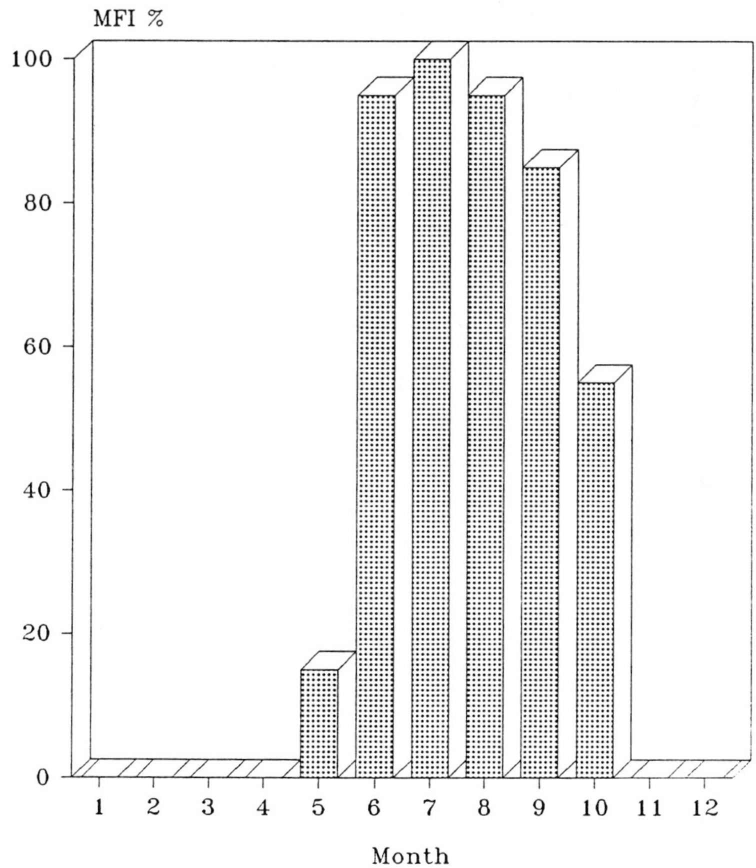

Datei: Kew3b

Fig 2a: Yearly Fruiting Periods FP of
Polyporus brumalis

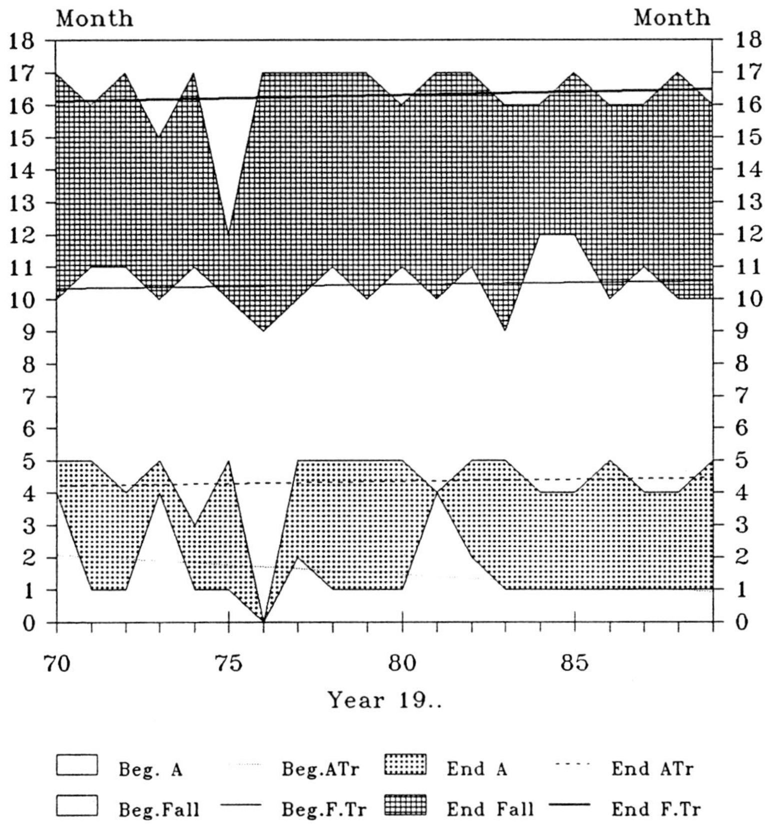

Datei: Kew5a

fructification was not possible, because for instance in the summer holiday time the observations were insufficient. The trends are the linear regression lines of the beginning or the end of all Yearly Fruiting Periods in the twenty years range. The area between the two trend lines documents the mean fruiting period per year. For *Amanita vaginata* as a non-highly specific ectomycorrhiza-forming species, a slight widening of the yearly fruiting period from 4 to 5.5 months is seen.

Now, a new term must be explained, the so called "Monthly Fructification Incidence", MFI. For example, *Amanita vaginata* was found in each of the 20 years in September, but only in 18 of the 20 years in October (Fig. 1b). The MFI is defined as the incidence in percentage of the fruiting in a selected month, referred to all years of the investigation. The MFI as probability of the Monthly Fructification for September is therefore 100%, the

Fig. 2b: Monthly Fructification Incidence MFI of Polyporus brumalis

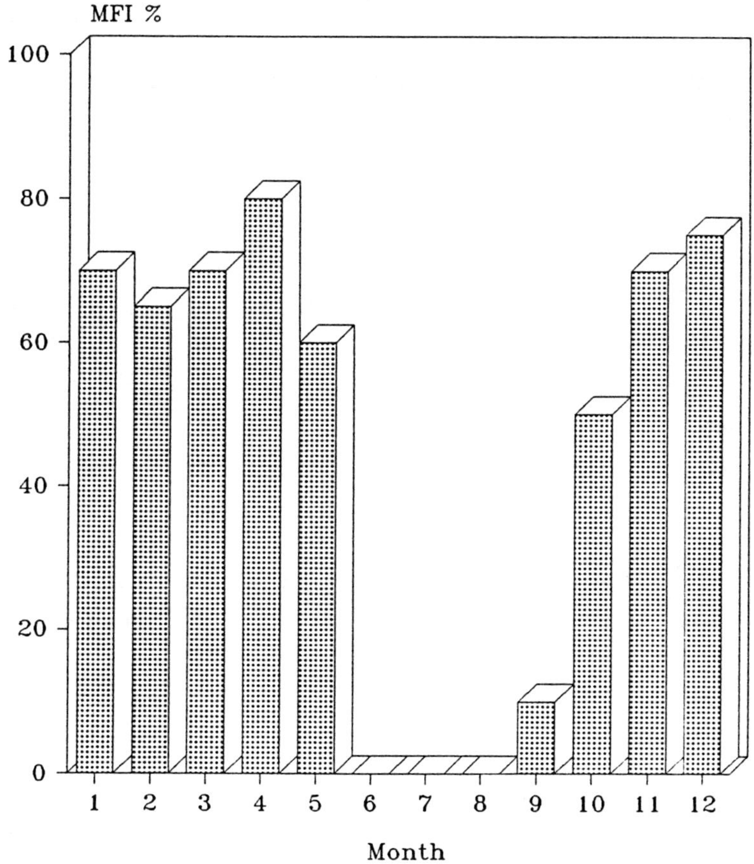

Datei: Kew5b

Fig. 3a: Yearly Fruiting Periods FP of
Polyporus lepideus

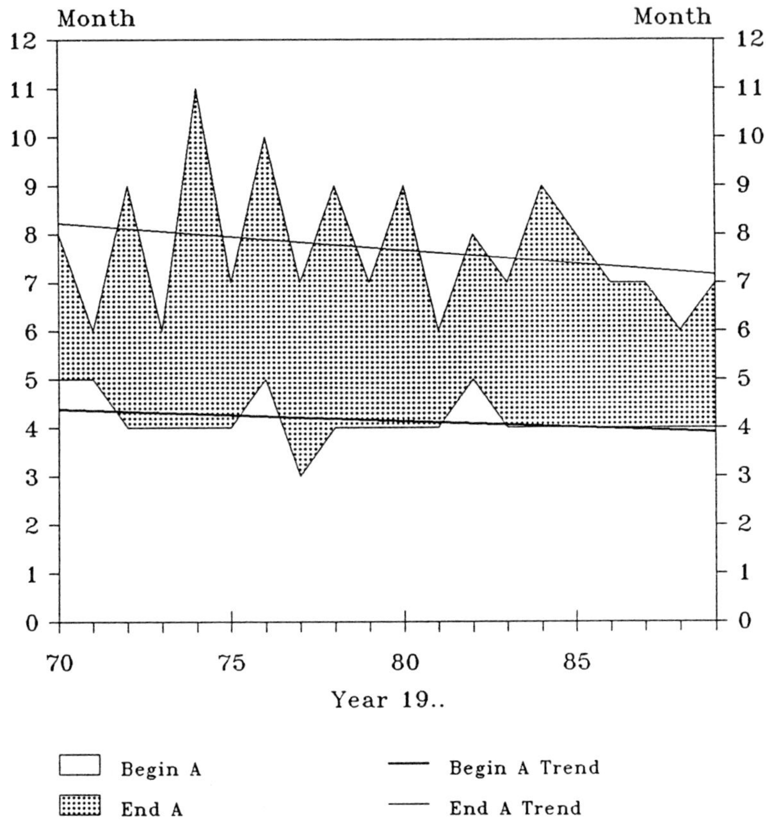

Datei: Kew6b

MFI for October 90%. This MFI data are a good complementation to the data of Monthly Fruiting Periods, showing the whole range of fructification in the 20 years period. *Amanita vaginata* shows a 100%-maximum of the Monthly Fructification Incidence in the case of July. The following examples show different types of changes or variation in width and position of fruiting periods of agarics from different ecological groups.

In many cases of fungi, which are fruiting over a year, from autumn to spring, the graphs show the fruiting periods beyond December as months numbered 13 for January, 14 for February and so on for the following year: see the example *Polyporus brumalis* (Pers.) Fr. (Fig. 2a), a lignicolous saprotroph on 12 different deciduous wood species in Saarland. The Yearly Fruiting Period, ranging from October to May, does not show any

Fig. 3b: Monthly Fructification Incidence MFI of Polyporus lepideus

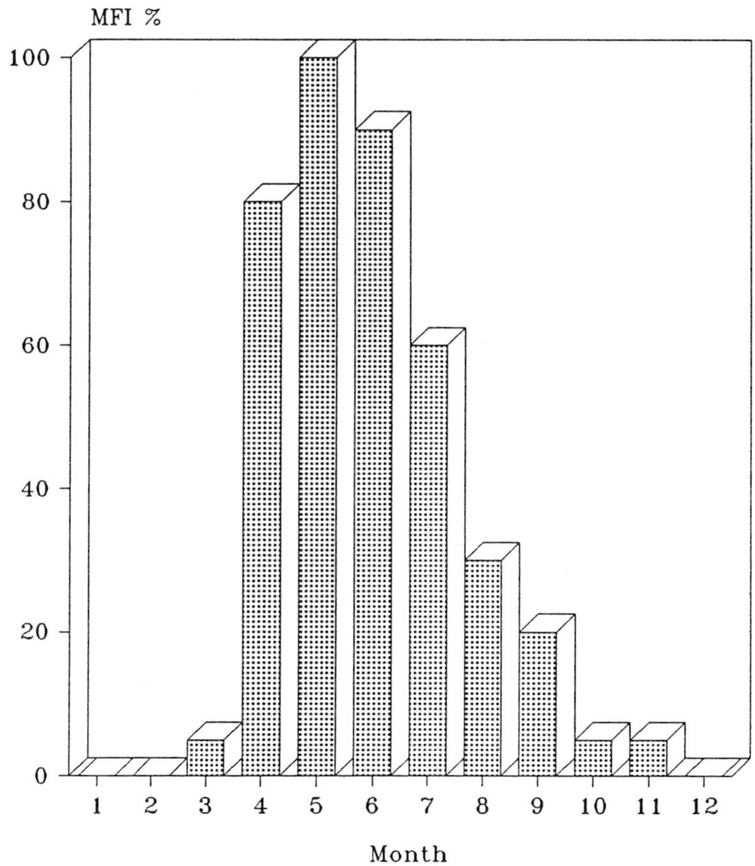

Datei: Kew6a

Fig. 4: Monthly Fructification Incidence MFI of Polyporus lepideus, P. brumalis

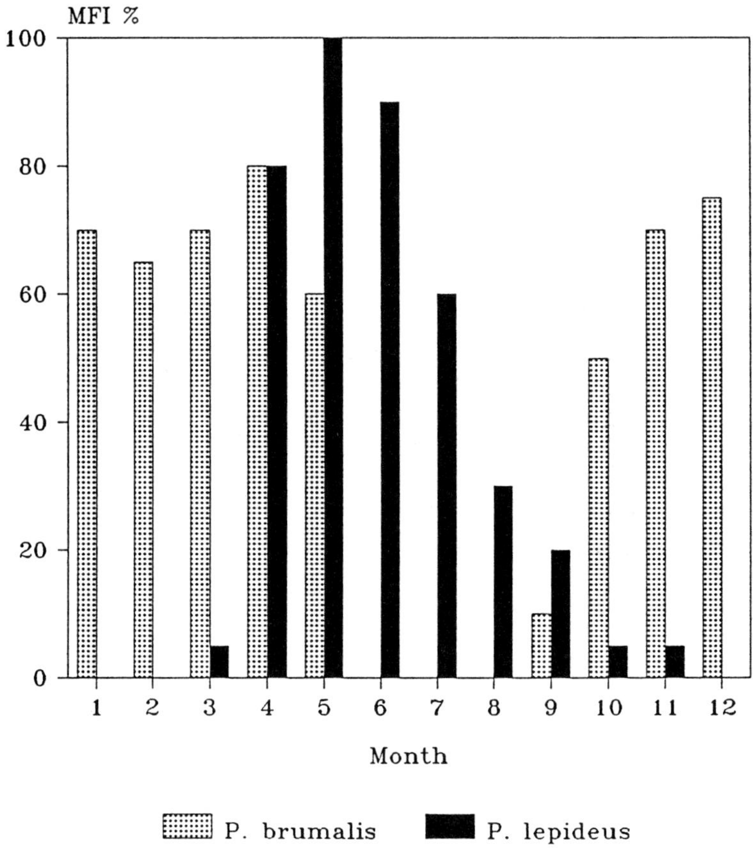

Datei: Kew6c

changing trend. Also in this example an often oscillating beginning and end of yearly fruiting periods is seen, dependent in particular on weather conditions in the different years. Reduced fruiting in dry summers or autumns, as in the years 1976 or 1981, is not yet fixed in the following normal year. But statistically, no significant changes in the 20 years-range are indicated, the trend lines showing a slope zero. Only if the trend lines show slopes significantly greater or lower than zero, is a change of fruiting periods significant. When considering the Monthly Fructification Incidence, two significant fruiting maxima in April and December are seen (Fig. 2b).

In contrast the situation of *Polyporus lepideus* Fr., also a lignicolous saprotroph with a comparable substrate spectrum, is documented in Fig. 3a. The mean fruiting period ranges from April to August and shows a significant shift in the last 20 years from the late summer to mid summer, combined with a shortening of around one month.

Fig. 5a: Yearly Fruiting Periods FP of
Russula fellea

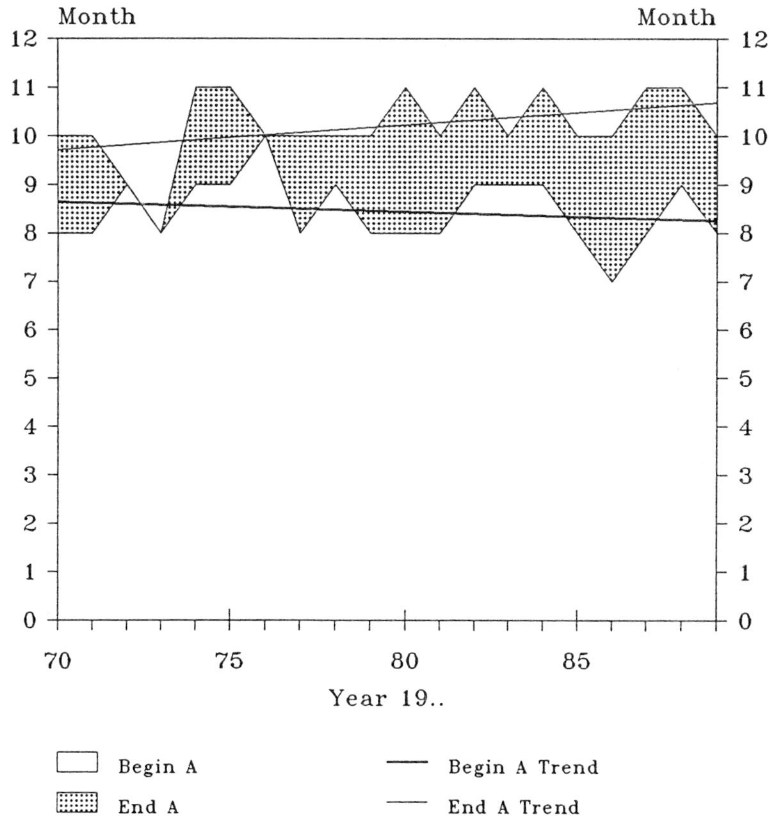

Datei: Kew7b

Fig. 5b: Monthly Fructification Incidence MFI of Russula fellea

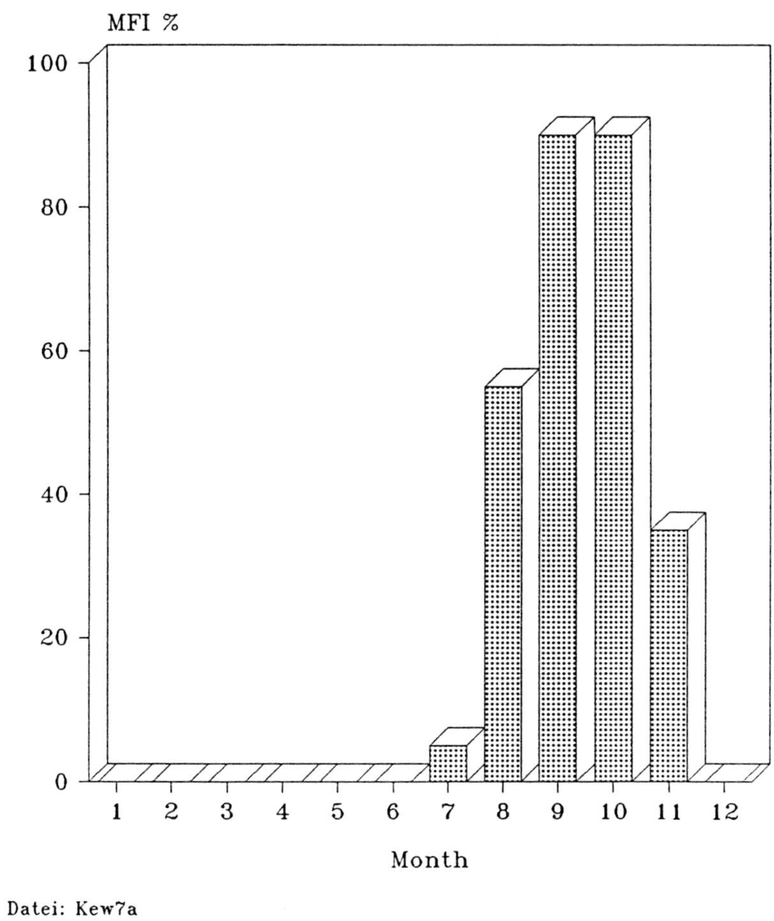

One maximum and a Monthly Fructification Incidence of 100% only in May is seen, with a tailing up to November (Fig. 3b). In Fig. 4, the Monthly Fructification Incidence of the two *Polyporus* species in one graph show the difference of both species and the overlapping parts in the fruiting periods.

Ectomycorrhizal fungi respond differently. The pattern for *Russula fellea* (Fr.: Fr.) Fr., an ectomycorrhizal symbiont mainly on *Fagus* and very seldom of *Carpinus* or *Picea* in Germany, is demonstrated in Fig. 5a. The small Yearly Fruiting Period is widened by around one month, i.e. 50% in 20 years. In no month did this fungus reach a Monthly Fructification Incidence of 100% (Fig. 5b).

Fig. 6a: Yearly Fruiting Periods FP of Russula delica

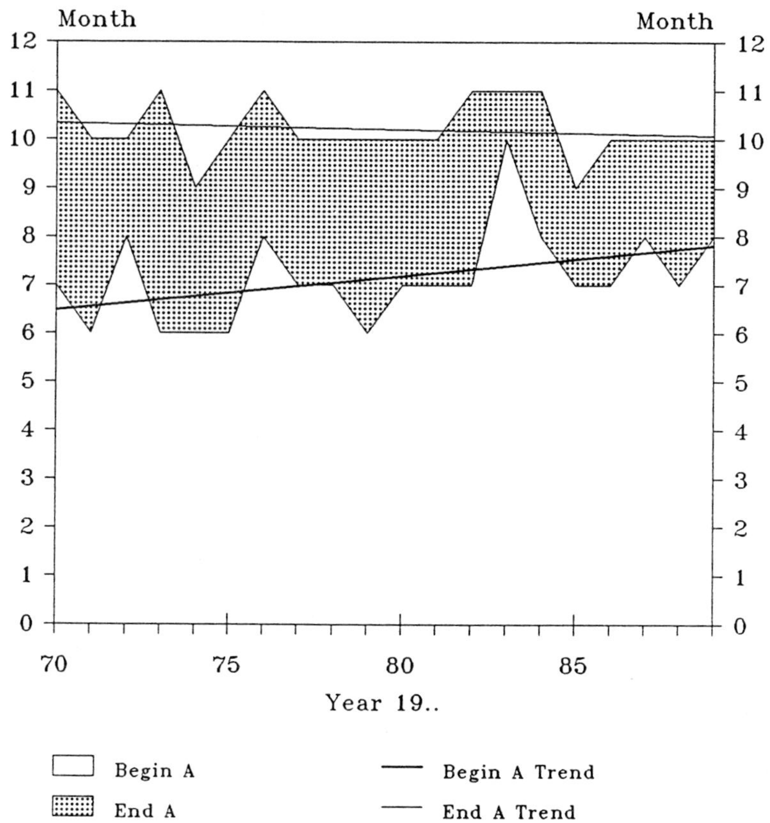

Datei: Kew8b

Fig. 6b: Monthly Fructification Incidence MFI of Russula delica

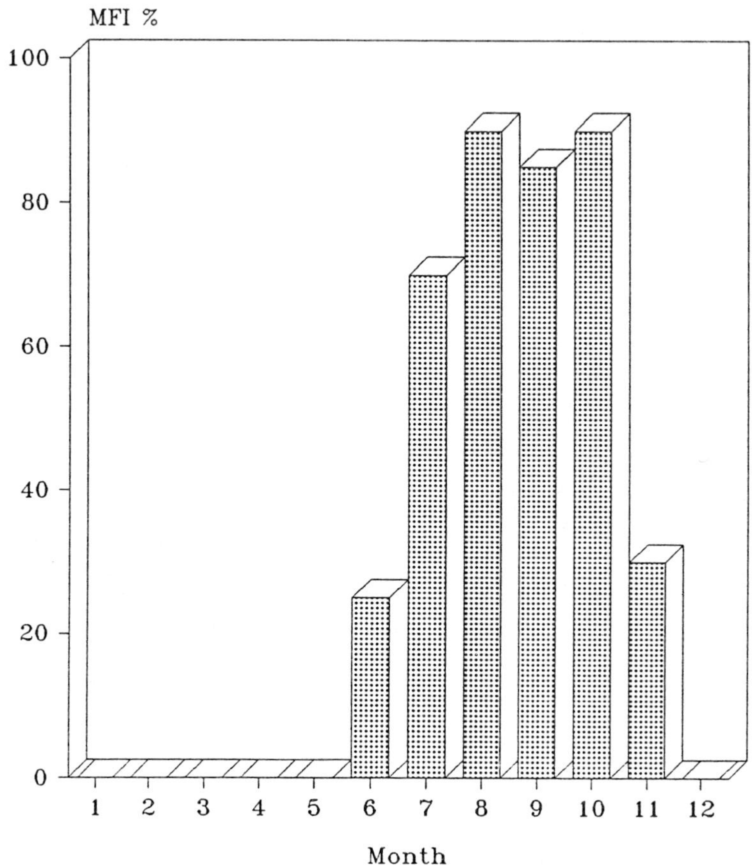

Datei: Kew8a

In contrast, *Russula delica* Fr, an ectomycorrhizal symbiont of *Fagus* and *Quercus* and possibly also *Carpinus,* a significant contracting trend of the Yearly Fruiting Period around two months is found (Fig. 6a). The Monthly Fructification Incidence shows a broad maximum from August to October (Fig. 6b).

In Fig. 7b the situation for a litter decomposing species, *Lepista nuda* (Bull.: Fr.) Cooke, is shown. This species is able to fruit over the whole year, with a lower Monthly Fructification Incidence maximum in May and a main 100%-maximum in November. In the last 20 years, the yearly fruiting periods of this species (Fig. 7a) show a significant trend to smaller periods (more than 50% reduction), the commencement of fruiting has shifted by about 1.5 month to summer whilst the end of fruiting goes back from spring to winter. In Figs. 8a and 8b, the data for one of the most widespread fungus species of all with the largest Yearly Fruiting Period, *Tubaria furfuracea* (Pers.: Fr.) Gillet, is shown. It is one of the few species, together with *Stropharia aurantiaca* (Cooke) P.D. Orton, with enlarging fructifications in the last years, caused by the increased use of shredded woody material in gardens and parks.

Fig. 7a: Yearly Fruiting Periods FP of
Lepista nuda

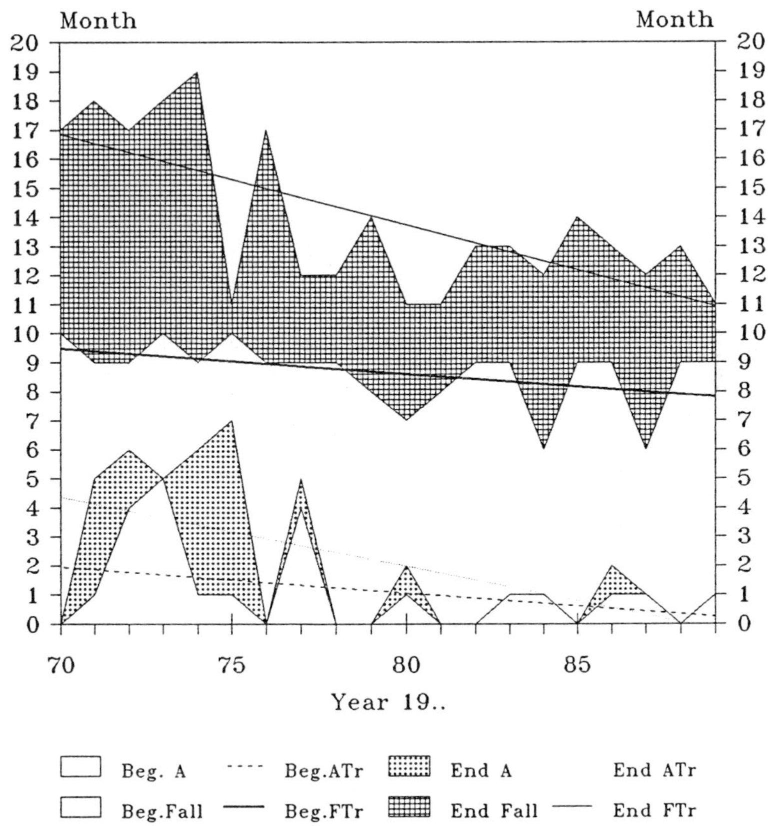

Datei: Kew9a

Fig. 7b: Monthly Fructification Incidence MFI of Lepista nuda

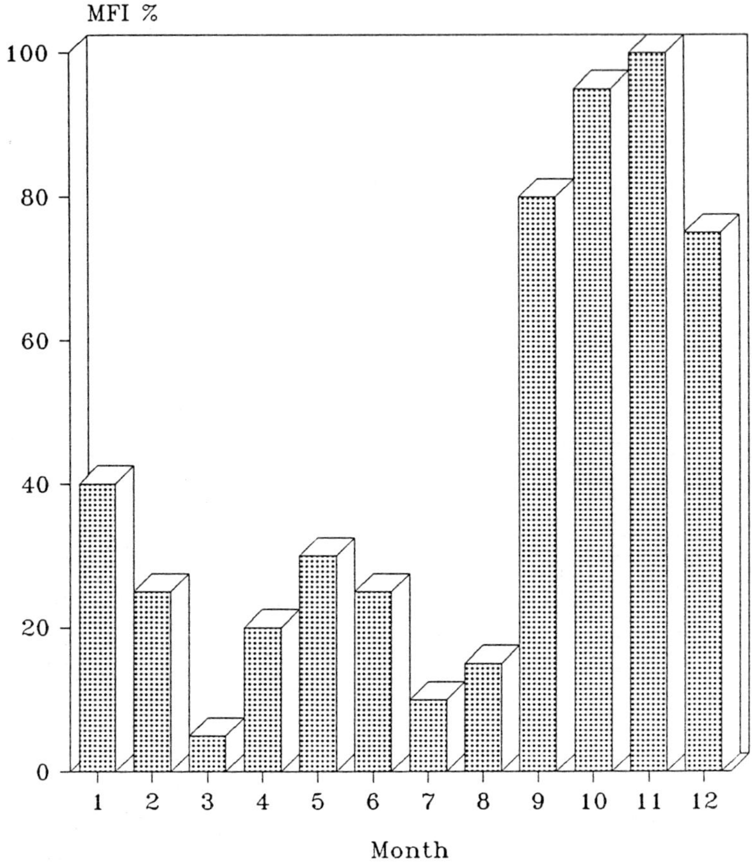

Datei: Kew9b

Consolidating the facts for the 100 investigated widespread fungus species (Table 1), the monthly numbers are rounded to 0.5 month) a series of generalizations emerge. Thus only in 18% of the investigated species no significant changes in their Yearly Fruiting Periods in the last 20 years were recorded. However in 82%, the fruiting periods are shifted and/or shortened (41%) or enlarged (34%). A seasonal shift of the fruiting period from autumn to winter or from summer to autumn can be caused by a lower availability of photosynthetic products from their plant symbionts to the mycorrhizal fungi in the growing season. A shortening of the fruiting period documents in general a deterioration of life conditions — also of the woody plant symbionts of mycorrhizal species. In the case of saproptrophs, changing biochemical conditions or characters in stressed wood, litter or soil, e.g. acidification, could be a reason for this fact.

On the other hand, the enlarging of fruiting periods — found in 34% of the investigated species — is in no case combined with an increase in populations or fructifications, that means with no improvement of environmental conditions, and is probably caused by the raising observation intensity in the last 5 to 10 years.

For each of the fungus species, the Monthly Fructification Incidence MFI in the 20 years-period is also mentioned. Only 60% of the most widespread fungi species show a MFI of 100% in one or more months, that means they are detected constantly in each year in the same months. In Table 2, the fruiting situation of the investigated fungus species is

Fig. 8a: Yearly Fruiting Periods FP of Tubaria furfuracea

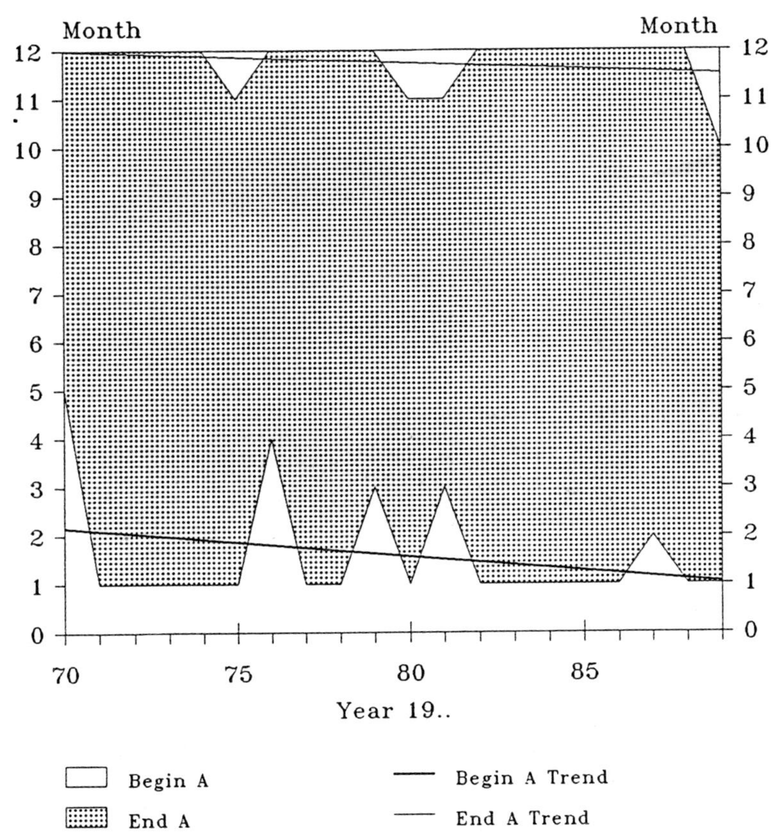

Datei: Kew12b

Fig. 8b: Monthly Fructification Incidence of Tubaria furfuracea

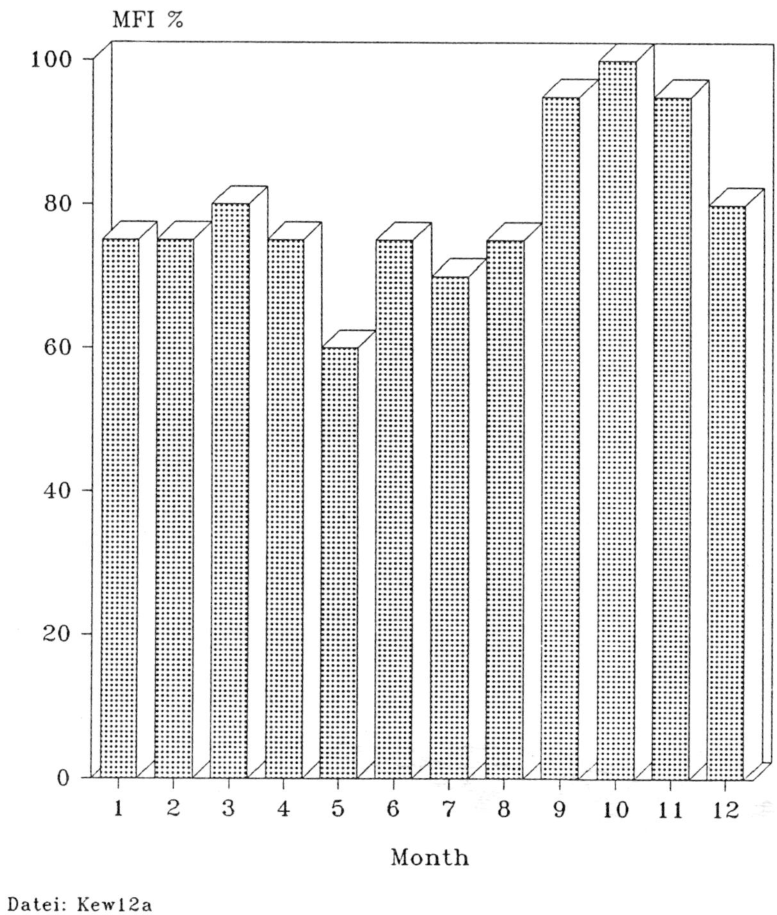

documented in relation to their ecological function. In the case of ectomycorrhizal fungi, 80% show declining populations, not one of them increasing. This is highly significant for those associated with *Fagus*, *Quercus* and *Picea*, but not in the case of *Betula*. Further, 34% of the saprotrophic species show declining populations, and here, the litter decomposing species are the most sensitive ones.

Concerning the Yearly Fruiting Periods, only 20% of the ectomycorrhizal Agaricales show no change, 86% show a shift and 76% a change in the amplitude of the shift. In the case of saprotrophic fungi, the litter decomposing species show a comparable situation.

TABLE 1. Yearly Fruiting Periods and their Changes of 100 Species of Agaricales s.l. between 1970 and 1989 in the Saarland Region

Ecology Type: s = saprotrophic; p = parasitic; m = Ectomycorrhiza-forming
Substrate: S = soil; C = coprophilous; L = Litter; W = Wood; D = deciduous wood; N = coniferous wood; P = pioneer wood; G. = grass; M = moss
Month: as number (January = 1; February = 2; .... December = 12)
Change: [a] positive numbers represents widening; negative numbers a contraction in months; [b] positive numbers represents an earlier, negative numbers a later begin of fruiting in months; [c] positive number represents a later, negative numbers an earlier end of fruiting in months

Fungus Species	Ecology Type	Substrate	Range [month] 1970	Range [month] 1989	Range Change[a]	Yearly Fruiting Periods [month] Begin in month 1970	Begin 1989	Begin Change[b]	End in month 1970	End 1989	End Change[c]
Agaricus abruptibulbus	s	S	5.5	5.0	−0.5	6.0	7.0	−1.0	10.5	11.0	+0.5
Agrocybe praecox	s	S	3.0	3.0	0	5.0	5.0	0	7.0	7.0	0
Amanita citrina	m	DN	4.5	3.5	−1.0	8.0	8.5	−0.5	11.5	11.0	−0.5
Amanita muscaria	m	DNP	4.0	3.5	−0.5	7.5	8.5	−1.0	11.0	11.5	+0.5
Amanita rubescens	m	DNP	5.5	5.5	0	6.0	6.0	0	10.5	10.5	0
Amanita spissa	m	DN	4.0	5.5	+1.5	6.5	6.0	+0.5	9.5	10.5	+1.0
Amanita vaginata	m	DN	4.0	5.5	+1.5	6.0	5.5	+0.5	9.0	10.0	+1.0
Armillariella mellea s.l.	ps	DNP	6.5	4.0	−2.5	6.5	7.5	−1.0	12.0	10.5	−1.5
Bolbitius vitellinus	s	SC	6.0	7.0	+1.0	5.0	5.0	0	10.0	11.0	+1.0
Boletus edulis	m	DN	4.5	3.5	−1.0	7.5	8.5	−1.0	10.5	10.5	0
Clitocybe gibba	s	LS	6.0	3.5	−2.5	5.5	7.5	−2.0	10.5	10.0	−0.5
Clitocybe suaveolens	s	LS	6.5	7.0	+0.5	9.0	9.0	0	3.0	3.5	+0.5
Collybia dryophila	s	LS	6.5	7.0	+0.5	5.0	5.0	0	10.5	11.0	+0.5
Collybia peronata	s	LS	5.0	5.0	0	7.0	6.5	+0.5	11.0	10.5	+0.5
Coprinus atramentarius	s	S	8.0	8.0	0	4.5	4.5	0	11.5	11.5	0

TABLE 1. Continued.

Fungus Species	Ecology	Substrate	Yearly Fruiting Periods [month]								
			Range [month]			Begin in month			End in month		
	Type		1970	1989	Change[a]	1970	1989	Change[b]	1970	1989	Change[c]
Coprinus comatus	s	S	5.5	5.5	0	6.5	6.5	0	11.0	11.0	0
Coprinus micaceus	s	WD	9.0	10.0	+1.0	4.0	3.0	+1.0	11.5	11.5	0
Coprinus plicatilis	s	S	7.0	6.0	–1.0	5.0	5.5	–0.5	11.0	10.5	–0.5
Gymnopilus penetrans	s	WDN	5.0	6.0	+1.0	8.0	6.5	+1.5	12.0	11.5	–0.5
Hebeloma crustuliniforme	m	DNP	4.0	3.0	–1.0	8.5	8.5	0	11.5	10.5	–1.0
Hebeloma mesophaeum	m	DNP	4.0	4.5	+0.5	9.0	8.0	+1.0	12.0	11.5	–0.5
Hygrophoropsis aurantiaca	s	LSWN	5.0	4.0	–1.0	7.5	8.5	–1.0	11.5	11.5	0
Hygrophorus cossus	m	Fagus	4.0	4.0	0	8.0	8.0	0	11.0	11.0	0
Hypholoma capnoides	s	WN	9.5	8.5	–1.0	9.5	8.5	+1.0	6.0	4.0	–2.0
Hypholoma fasciculare	s	WDNP	9.5	11.0	+1.5	3.5	1.5	+2.0	12.0	11.5	–0.5
Hypholoma sublateritium	s	WDN	7.5	7.0	–0.5	9.0	9.0	0	3.5	3.0	–0.5
Inocybe dulcamara	m	PDN	6.0	4.5	–1.5	5.5	6.0	–0.5	10.0	9.0	–1.0
Inocybe fastigiata	m	DNP	6.0	5.5	–0.5	5.5	5.5	0	10.5	10.0	–0.5
Inocybe geophylla	m	DN	6.0	4.0	–2.0	6.5	8.0	–1.5	11.5	11.0	–0.5
Inocybe maculata	m	DN	4.5	3.5	–1.0	6.5	8.0	–1.5	10.0	10.5	–0.5
Kühneromyces mutabilis	s	DNP	9.0	10.5	+1.5	4.0	2.0	+2.0	12.0	11.5	–0.5
Laccaria amethystina	m	DN	3.5	4.0	+0.5	8.5	7.5	+1.0	11.0	10.5	–0.5
Laccaria laccata	m	DN	6.0	5.5	–0.5	6.5	6.5	0	11.5	11.0	–0.5
Laccaria proxima	m	DN	4.5	5.0	+0.5	7.5	7.5	0	11.0	11.5	+0.5
Lactarius blennius	m	Fagus	3.5	4.5	+1.0	8.0	7.5	+0.5	10.5	11.0	+0.5
Lactarius deterrimus	m	Picea	4.0	3.0	–1.0	7.5	8.5	–1.0	10.5	10.5	0
Lactarius necator	m	NPD	5.0	3.0	–2.0	7.0	8.5	–1.5	11.0	10.5	–0.5
Lactarius quietus	m	D	3.5	5.0	+1.5	8.0	7.0	+1.0	10.5	11.0	+0.5

TABLE 1. Continued.

Fungus Species	Ecology Type	Substrate	Yearly Fruiting Periods [month] Range [month] 1970	1989	Change[a]	Begin in month 1970	1989	Change[b]	End in month 1970	1989	Change[c]
Lactarius vellereus	m	DN	5.0	3.0	–2.0	7.0	8.0	–1.0	11.0	10.0	–1.0
Leccinum scabrum	m	Betula	4.5	5.5	+1.0	7.0	6.0	+1.0	10.5	10.5	0
Lepiota cristata	s	S	6.0	4.5	–1.5	6.5	7.0	–0.5	11.5	10.5	–1.0
Lepista inversa	s	LND	7.0	6.0	–1.0	6.5	7.5	–1.0	0.5	0.5	0
Lepista nebularis	s	LDN	3.0	3.5	+0.5	9.5	9.0	+0.5	11.5	11.5	0
Lepista nuda	s	LDN	8.5	4.0	–4.5	9.5	8.0	+1.5	5.0	11.0	–6.0
Macrolepiota procera	s	SL	3.5	3.5	0	8.0	8.0	0	10.5	10.5	0
Marasmiellus ramealis	s	W	5.0	5.0	0	6.5	6.5	0	10.5	10.5	0
Marasmius oreades	sp	GS	7.5	6.5	–1.0	5.0	5.0	0	11.5	10.5	–1.0
Marasmius rotula	s	SL	6.0	4.0	–2.0	6.0	6.5	–0.5	11.0	9.5	–1.5
Melanoleuca melaleuca	s	SL	6.0	3.0	–3.0	7.0	8.0	–1.0	12.0	10.0	–2.0
Micromphale perforans	s	LN	7.5	5.5	–2.0	5.0	6.5	–1.5	11.5	11.0	–0.5
Mycena epipterygia	s	SW	5.0	3.0	–2.0	8.5	9.5	–1.0	0.5	11.5	–1.0
Mycena galericulata	s	W	7.5	8.5	+1.0	6.0	5.0	+1.0	0.5	0.5	0
Mycena galopoda	s	SLND	6.5	6.5	0	6.0	6.0	0	11.5	11.5	0
Mycena pura	m	DN	6.5	5.5	–1.0	6.0	6.5	–0.5	11.5	11.0	–0.5
Mycena sanguinolenta	s	LND	5.5	5.5	0	6.0	6.0	0	10.5	10.5	0
Oudemansiella platyphylla	s	SWDN	6.0	6.0	0	5.5	5.0	+0.5	10.5	10.0	–0.5
Oudemansiella radicata	s	WD	6.0	5.5	–0.5	6.0	6.0	0	11.0	10.5	–0.5
Panaeolina foenisecii	s	GS	5.0	5.0	0	6.0	5.5	+0.5	10.0	9.5	–0.5
Panellus stypticus	s	W	10.0	8.0	–2.0	8.0	8.0	0	5.0	3.0	–2.0
Paxillus involutus	m	PND	6.0	6.5	+0.5	6.5	5.5	+1.0	11.5	11.0	–0.5
Pholiota lenta	s	WDN	8.0	6.5	–1.5	9.0	8.0	+1.0	4.0	1.5	–2.5
Pluteus atricapillus	s	WDN	8.0	8.0	0	4.5	4.5	0	11.0	11.5	0

TABLE 1. Continued.

Fungus Species	Ecology Type	Substrate	Range [month] 1970	Range [month] 1989	Range [month] Change[a]	Yearly Fruiting Periods [month] Begin in month 1970	Begin in month 1989	Begin in month Change[b]	End in month 1970	End in month 1989	End in month Change[c]
Pluteus phlebophorus	s	WD	5.0	5.0	0	5.5	6.0	–0.5	9.5	10.0	+0.5
Polyporus brumalis	s	WD	6.5	7.0	+0.5	10.5	10.5	0	4.0	4.5	+0.5
Polyporus lepideus	s	WD	5.0	4.0	–1.0	4.5	4.0	+0.5	8.5	7.0	–1.5
Polyporus varius	s	WD	4.5	6.5	+2.0	6.0	5.0	+1.0	9.5	10.5	+1.0
Psathyrella candolleana	s	SW	5.0	5.5	+0.5	5.5	5.5	0	9.5	10.0	+0.5
Psathyrella hydrophila	s	WDN	4.5	5.0	+0.5	8.5	8.0	+0.5	12.0	12.0	0
Psathyrella velutina	s	S	6.0	5.0	–1.0	5.5	6.5	–1.0	10.5	10.5	0
Rhodophyllus staurosporus	s?	SLN	6.5	6.0	–0.5	5.0	5.5	–0.5	10.5	10.5	0
Rickenella fibula	sp	M	5.0	6.5	+1.5	6.5	6.0	+0.5	10.5	11.5	+1.0
Russula aeruginea	m	PN	5.0	5.5	+0.5	6.5	6.0	+0.5	10.5	10.5	0
Russula chamaeleontina	m	D	5.0	5.5	0	6.0	6.0	0	10.0	10.0	0
Russula cyanoxantha	m	DNP	4.0	5.0	+1.0	6.5	6.0	+0.5	9.5	10.0	+0.5
Russula delica	m	D	5.0	3.0	–2.0	6.5	8.0	–1.5	10.5	10.0	–0.5
Russula exalbicans	m	Betula	5.0	6.0	+1.0	6.0	5.5	+0.5	10.0	10.5	+0.5
Russula fellea	m	DN	2.0	3.0	+1.0	8.5	8.5	0	9.5	10.5	+1.0
Russula foetens	m	DNP	3.5	3.5	0	7.0	7.0	0	9.5	9.5	0
Russula fragilis	m	DN	3.5	3.5	0	8.0	8.5	–0.5	10.5	11.0	+0.5
Russula integra	m	N	5.0	3.5	–1.5	7.0	8.0	–1.0	11.0	10.5	–0.5
Russula lepida	m	Fagus	3.0	3.5	+0.5	7.5	7.5	0	9.5	10.0	+0.5
Russula nauseosa	m	Picea	6.0	6.5	+0.5	6.0	5.5	+0.5	11.0	11.0	0
Russula nigricans	m	DP	3.5	4.0	+0.5	7.0	7.5	–0.5	9.5	10.5	+1.0
Russula ochroleuca	m	DN	4.5	4.5	0	7.5	8.0	–0.5	11.0	11.5	+0.5
Russula parazurea	m	DNP	4.0	5.5	+1.5	6.5	6.5	0	9.5	11.0	+1.5

TABLE 1. Continued.

Fungus Species	Ecology Type	Substrate	Yearly Fruiting Periods [month] Range [month] 1970	1989	Change[a]	Begin in month 1970	1989	Change[b]	End in month 1970	1989	Change[c]
Russula pectinatoides	m	DP	4.0	5.0	+1.0	6.5	6.0	+0.5	9.5	10.0	+0.5
Russula puellaris	m	DN	5.0	5.0	0	6.5	6.5	0	10.5	10.5	0
Russula rosea	m	D	4.0	3.5	–0.5	6.5	7.5	–1.0	9.5	10.0	+0.5
Russula vesca	m	DN	4.0	4.5	+0.5	6.5	6.0	+.05	9.5	9.5	0
Russula violeipes	m	D.N?	4.0	4.0	0	6.5	6.5	0	9.5	9.5	0
Russula virescens	m	D	3.5	2.5	–1.0	6.5	7.5	–1.0	9.0	9.0	0
Stropharia aeruginosa	s	SL.W?	4.5	4.0	–0.5	8.5	8.0	+0.5	12.0	11.0	–1.0
Suillus grevillei	m	Larix	5.0	5.0	0	6.0	6.5	–0.5	10.0	10.5	+0.5
Tricholoma scalpturatum	m	DP	8.0	6.5	–1.5	5.5	5.5	0	0.5	11.5	–1.5
Tricholoma sulphureum	m	Fagus	3.0	3.0	0	8.5	8.5	0	10.5	10.5	0
Tricholomopsis rutilans	s	WN	4.0	3.5	–0.5	8.0	8.5	–0.5	11.0	11.0	0
Tubaria furfuracea	s	SW	11.0	11.5	+0.5	2.0	1.0	+1.0	12.0	11.5	–0.5
Xerocomus badius	m	ND	5.0	5.0	0	7.0	7.0	0	11.0	11.0	0
Xerocomus chrysenteron	m	DN	6.0	6.0	0	6.0	6.0	0	11.0	11.0	0
Xerocomus subtomentosus	m	DN	5.5	5.5	0	6.0	6.0	0	10.5	10.5	0

TABLE 2 Changes in Fruiting Periods of Special Ecological Groups of 100 Agaricales Species

Taxon/Ecol.Gr.	Substrate	Number	Populations damaged		Populations constant		Fruiting Periods Pos.+ Range constant		Fruiting Periods Pos.– Range		Fruiting Periods Range–Change		Fruiting Periods Pos. or Range Change	
			Number	%	Number	%	Number	%	Number	%	Number	%	Number	%
Agaricales	all	100	58	58	42	42	18	18	82	82	75	75	82	82
Mycorr	all	50	40	80	10	20	10	20	43	86	38	76	40	80
Mycorr	Fagus	4	4	100	0	0	2	50	2	50	2	50	2	50
Mycorr	Quercus	3	3	100	0	0	1	33	2	67	2	67	2	67
Mycorr	Picea	3	2	67	1	33	1	1	2	67	2	67	2	67
Mycorr	Betula	2	0	0	2	100	0	0	2	100	2	100	2	100
Russula/Mycorr	all	20	14	70	6	30	4	20	16	80	14	70	16	80
Sapro	all	50	17	34	33	66	–	–	–	–	–	–	–	–
Sapro	Soil	21	6	29	15	71	4	19	9	43	8	38	17	81
Sapro	Litter	7	5	71	2	29	3	43	4	86	4	86	4	57
Sapro	Wood	17	5	29	12	71	2	12	11	65	4	24	15	8

Conclusion

Summarizing the results of the investigations on fruiting period changes of selected, widespread Agaricales species, only 18% of the 100 species show no significant change. In 82% of species, the periods are shifted and/or shortened or enlarged, yet in the last case the change is not combined with any increase of fructification. That means, if widespread non sensitive species show in many cases changes in their fruiting behaviour, the Yearly Fruiting Period may be a useful parameter to indicate changes in the environmental conditions of other agaric species.

Acknowledgements

I thank very much Dr D Pegler, Royal Botanic Gardens, Kew, for the kind invitation to present this paper on the XIth Congress of European Mycologists in Kew 1992. I am also indebted to Dr Roy Watling, Royal Botanic Garden, Edinburgh, for the critical reading of the manuscript.

References

Arnolds, E. (1991). Decline of ectomycorrhizal fungi in Europe. *Agric. Eco-Syst. Environm.* **35**: 209–244.

Derbsch, H. and Schmitt, J.A. (1984). Atlas der Pilze des Saarlandes, Teil 1: Verbreitung und Gefährdung. 535 pp. Arbeitsgemeinschaft für tier- und pflanzengeographische Heimatforschung im Saarland e.V., Saarbrücken. (Minister für Umwelt, Raumordnung und Bauwesen des Saarlandes und die DELATTINIA (editors). Natur und Landschaft im Saarland, Sonderband 2).

Derbsch, H. and Schmitt, J.A. (1987). Atlas der Pilze des Saarlandes, Teil 2: Nachweise, Ökologie, Vorkommen und Beschreibungen. 816 pp. Arbeitsgemeinschaft für tier- und pflanzengeographische Heimatforschung im Saarland e.V., Saarbrücken. (Minister für Umwelt des Saarlandes und die DELATTINIA (editors). Natur und Landschaft im Saarland. Sonderband 3).

Schmitt, J.A. (1990–1991). Bestand und Dynamik der Pilzfloren ausgewählter Biotope des Saarlandes nach Langzeitbeobachtung (mindestens 15-jähriger Beobachtungszeitraum), Teile 1–6. 1500 pp. Saarbrücken. Project supported by the Ministry of Environment in Saarland, Saarbrücken.

Ryvarden, L. (1993). Distribution of aphyllophoroid fungi in the Taiga region of Fennoscandia. In D.N. Pegler, L. Boddy, B. Ing & P.M. Kirk (Editors). Fungi of Europe: Investigation, Recording and Conservation, pp. 71–76 Royal Botanic Gardens, Kew.

6. DISTRIBUTION OF APHYLLOPHOROID FUNGI IN THE TAIGA REGION OF FENNOSCANDIA

L. RYVARDEN

Department of Botany, University of Oslo, P.O. Box 1045, Blindern, N-0316 Oslo, Norway

Abstract

The distribution of the aphyllophoroid basidiomycetes found in the most northern coniferous forest of Eurasia is considered. Climatic, spore dispersal, host migration, and growth requirements are assessed for the species present.

The word 'taiga' is Russian and means forest, but in the phytogeographical sense it means the most northern, boreal coniferous forest in Eurasia. It stretches from Eastern Siberia to the interior part of the Fennoscandian peninsula and is dominated by *Pinus* spp. and *Picea abies*. The taiga area in Fennoscandia is shown on Fig. 1. The area has a distinct flora and fauna, but emphasis here will be placed on some wood inhabiting basidiomycetes only. The climate is strongly continental and is characterized by cold, dry winters and warm, dry summers. Typically, many places within this region have average temperatures of about –15°C in January–February and +15–17°C in July. No other region in Fennoscandia has such a high amplitude in average temperatures. Karasjok in Finnmark, Norway, for example, has the Norwegian record for both highest and lowest temperatures, +35°C and –51°C respectively. The precipitation is low in the area, averaging only 300–400 mm per year.

Whilst trees like *Pinus sylvestris*, *Betula tortuosa* and *Salix caprea* immigrated rather quickly following the last glaciation, *Picea abies* is a late, eastern, immigrant. Fig. 2 shows the approximate dates for its advance in Norway. Although the trees mentioned above have a very wide distribution in Fennoscandia, many fungi connected with them are restricted to the taiga area as shown on Fig. 1. Thus, in addition to their host relationship, some other factor(s) must be restricting their distribution.

First, are the areas of distribution shown on the maps true, or do they only reflect collection records? Even if there are fairly large areas of Fennoscandia which have not been visited by mycologists, the taiga area is well collected. This is mainly due to the existence of many forests reserves in the area. All the mapped species have rather conspicuous basidiomes and over the years have especially been looked for in suitable localities and forest reserves outside the taiga area. Thus, the maps are reliable as a basis for the following discussion. Further, all the mapped species are distinct and without a disputable taxonomic status. Figs. 3 to 5 show the distribution in Fennoscandia of *Laurilia sulcata* (Burt) Pouzar, *Antrodia albobrunnea* (Rom.) Ryv., and *Phlebia centrifuga* Karsten, (fig. 3), *Asterodon furruginosum* Pat. and *Columnocystis abietina* (Pers.: Fr.) Pouzar (Fig. 4).

As fungi in general are assumed to have wind-dispersed spores it is difficult to explain the distributional patterns shown on the figures in terms of insufficient establishment time following host immigration to the area. Many species with the same type of small, thin-walled, hyaline spores, have been able to migrate concurrently with their hosts wherever they occur in Fennoscandia. Thus, historical reasons cannot adequately explain the distribution of the taiga species.

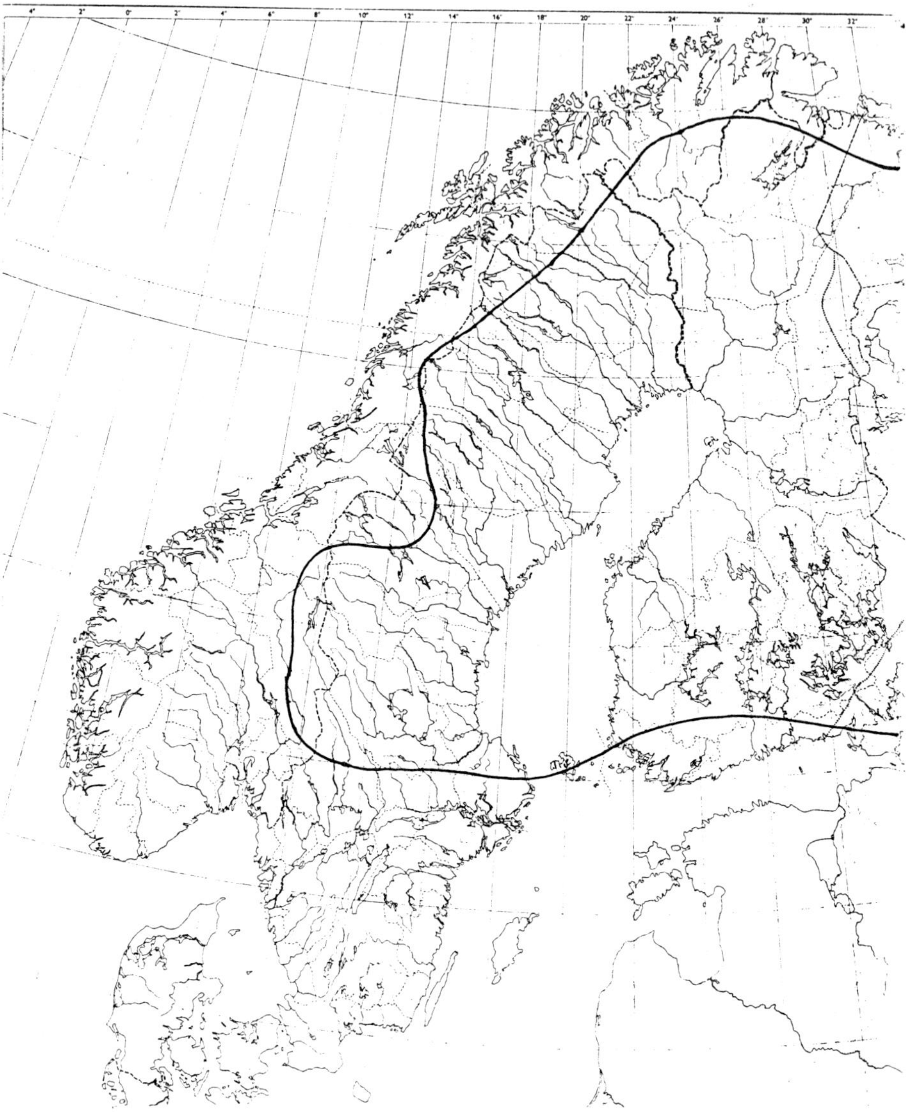

FIG. 1. The taiga area in Fennoscandia (see text for explanation).

It is also difficult to understand the particular climatic factors in the continental areas that should favour the taiga species. As mentioned above, summers are hot and dry, conditions known to be unfavourable to fungal growth in general. The winter is very cold, lasting for long periods and must be a difficult period for the mycelia inside the frozen logs. Further, the snow cover is thin, thus, providing little physical barrier between the lying logs and the ground. From the point of view of climatic factors, it is difficult to understand why these fungi have not spread into the lowland areas with higher summer and winter temperatures and precipitation.

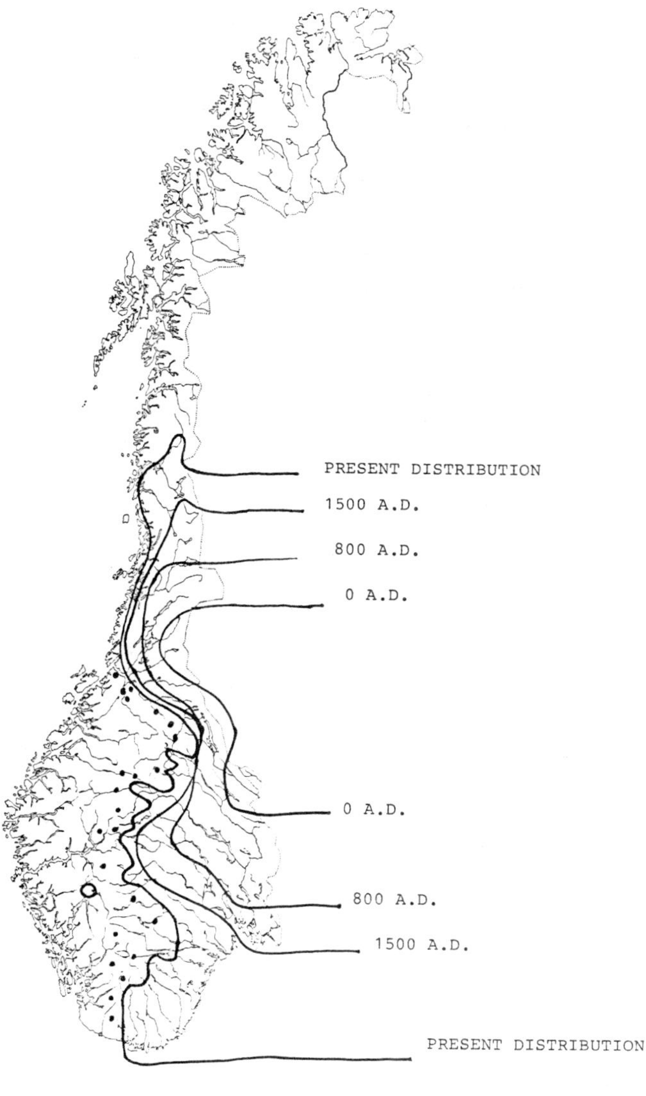

FIG. 2. The advance of *Picea abies* in Norway after the last glaciation.

Do then the taiga species, in contrast to other spruce restricted species, have special features that make them able to survive the rather harsh conditions of this area? We do not know. However, as far as I know, there are no species which are more or less restricted to spruce, that do not also occur in the taiga area.

Since we can also clearly reject the climatic factors as an to explanation of the distribution of the taiga species, we are left with age of the hosts and competitive force of the taiga species as the alternative explanation.

For the trees, the short and dry summer provides slow growth and many trees reach an age of 300 years or more. In the lowlands of Fennoscandia, 60–80 years are sufficient to mature a tree for logging. After that, stagnation sets in and within the next 100 years either storms or insects and/or fungi will have killed almost all the trees. Thus, if a fungus has a

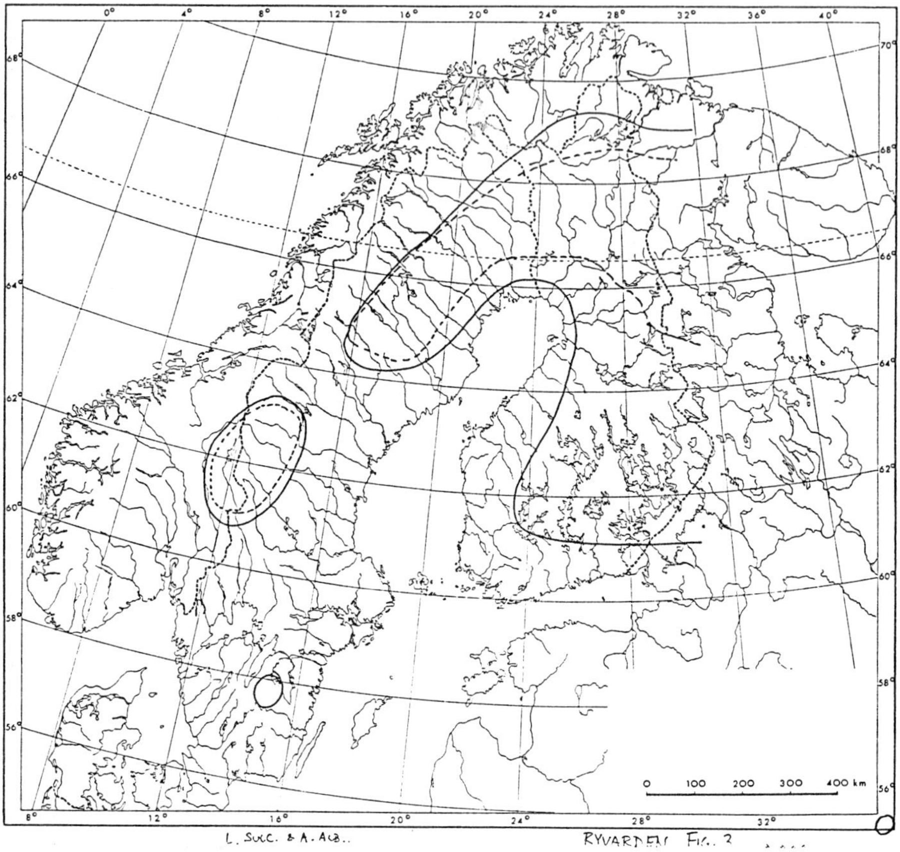

FIG. 3. Approximate distribution in Fennoscandia of *Laurilia sulcata* (dotted line) and *Antrodia albobrunnea* (whole line).

slow growth rate, needing some 200 years to accumulate sufficient energy to produce a basidiome, it may prove difficult for it to find suitable hosts in such areas. More aggressive species with faster growth rates will take over.

In the taiga area, such species will compete with the taiga species although the latter will produce less basidiomes than the faster growing species. This is reflected in the fact that taiga species are rather rare even where they occur. In the last hundred years the lowland forests of Fennoscandia have been logged at least once, and in many cases several times over. Consequently, larger and older potential hosts for the slow growing species have continuously been removed from the forest. Thus, the economical development of the forests has pushed these slow growing species back to the continental areas, i.e. the taiga area.

Picea was not the only tree coming to Fennoscandia from the east after the glaciation. *Salix caprea* was another. With its tiny, hairy seeds it has a very effective dispersal and is known from every forested area in Fennoscandia. *Salix caprea* is the specific host to some species and one of the most interesting of these is *Haploporus odorus* (Somf.: Fr.) Bond. & Singer. The distribution of this species is shown on Fig. 5 and shows a typical taiga species

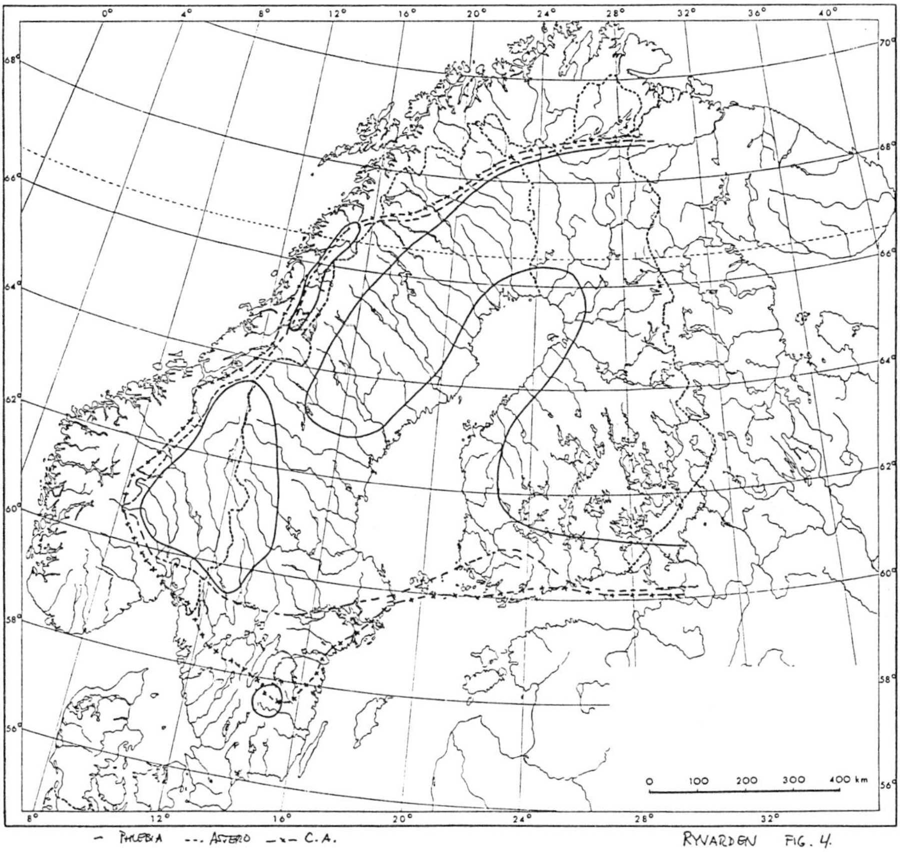

FIG. 4. Approximate distribution in Fennoscandia of *Phlebia centrifuga* (whole line), *Asterodon ferruginosus* (dotted line) and *Columnocystis abietina* (dots and crosses).

distribution pattern. The basidiomes of this polypore have a strong odour, often so strong that it is detected long before the basidiome is seen. Although *Salix caprea* is a tree of no commercial importance it may take hundreds of years for a tree to mature in the continental taiga area. This may be necessary for *Haploporus odorus* to accumulate sufficient energy to produce a basidiome. In the lowland and the more humid areas *Salix caprea* grows and degrades faster, thus excluding *Haploporus*.

It may be speculated whether some type of insect dispersal is involved. The strong scent; is it only accidental or has it been established as a result of selective pressure? It may be recalled that *Haploporus odorus* is one of the very few polypores in Fennoscandia with ornamented spores. This could be an adaptation to insect dispersal. If the dispersing insect for some unknown reasons is restricted to *Salix caprea* in the taiga area, this will also be the case for *Haploporus odorus*. Among higher plants such cases are known. In Fennoscandia, *Aconitum septentrionalis* for example, is restricted within the distribution area of the only bumble bee that is able to pollinate it. Outside that area, *Aconitum* grows excellently in gardens etc, but does not produce seeds.

No experiments have been performed to see whether the taiga species on *Picea* or *Salix* have slower growth rates than those of the more widely distributed species restricted to the same hosts. However, investigations have just recently been started to address this problem and the future will give some clues to its solution.

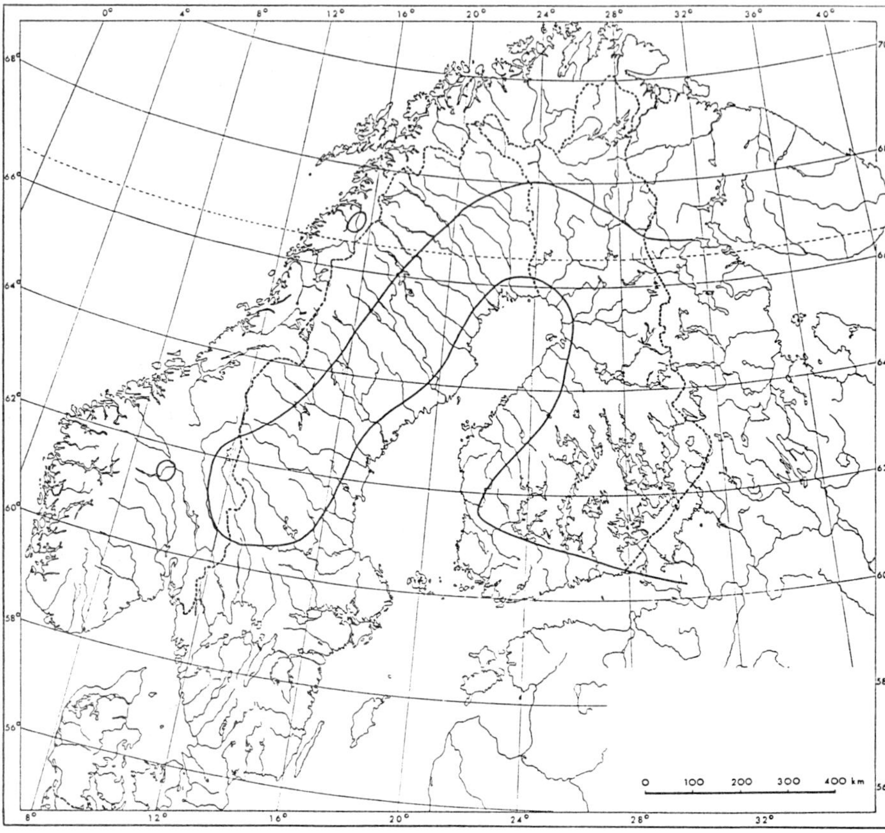

FIG. 5. Approximate distribution in Fennoscandia of *Haploporus odorus*.

Acknowledgements

Drs. T. Niemalä and H. Kotiranta, Helsinki, Finland, have given valuable information about the distribution of the taiga species in Finland, and their help is acknowledged.

References

Eriksson, J. and Strid, A. (1969). Studies in the Aphyllophorales (Basidiomycetes) of Northern Finland. *Ann. Univ. Turku., A.* **40**: 112–158.

Hafsten, U. (1991). Granskogens historie i Norge under opprulling. *Blyttia* **49**: 171–181.

Niemelä, T. (1971). On Fennoscandia polypores: I. *Haploporus odorus* (Sommerf.) Bond. & Sing. *Ann. Bot. Fenn.* **8**: 237–244.

Perini, C., Barluzzi, C. & De Dominicis, V. (1993). Fungal communities in mediterranean and submediterranean woodlands. In D.N. Pegler, L. Boddy, B. Ing & P.M. Kirk (Editors). Fungi of Europe: Investigation, Recording and Conservation, pp. 77–92. Royal Botanic Gardens, Kew.

7. FUNGAL COMMUNITIES IN MEDITERRANEAN & SUBMEDITERRANEAN WOODLANDS

C. PERINI, C. BARLUZZI & V. DE DOMINICIS

Dipartimento di Biologia Ambientale, Via P.A. Mattioli, 4 — 53100 Siena, Italy

Abstract

A summary of data based on several mycocoenological studies, carried out in a belt extending from the Tyrrhenian coast to the foot of the Central Apennines, is presented. The studies, concerning macromycetes, have been carried out in mediterranean and submediterranean communities; juniper scrubs on sandy dunes and evergreen oak woods of the Maremma coast; evergreen oak woods, heathlands and chestnut coppices of the hills; chestnut coppices and fir woods of submontane areas. Association between about 700 fungal species, collected in 5.5 hectares, and the seven phytocoenoses are analysed. The species diversity is noteworthy: 14.4% are species widely distributed in the phytocoenoses and 54.9% are associated with only one type of vegetation community.

Introduction

Mycocoenological research, concerning macrofungi, were carried out over a period of about 15 years in the Mediterranean belt of central southern Tuscany. The observations for the main types of woodland vegetation going from the Tyrrhenian sea to the foot of the Apennines, are summarized in this paper.

There were several early studies of the fungal flora of central southern Tuscany (A. Tassi, 1862; Valenti-Sereni, 1868; Arcangeli, 1899; F. Tassi, 1896, 1897, 1906). No studies were published for sixty years, until Carla Barluzzi and her collaborators first examined floristic aspects (Barluzzi & Bianciardi, 1969, 1970; Barluzzi et al., 1978, 1980, 1983a, 1985) and then took a mycocoenological approach (De Dominicis & Barluzzi, 1983; Barluzzi et al., 1983b, 1984, 1986, 1990, 1992; Perini et al., 1989).

Study Sites

Climate and Geography

The study area is orographically varied, ranging from sandy or rocky Mediterranean coast, through alluvial plains and groups of hills and mountains, to the Central Apennines.

The climate is typically Mediterranean on the coast and becomes submediterranean, tending to continental inland and towards higher elevations. On the coast the summers are hot and the drought is quite marked; the winters are mild with adequate rain. Inland, the summer drought is much less severe, the winters are relatively hard and the temperature often falls below freezing point (Fig. 1). The coastal climatic conditions are optimal for the development of *Quercetum ilicis*, while inland conditions are more favourable for the growth of deciduous oak (*Quercus* spp.) forests at lower altitudes (0–1000 m) and beech (*Fagus* spp.) woods above 1000 m.

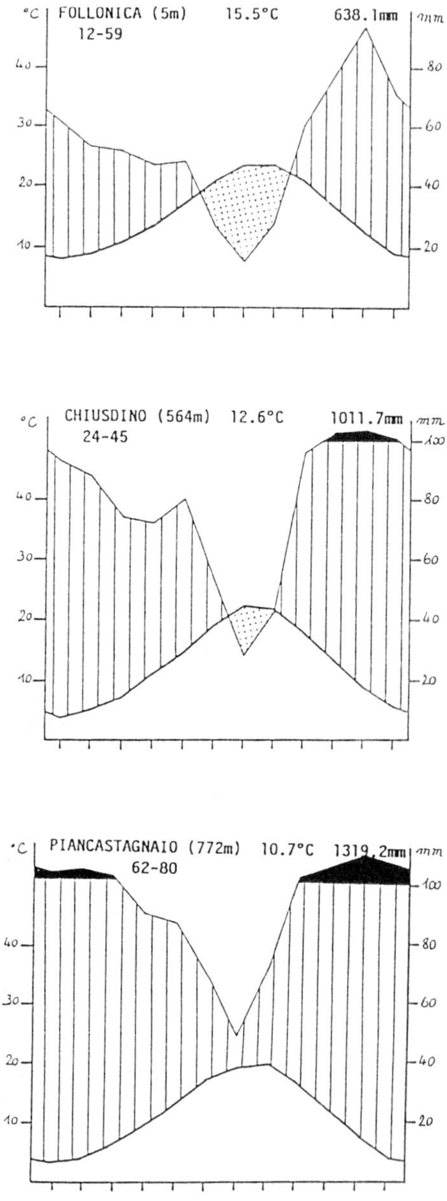

FIG. 1. Thermopluviometric diagrams showing the climatic changes from the coast to the inland. The diagrams constructed according to Walter and Lieth (1960). At the top of the diagram are reported: the name and the altitude of the thermopluviometric station, the mean annual temperature and rainfall, the observation years.

a. Juniper Scrubs

The juniper scrubs were situated on the double coastal dunes between the Tyrrhenian sea and the Burano Lake in the Burano Natural Reserve. The protected area, established in 1980 by the World Wildlife Fund, is located in southern Tuscany, 50 km from Grosseto (Fig. 2). The more consolidated dunes, up to 9.6 metres high, are covered by a discontinuous "macchia" of variable height, in which *Juniperus oxycedrus* L. ssp. *macrocarpa* (S. & S.) Ball, *J. phoenicea* L. and *Quercus ilex* L. dominate together with *Phillyrea angustifolia* L., *Myrtus communis* L. and *Pistacia lentiscus* L. This is the most natural of the vegetation types studied, and is probably the only area of the "Maremma" coast not colonized by umbrella pines (*Pinus pinea* L.). The latter was widely dispersed during reclamation in the last century. The geological substratum is principally formed by yellow, sometimes red, sandies (Carta Geologica d'Italia, foglio 135). On the basis of the phytosociological map 1:2000 (Pedrotti et al., 1975) the eight plots chosen in this area (about 900 sqm) belong to the *Juniperetum macrocarpae phoenicae* and *Oleo-Lentiscetum*. The mycocoenological observations were started in 1988 and are still in progress.

b. Evergreen Oak Woods

The distribution of *Quercus ilex* L., the dominant species of Mediterranean maquis, is not limited to the coast and coastal hills but reaches inland wherever environmental conditions permit. *Q. ilex* woods are generally cut every twenty years or less, and show a compact cover of low trees and tall scrubs. The characteristic species, besides *Q. ilex*, are: *Arbutus unedo* L., *Phillyrea angustifolia* L., *P. latifolia* L., *Rubia peregrina* L., *Smilax aspera* L., *Viburnum tinus* L., etc. The herbaceous layer is poor and the ground is sometimes covered with *Hedera helix* L., *Cyclamen repandum* S. & S. and *C. hederifolium* Aiton.

1. *Evergreen oak woods of the Maremman coast.* Mycocoenological research was carried out from 1981 to 1984 in five plots (10,000 m^2) situated along and near the Grosseto Maremman coastline (Fig. 2; Perini et al., 1989). The plots have not been cut for 30 years and therefore are both aged coppices and one of the most natural wooded areas in the study. Phytosociologically, these evergreen forests can be assigned to the *Viburno-Quercetum ilicis* (Br.-Bl. 1936) Rivas-Martinez 1975 subass. *ornetosum* Allier & Lacoste 1979. This subassociation is a transition between the *Viburno-Quercetum ilicis* of the French Mediterranean coast and the *Orno-Quercetum ilicis* of Dalmazia.

2. *Evergreen oak woods of the Sienese hills.* Five plots (10,000 m^2), located in the hills near Siena, (50 km from the Tyrrhenian sea, Fig. 2), were examined from 1977 to 1979 (De Dominicis & Barluzzi, 1983). These *Q. ilex* groves are situated on slopes where the scanty soil depth and rapid drainage of water create xerophilous conditions, and are characterized by the presence of numerous species belonging to the *Querco-Fagetea* (in particular *Crataegus monogyna* Jacq., *Fraxinus ornus* L., *Quercus pubescens* Willd., *Sorbus torminalis* (L.) Crantz, etc.). These forests are intermediate between the mediterranean evergreen sclerophyllous forests and the deciduous oak woods (*Quercetalia pubescentis* Br.-Bl. (1931) 1932) growing at the next higher level. Phytosociologically, these woods have been attributed to the *Quercion ilicis* s.l. without indicating a definitive association.

c. *Calluna vulgaris* heathlands

Calluna vulgaris heathlands, common throughout Atlantic and sub-Atlantic Europe, are at the southeastern limits of their distribution in Tuscany. They consist of dense scrubs, varying in height from 50 to 150 cm, and are physionomically dominated by *Ericaceae* (*Calluna vulgaris* (L.) Hull., *Erica scoparia* L. and *Arbutus unedo* L.). Pines (*Pinus pinaster* Aiton) introduced as a forestry species, spread spontaneously and now form a dense canopy. These heathlands (*Tuberario lignosae-Callunetum* De Dominicis & Casini 1979) occur on acid soils and arise from the degradation of acidophile oak woods with *Quercus cerris* L. and *Castanea sativa* Miller.

The mycocoenological research was performed, from 1978 to 1980, in 6 plots (12,000 m^2) located on the basin of the Merse river in the oriental part of the "Colline Metallifere" (South of Siena, Fig. 2) (Barluzzi et al., 1986).

d. *Castanea sativa* coppices

The central-southern Tuscan chestnut coppices mostly descend from the transformation of *Endothia*-infested chestnut groves. Mixed coppices with oaks and chestnuts (*Castanea sativa* Miller), on siliceous substrata, are not uncommon and are probably the most natural communities from which chestnut woods were created by elimination of other tree species.

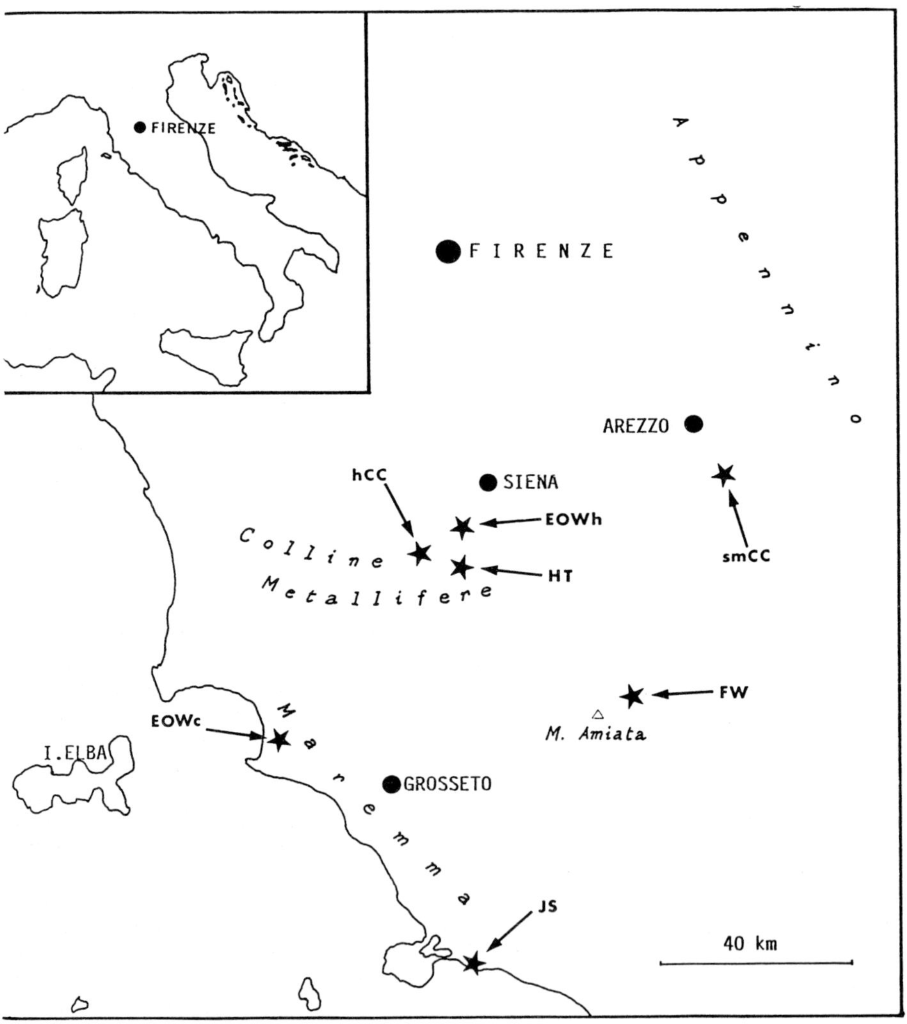

FIG. 2. Map of the vegetational communities studied. **JS** = juniper scrubs; **EOWc** = evergreen oak woods of the Maremma coast; **EOWh** = evergreen oak woods of the Sienese hills; **Ht** = *Calluna vulgaris* heathlands; **hCC** = hill chestnut coppices; **smCC** = submontane chestnut coppices; **FW** = fir woods.

The coenoses studied are characterized by silicicolous species of the *Quercetalia robori-petraeae* Tx. 1931: *C. sativa*, *Teucrium scorodonia* L., *Lathyrus montanus* Bernh., *Luzula forsteri* (Sm.) DC., *Pteridium aquilinum* (L.) Kühn, etc. These species decrease with altitude while the central European and montane species of the *Carpinion betuli* and *Fagetalia sylvaticae* increase: e.g. *Primula vulgaris* Hudson, *Symphytum tuberosum* L., *Melica uniflora* Retz. and *Fagus sylvatica* L. Notwithstanding several phytosociological studies, the syntaxonomy of these chestnut groves is still puzzling. According to Bergman *et al.* (1988) two associations can be distinguished:

1. *Hill chestnut coppices*: below five hundred metres *Rubia peregrina-Quercus cerris* communities Pignatti & Pignatti 1968 are found.

2. *Submontane chestnut coppices*: at higher altitudes the chestnut woods belong to the *Physospermo-Quercetum petraeae* Oberdorfer & Hofmann 1967.

Mycocoenological studies (from 1979 to 1982) include nine chestnut coppices (18,000 m^2) located in a belt from the "Colline Metallifere" to the foot of the Apennines (Fig. 2) (Barluzzi et al., 1992).

e. Fir Woods of Monte Amiata

Abies alba Miller, even if favoured by forestry, generally finds its ecological optimum at high altitudes in the Apennines. The importance of the fir woods of "Monte Amiata", an isolated pre-Apennine massif, lies in the fact that they are low-altitude relics of autochthonous *Abies alba.*

According to De Dominicis et al. (in press) the phytocoenoses studied belong to the *Aceri (obusati)-Quercetum cerris* Ubaldi & Speranza 1982 of the *Laburno-Ostryon* Ubaldi 1980 alliance, a central southern european vicariant of the *Carpinion betuli* Oberdorfer 1953. Besides *A. alba*, *Quercus cerris* L. and *Acer obtusatum* W. & K., this type of forest also contains *C. sativa*, *Carpinus betulus* L. and *Fagus sylvatica* L.

Mycocoenological research, which started in 1986 and is continuing, has been performed in four plots (about 2500 m^2, Fig. 2). The geological substratum is formed by sandstones of the type "Pietraforte" (Carta Geologica d'Italia, foglio 129).

Results and Discussion

Ninety-four fungal species were found in the seven vegetation communities. (Table 1). Fifteen species were common to six communities out of the seven, 28 to five and 51 to four. Only 10 species were found in juniper scrubs and their presence was quite low. Most of the fungi found in the seven communities have a wide ecological range, e.g.: *Amanita pantherina* DC: Fr., *A. rubescens* (Pers.: Fr.) S.F. Gray, *Cortinarius torvus* (Bull.: Fr.) Fr., *C. trivialis* Lge, *M. pura* (Pers.) Kumm., *Russula cyanoxantha* Schaeff.: Fr. etc. *Lactarius chrysorrheus* Fr., considered to be mycorrhizal on *Quercus* spp. (Trappe, 1962) and characteristic of acid oak forests (Bohus & Babos, 1967; Jansen, 1981; Arnolds, 1984), were not found in submontane chestnut coppices even if *Quercus cerris* L. and *Quercus pubescens* Willd. were present.

Seventy-eight species were found in three out of seven phytocoenoses, (Table 1), and are also wideranging, e.g. *Lactarius vellereus* (Fr.) Fr., *Marasmius splachnoides* Fr., *Mycena polygramma* (Bull.: Fr.) S.F. Gray. Two fungi, *Cortinarius calochrous* Fr. and *C. sodagnitus* R. Hry., can be considered characteristic species of the Mediterranean alliance *Quercion ilicis*, while the thermophilous *Leucopaxillus gentianus* (Quél.) Kotl. distinguishes the *Viburno-Quercetum ilicis ornetosum* of the coast. A large group of species ranging from *Russula rosacea* Pers. ex S.F. Gray to *Otidea cochleata* (L. ex St.-Am.) Fuck. occur in the coastal evergreen oak woods and in all chestnut coppices, and distinguish broad-leaved woodland from juniper scrubs, heathlands and fir woods. The reason for their absence from the evergreen oak woods of the Siena hills is unclear.

One hundred and twenty three species occur in two out of the seven vegetational communities (Table 1). Fungi (group **A**) occurring in Juniper scrubs and coastal *Quercus ilex* woods are mainly thermo-xerophilous, e.g. *Rhodocybe truncata* (Schaeff.: Fr.) Sing., *Agaricus praeclaresquamosus* Freem. var. *praeclaresquamosus* Freem., *Hydnocystis piligera* Tul. and *Flammulaster carpophiloides* (Kühn.) Watl.

The species of group **B** found in both the juniper scrubs and fir woods (which have vastly different climates) may be linked to the presence of resins. It emerges from the literature, however, that only *Ramaria flaccida* (Fr.) Bourd. and *Hemimycena gracilis* (Quél.) Sing. are exclusively linked to conifers.

The nineteen species of group **C** present in both types of evergreen oak woods may be regarded as characteristic of these woods in the Mediterranean. *Krombholziella lepida* (Bouch.) Alessio, *Hygrophorus russula* (Schaeff.: Fr.) Quél. and *Hygrophorus dichrous* Kühn. & Romagn. var. *fuscovinosus* Bon were just considered as characteristic taxa in a previous report (Perini et al., 1989).

In group **D** (Sienese *Quercus ilex* woods and heathlands) species such as *Lactarius deliciosus* (L.: Fr.) S.F. Gray, *Suillus granulatus* (L.: Fr.) Kuntze, *S. boudieri* (Quél.) Watl. and *Mycena seynii* Quél. are linked to incidental presence of *Pinus pinea* L. in one plot and *Pinus pinaster* Aiton near two others. Since pines are foreign in these forests, such fungi can be regarded as "alien" *sensu* Barkman (1976). *Cantharellus lutescens* Pers.: Fr., a montane species characteristic of the Belgian beech woods (Heinemann & Darimont, 1956; Darimont, 1973) and differential of *Galio-Abietetum* from other fir woods in the Carpathians (Bujakiewicz, 1982), can be considered a differential species of Tuscan heathlands. Another differential taxon is *Scleroderma citrinum* Pers., also reported from European moorlands (Runge, 1958, 1960; Bon & Gehu, 1973).

A species strictly linked to conifers is *Tricholomopsis rutilans* (Schaeff.: Fr.) Sing. found in heathlands and fir woods (group **E**). On a local scale the species of group **F** are characteristic of chestnut coppices (e.g. *Fistulina hepatica* Schaeff.: Fr., *Rutstroemia echinophila* (Bull.: Fr.) Quél. and *Mycena inclinata* (Fr.) Quél.).

Sixty-one species out of 345 found only in single phytocoenoses, have a class of presence of >40% (Table 1). Six species are differential of juniper scrubs e.g., *Geastrum pseudolimbatum* Hóllos and *G. saccatum* (Fr.) E. Fischer, which are new taxa for the Italian flora (Calonge *et al.*, 1991). Among the 43 sporadic species of juniper scrubs, the presence of *Macrolepiota phaeodisca* Bellù, *Scleroderma meridionale* Demoul. & Malenç., *Agaricus menieri* Bon and *Hydropus mediterraneus* Pacioni & Lalli is noteworthy; they reflect the xerothermic conditions and sandy substrate. The presence of a tropical lignicolous species, *Perenniporia ochroleuca* (Berk.) Ryv., observed previously on certain Italian islands in the Tyrrhenian sea (Bernicchia, 1990) is interesting. In relatively cooler and moister habitats, sheltered from winds and shaded by scrubs, species such as *Hygrocybe conicoides* (Orton) Orton & Watl. and *Arrhenia spathulata* (Fr.) Redhead are found in spring and autumn.

One hundred and four fungi were found only in the *Quercus ilex* woods of the Maremma. Some of these fungi, which have also been reported in Mediterranean environments of France, Spain, Morocco and Italy, are more thermophilous and/or xerophilous species of warmer evergreen oak woods, e.g. *Clavariadelphus flavoimmaturus* Pet., *Hymenoscyphus fructigenus* (Bull. ex Mèr.) S.F. Gray, *Mycena lenta* (R. Mre) Kühn. & Romagn with a high class of presence, and *Cortinarius ionochlorus* R. Maire, *Lepiota rhodorrhiza* Romagn. & Locq. ex Orton, *Mycena algeriensis* R. Maire, *Russula seperina* Dup. with a low class of presence (Perini et al., 1989).

Among 41 species found only in the heathlands many are associated with pines. *Lyophyllum loricatum* (Fr.) Kühn. and *Amanita citrina* Schaeff.: Fr. var. *alba* Price are differential species; the former has been reported (Bon & Gehu, 1973) from *Fago-Quercetum* and has been a common species of deciduous woods (Moser, 1978). The latter is an acidophilous species of the North European deciduous forests (Thoen, 1970, 1971; Darimont, 1973).

Eighty five macrofungi were found only in fir woods, of which 28 have a high class of presence. Most species are linked to montane and submontane forests of the alliance *Fagion sylvaticae.* Thirteen species have been reported from coenoses with *Abies alba* Miller of Czechoslovakia (Smarda, 1973) and Poland (Bukakiewicz, 1982), namely: *Clitocybe vibecina* (Fr.) Quél., *Cortinarius flexipes* Fr., *Galerina stylifera* (Atk.) Smith & Sing., *Lachnellula subtilissima* (Cke.) Dennis, *Lactarius salmonicolor* Heim & Lecl., *Mycena amicta* (Fr.) Quél., *M. erubescens* Höhn., *M. flavescens* Vel., *M. zephyrus* (Fr.) ss. Höhnel, *Pseudohydnum gelatinosum* (Scop.: Fr.) Karst., *Rhodocybe nitellina* (Fr.) Sing., *Russula laurocerasi* Melz. and *R. puellaris* Fr. According to Kost & Haas (1989) *Amylostereum chailletii* (Pers.) Boudin, *Lachnellula subtilissima* (Cke.) Dennis, *Lactarius salmonicolor* Heim & Lecl., *Oudemansiella melanotricha* (Dörfelt) Mos. and *Russula cavipes* Britz. (ss. Heim.) are strictly associated with *A. alba.*

Conclusions

During these mycocoenological studies 654 fungal species have been collected in less than 5.5 hectares, showing the high species diversity of this area of the Mediterranean. The plant communities studied are well diversified in regard to the respective mycological population. In fact, 14.4% of species are widely distributed in the phytocoenoses and 54.9% associated with only one type of vegetation community.

The *Quercus ilex* woods of the Maremma coast have the most abundant mycoflora with 309 species, followed by the hill chestnut coppices (194 species). The juniper scrubs have the fewest fungi (83 species), but the area studied was smaller (900 m^2 compared with 10,000 m^2) and the studies are still in progress. However, it is unlikely that many more species will be found in such a hostile environment. It is worth adding that the plant communities with the richest and the poorest mycoflora are those least affected by man.

References

Arcangeli, G. (1899). Una rapida escursione a Moncioni ed a Brolio. *Nuovo Giorn. Bot. Ital., n.s.* **6**: 38–50.

Arnolds, E. (1984). Standard list of macrofungi in the Netherlands. *Coolia* **26**: 1–357.

Barkman, J.J. (1976). Algemene inleiding tot der oecologie en sociologie van macrofungi. *Coolia* **19**: 57–66.

Barluzzi, C. and Bianciardi, V. (1969). Studies on the macromycetes of the wood of Lecceto (Siena). *Giorn. Bot. Ital.* **103**(6): 599.

Barluzzi, C. and Bianciardi, V. (1970). Macromiceti raccolti nella provincia di Siena. *Inform. Bot. Ital.* **2**: 21–41.

Barluzzi, C., Mariani, M. and Appella, R. (1978). Macromiceti nuovi per il Senese. *Inform. Bot. Ital.* **10**: 229–238.

Barluzzi, C., De Dominicis, V. and Marchetti, P. (1980). Macromiceti nuovi per il Senese e per il Grossetano. *Inform. Bot. Ital.* **12**: 61–70.

Barluzzi, C., De Dominicis, V., Marchetti, P. and Perini, C. (1983a). Macromiceti nuovi per le province di Arezzo, Siena e Grosseto. *Micol. Ital.* **12**: 37–54.

Barluzzi, C., De Dominicis, V. and Perini, C. (1983b). Primi risultati di indagini micologiche sulle leccete del litorale Grossetano. *Micol. Ital.* **12**: 29–35.

Barluzzi, C., Perini, C. and De Dominicis, V. (1984). Comparazione preliminare tra micocenosi di leccete e di castagneti della Toscana centro-meridionale. *Giorn. Bot. Ital.* **118** (suppl. 2): 14–16.

Barluzzi, C., Perini, C., De Dominicis, V. and Marchetti, P. (1985). Contributo alla conoscenza della flora micologica delle province di Arezzo, Siena e Grosseto. *Micol. Ital.* **14**: 41–52.

Barluzzi, C., Perini, C. and De Dominicis, V. (1986). Ricerche geobotaniche in Val di Merse (Toscana meridionale). II. Micocenologia delle lande a calluna. *Micol. Ital.* **15**: 39–48.

Barluzzi, C., Perini, C., De Dominicis, V. and Barkman, J.J. (1990). Studi micocenologici in castagneti della Toscana centro meridionale. *Not. Soc. Ital. Fitosociol.* **23**: 189–196.
Barluzzi, C., Perini, C. and De Dominicis, V. (1992). Coenological research on macrofungi in chestnut coppices of Tuscany. *Phytocoenologia* **20**: 449–465.
Bergman, B., Stortelder, A. and Westhoff, V. (1988). Flora e vegetazione. In B. Pedroli, G.M. Vos, H. Dijkstra, and R. Rossi, Studio degli effetti ambientali della diga sul torrente Farma. pp. 174–188. Marsilio eds., Venezia.
Bernicchia, A. (1990). Polyporaceae s.l. in Italia. 594 pp. Ist. Pat. Veg. Bologna.
Bohus, G. and Babos, M. (1967). Mycocoenological investigation of acidophilous deciduous forets in Hungary. *Bot. Jahrb. Syst.* **87**: 304–360.
Bon, M. and Géhu, J.M. (1973). Unités supérieures de végétation et récoltes mycologiques. *Doc. Mycol.* **6**: 1–40.
Bujakiewicz, A. (1982). Fungi of Mt. Babia Gora. III. Indicating value of macromycetes in the forest associations. *Acta Mycol.* **18**: 3–44.
Calonge, F.D., De Dominicis, V., Barluzzi, C. and Perini, C. (1991). *Geastrum pseudolimbatum* Hollós and *G. saccatum* (Fr.) E. Fischer (Gasteromycetes) in Italy. *Bol. Soc. Micol. Madrid* **15**: 177–181.
Carta Geologica d'Italia. Scala 1:100,000, fogli 129, 135.
Darimont, F. (1983). Recherches mycologiques dans les forêts de haute Belgique. *Inst. Rech. Sci. Nat. Belg.* 170–220.
De Dominicis, V. and Barluzzi, C. (1983). Coenological research on macrofungi in evergreen oak woods in the hills near Siena (Italy). *Vegetatio.* **54**: 177–187.
De Dominicis, V., Loppi, S., Chiarucci, A., Mariotti, M.G., Perini, C. and Angiolini, C. (in press.) Woods with *Abies alba* of Mt. Amiata (Central Italy). *Doc. Phytosoc.*, in press.
Heinemann, P. and Darimont, F. (1956). Premières indications sur les relations entre les groupements végétaux et les champignons en Belgique. *Naturalistes Belges* **37**: 141–155.
Jansen, A.F. (1981). The vegetation and macrofungi of acid oak woods in the North East Netherlands. Thesis 131 pp. Wageningen.
Kost, G. and Haas, H. (1989). Ein Beitrag zur Kenntnis der Vergesellschaftung Höherer Pilze in einigen süddeutschen Waldgesellschaften. In Mykologische u Ökologische Untersuchungen in Waldschutzgebieten, Freiburg *Waldschutzgebiete* **4**: 9–182.
Moser, M. (1978). Die Röhrlinge und Blätterpilze. Fischer, Stuttgart and New York.
Pedrotti, F., Orsomando, E. and Cortini-Pedrotti, C. (1975). Carta della vegetazione del Lago di Burano e della Duna di Capalbio (Grosseto). Litografia Artistica Cartografica, Firenze.
Perini, C., Barluzzi, C. and De Dominicis, V. (1989). Mycocoenological research on evergreen oak woods in the hills adjacent the Maremma coastline (NW of Grosseto, Italy). *Phytocoenologia.* **17**: 289–306.
Runge, A. (1958). Zur Pilzflora des Naturschutzgebietes "Gelmer Heide". *Natur & Heimat* **18**: 1–8.
Runge, A. (1960). Pilzökologische und soziologische Untersuchungen in den Bockholter Bergen bei. Münster *Abh. Landesmus. Naturk. Münster.* **22**: 3–21.
Smarda, F. (1973). Die Pilzgesellschaften einiger Fichtenwälder Mährens. *Přir. Práce Ústavů Ceskoslov. Akad. Věd Brně* **7**: 1–44.
Tassi, A. (1862). Regno vegetabile. In Siena e il suo territorio. pp. XLI–CII Tip. R. Ist. Sordo-Muti L. Lazzeri, Siena.
Tassi, F. (1896). Micologia della provincia Senese. Imenomiceti. *Atti Reale Accad. Fisiocrit. Siena, ser. IV.* **8**: 321–337.
Tassi, F. (1897). Micologia della provincia Senese. Imenomiceti. *Atti Reale Accad. Fisiocriti Siena, ser. IV.* **9**: 113–147.
Tassi, F. (1906). Elenco generale dei funghi della provincia Senese rinvenuti fino a tutto l'anno 1905. *Bull. Lab. Orto Bot. Reale Univ. Siena* **8**: 223–254.
Thoen, D. (1970–1971). Etude mycosociologique de quelques associations forestières des district Picardo-Brabançon, mosan et ardennais de Belgique. *Bull. Inst. Agron. Etat Gembloux n.s.* **5**: 309–326; **6**: 215–243.
Trappe, J.M. (1962). Fungus associates of ectotrophic mycorrhizae. *Bot. Rev.* **21**: 538–606.
Valenti-Serini, F. (1868). Funghi sospetti e velenosi della provincia Senese. Torino.
Walter, H. and Leith, H. (1960). Klimadagram Weltatlas. Fischer, Jena.

TABLE 1. Presence of fungi in vegetational coenoses (arranged in order of increasing altitude).

Roman numerals indicate the class of presence (percentage of relevés): I, 1–20%; II, 21–40%; III, 41–60%; IV, 61–80%; V, 81–100%. Fungi found in only one type of vegetation with a presence of <40% are considered as 'sporadic species' and are not listed.

	JS	EOWc	EOWh	HT	hCC	smCC	FW
Average altitude (m)	8	119	333	421	498	785	810
Total number of species	83	309	177	141	194	183	184
Hebeloma crustuliniforme (Bull.: Fr.) Quél.	I	IV	III	IV	I	III	
Xerocomus chrysenteron (Bull. ex St. Am.) Quél.	III	IV	I		I	II	II
Mycena filopes (Bull.: Fr.) Kumm.	III	I		I	I	III	III
Clitocybe phaeophtalma (Pers.) Kuyp.	III	II		I	I	II	II
Laccaria laccata (Scop.: Fr.) Bk. & Br.		V	IV	V	IV	V	IV
Cortinarius trivialis Lge.		V	IV	I	III	III	III
Mycena galopoda (Pers.: Fr.) Kumm.		V	III	V	V	V	V
Mycena pura (Pers.) Kumm.		V	III	I	II	V	V
Cortinarius elatior Fr.		V	I	III	IV	II	V
Stereum hirsutum (Willd.: Fr.) S.F. Gray		IV	II	III	V	V	II
Tricholoma saponaceum (Fr.) Kumm.		IV	II	III	II	III	IV
Cantharellus cibarius Fr.		IV	I	II	V	V	II
Hydnum repandum L.: Fr.		II	II	II	V	III	II
Lyophyllum infumatum (Bres.) Kühn.		II	II	II	IV	II	II
Amanita rubescens (Pers.: Fr.) S.F. Gray		II	I	III	V	V	III
Phellinus torulosus (Pers.) Bourd. & Galz.	II	V	IV		I	II	
Lepista nuda (Bull.: Fr.) Cke.	II	V	III		II	III	
Crepidotus variabilis (Pers.: Fr.) Kumm.	II	II	I		IV	III	
Hebeloma sinapizans (Paul.: Fr.) Gill.		V	IV	I	II	III	
Russula fragilis (Pers.: Fr.) Fr.		V	III	V	IV	IV	
Amanita pantherina (DC.: Fr.) Secr.		V	III	II	III	IV	
Hydnum rufescens Pers.: Fr.		V	III	II	II	IV	
Amanita vaginata (Bull.: Fr.) Quél.		V	II	I	IV	IV	
Tricholoma ustaloides Romagn.		IV	V	II	III	III	
Trametes versicolor (L.: Fr.) Pilat		III	IV	II	II	II	
Dermocybe cinnamofulva R.Hry.		I	III	III	I	IV	
Mycena galericulata (Scop.: Fr.) S.F. Gray		I	III	II	III	II	
Lactarius chrysorrheus Fr.		V	IV	V	IV		II
Hygrophorus cossus (Sow. ex Bk.) Fr.		III	V	II	II		II
Lactarius subdulcis Bull.: Fr.		III	I	I	II		II
Cortinarius uraceus Fr.		II	III	I	III		III
Collybia dryophila (Bull.: Fr.) Kumm.		III	I	I		II	IV
Lycoperdon perlatum Pers. ex Pers.		V	I		III	IV	IV
Inocybe geophylla (Sow.: Fr.) Kumm.		IV	III		II	II	V
Cratherellus cornucopioides (L.: Fr.) Pers.		III	II		III	III	II
Lepiota clypeolaria (Bull.: Fr.) Kumm.		I	I		II	III	II
Mycena pelianthina (Fr.) Quél.		I	I		I	III	V
Lycoperdon pyriforme Schaeff. ex Pers.		V		IV	III	IV	III
Collybia butyracea (Bull.: Fr.) Quél.		V		I	I	II	IV
Clitocybe gibba (Pers.: Fr.) Kumm.		IV		I	I	IV	V
Mycena epipterygia (Scop.) S.F. Gray		I		I	II	III	V
Laccaria amethystina (Bolt. ex Hooker) Murr.		I		I	II	III	IV
Hypholoma fasciculare (Huds.: Fr.) Kumm.			I	III	I	IV	II
Armillariella mellea (Vahl.: Fr.) Kumm.	II	III	I		I		
Astraeus hygrometricus (Pers.) Morg.	II	II	I			III	
Marasmius androsaceus (L.: Fr.) Fr.	II		I		II		IV

TABLE 1 continued

	JS	EOWc	EOWh	HT	hCC	smCC	FW
Average altitude (m)	8	119	333	421	498	785	810
Total number of species	83	309	177	141	194	183	184
Hydnellum concrescens (Pers. ex Schw.) Banker		IV	IV	II	IV		
Phellodon niger (Fr.: Fr.) Karst.		IV	III	III	III		
Ganoderma lucidum (W. Curt.: Fr.) Karst.		III	II	I	I		
Russula albonigra Krombh.		III	I	I	II		
Tricholoma sejunctum (Sow.: Fr.) Quél.		II	I	II	I		
Ramaria flava (Fr.) Quél.		I	I	II	I		
Bovista plumbea Pers. ex Pers.		I	I	I	I		
Ramaria botrytis (Pers.: Fr.) Boud.		I	I	I	I		
Hygrocybe nigrescens (Quél.) Kühn.		V	II	II		II	
Macrolepiota procera (Scop.: Fr.) Sing.		V	I	I			II
Cortinarius anomalus (Fr.: Fr.) Fr.		I	II	I			II
Mycena rosea (Bull.) Sacc. & Dalla Costa		V	III		III	IV	
Russula chamaeleonthina (Fr.) Fr. ss. Romagn.		V	I		V	IV	
Coprinus plicatilis (Curt.: Fr.) Fr.		IV	II		IV	IV	
Hygrophorus nemoreus (Lasch) Fr.		IV	II		I	II	
Entoloma nidorosum (Fr.) Quél.		III	II		II	IV	
Entoloma sinuatum (Bull.: Fr.) Kumm.		III	II		I	III	
Cortinarius amoenolens R.Hry.		III	II		I	II	
Entoloma rhodopolium (Fr.) Kumm.		II	I		II	IV	
Russula atropurpurea Krbh.		I	III		I	II	
Marasmiellus ramealis (Bull.: Fr.) Sing.		I	I		III	II	
Lactarius uvidus Fr.		I	I		II	II	
Tricholoma argyraceum (Bull.: Fr.) Sacc.		II	II		I		II
Sarcoscypha coccinea (Scop. ex S.F.Gray)		III	I			III	II
Lepista nebularis (Fr.) Harmaja		II	I			V	II
Cortinarius duracinus Fr.		V		III	II	V	
Clavulina rugosa (Bull.: Fr.) Schroet.		V		I	I	III	
Amanita citrina (Schaeff.) S.F.Gray		IV		III	V	III	
Boletus edulis Bull.: Fr.		II		III	III	II	
Clavulina cinerea (Bull.: Fr.) Schroet.		II		II	I		
Russula delica Fr.		V			III	III	V
Lycoperdon atropurpureum Vitt.		V			I	III	III
Mycena vitilis (Fr.) Quél.		IV			III	III	IV
Amanita phalloides (Vaill.: Fr.) Secr.		IV			I	III	IV
Cortinarius torvus (Bull.: Fr.) Fr.		III			V	III	III
Russula cyanoxantha Schaeff.: Fr.		III			IV	V	III
Leotia lubrica Pers.		III			IV	IV	II
Lactarius pterosporus Romagn.		III			I	II	II
Mycena sanguinolenta (A.-S.: Fr.) Kumm.)		II			V	III	III
Xylaria hypoxylon (L. ex Hook.) Greville		II			I	IV	III
Collybia peronata (Bolt.: Fr.) Sing.		II			I	III	II
Cantharellus tubaeformis (Fr.) Quél.		I			V	III	II
Clitocybe odora (Bull.: Fr.) Kumm.		I			IV	IV	II
Boletus fetchneri Vel.		I			I	II	II
Cortinarius obtusus Fr.			II	III	III	II	
Tricholoma ustale (Fr.: Fr.) Kumm.			II	I	III	II	
Dermocybe semisanguinea (Fr.) Mos.			I	IV	I	II	
Rickenella fibula (Bull.: Fr.) Raith.				III	V	II	II
Agaricus silvaticus Schaeff.: Fr.	IV	III					II
Micromphale foetidum (Sow.: Fr.) Sing.	II	I					II
Hemimycena cephalotricha (Joss.) Sing.	I	III					II
Scleroderma verrucosum (Bull.) Pers.	II	II				IV	

TABLE 1 continued

	JS	EOWc	EOWh	HT	hCC	smCC	FW
Average altitude (m)	8	119	333	421	498	785	810
Total number of species	83	309	177	141	194	183	184
Cyathus striatus Huds. ex Pers.	I				I		II
Leucopaxillus gentianeus (Quél.) Kotl.		IV	I	I			
Russula luteotacta Rea		III	I	III			
Tricholoma flavovirens (Pers.: Fr.) Lund. & Nannf.		II	I	II			
Krombholziella nigrescens (Richon & Roze) Sutara		II	I	I			
Tremella lutescens (Pers. ex Pers.) Fr.		I	II	II			
Tricholoma atrosquamosum (Chev.) Sacc.		V	V		IV		
Macrolepiota mastoidea (Fr.) Sing.		V	I		IV		
Cortinarius sodagnitus R.Hry.		IV	IV		I		
Boletus aereus Bull.: Fr.		III	I		II		
Cortinarius trivialis Lge. var. squmosipes R.Hy.		III	I		I		
Lactarius vellereus (Fr.) Fr.		II	II		III		
Sarcodon cyrneus Maas G.		I	II		I		
Boletus calopus Fr.		I	I		II		
Cortinarius infractus (Pers.: Fr.) Fr.		V	V			II	
Cortinarius calochrous Fr.		V	IV			II	
Boletus luridus Schaeff.: Fr.		V	II			II	
Russula olivacea (Schaeff. ex Secr.) Fr.		II	I			II	
Tricholoma album (Schaeff.: Fr.) Quél.		I	IV			III	
Russula heterophylla (Fr.) Fr.		I	I			II	
Cortinarius citrinus Lge. ex Orton		I	I			II	
Tricholoma squarrulosum Bres.		V	I				II
Collybia marasmioides (Britz.) Brsky. & Stangl.		III	I				IV
Ramaria formosa (Pers.: Fr.) Quél.		I		I	III		
Amanita caesarea (Scop.: Fr.) Pers. ex Schw.		I		I	I		
Entoloma hirtipes (Schum.: Fr.) Mos.		IV		III		III	
Russula rosacea Pers. ex S.F.Gray		V			IV	III	
Russula vesca Fr.		V			IV	III	
Mycena metata (Fr.) Kumm.		V			III	II	
Xerocomus spadiceus (Fr.) Quél.		V			II	II	
Lactarius zonarius Fr.		IV			II	II	
Camarophyllus niveus (Scop.: Fr.) Karst.		IV			I	II	
Tricholoma sulphureum (Bull.: Fr.) Kumm.		III			II	V	
Armillariella tabescens (Scop.: Fr.) Sing.		III			I	II	
Lactarius atlanticus Bon		III			I	II	
Russula densifolia Secr. (ss. Romagn.)		III			I	II	
Russula aurata (With.) Fr.		II			III	III	
Mycena maculata Karst.		II			I	III	
Tricholoma bresadolianum Clem.		II			I	II	
Cortinarius crassus Fr. ss. Lge. non Fr.		II			I	II	
Oudemansiella longipes (Bull. ex St.-Am.) Mos.		I			IV	V	
Psathyrella hydrophila (Bull. ex Mer.) R. Mre.		I			III	V	
Clitopilus prunulus (Scop.: Fr.) Kumm.		I			III	IV	
Marasmius rotula (Scop.: Fr.) Fr.		I			III	III	
Russula romellii R.Mre.		I			I	IV	
Agaricus sylvicola (Vitt.) Sacc.		I			I	III	
Cortinarius trivialis Lge. var. *subolivascens* R.Hy.		I			I	III	
Lactarius camphoratus Bull.: Fr.		I			I	II	
Marasmius epiphyllus (Pers.: Fr.) Fr.		I			I	II	
Otidea cochleata (L. ex St.-Am.) Fuck.		I			I	II	
Cystolepiota sistrata (Fr.) Sing.		III			II		III
Crepidotus pubescens Bres.		III			I		II

TABLE 1 continued

		JS	EOWc	EOWh	HT	hCC	smCC	FW
Average altitude (m)		8	119	333	421	498	785	810
Total number of species		83	309	177	141	194	183	184
Inocybe sindonia (Fr.) Karst.			II				II	III
Marasmius lupuletorum (Weinm.) Fr.			I				III	II
Mycena vitrea (Fr.) Quél.			I				II	IV
Tricholoma scalpturatum (Fr.) Quél.				III	I	I		
Cortinarius castaneus (Bull.: Fr.) Fr.				I	I	I		
Tricholoma stans (Fr.) Sacc.				I	III			II
Hygrophorus chrysodon (Batsch) Fr.				I	I			III
Lactarius pergamenus (Swartz.: Fr.) Fr.				V		V	III	
Tricholoma acerbum (Bull.: Fr.) Quél.				V		III	IV	
Cortinarius rufoolivaceus Fr.				III		I	III	
Mycena polygramma (Bull.: Fr.) S.F.Gray				I		V	IV	
Mycena xantholeuca Kühn.				I		II	II	
Panellus stypticus (Bull.: Fr.) Karst.			I		I	IV		
Mycena pura (Pers.) Kumm. f. *alba* Gill.				I		I	III	
Cortinarius acutus Fr.					V	V	III	
Marasmius splachnoides Fr.					II	IV	II	
Tricholoma flavobrunneum (Fr.) Kumm.					III	I		II
Mycena oortiana Kühn. ex Hora					I	I		II
Galerina marginata (Fr.) Kühn.					IV		III	IV
Clitocybe suaveolens (Schum.: Fr.) Kumm.					II		II	III
Clavulina cristata (Fr.) Schroet.						III	III	V
Marasmius bulliardii Quél.						I	V	II
Rhodocybe truncata (Schaeff.: Fr.) Sing.	)	IV	II					
Agaricus praeclaresquamosus Freem. var. *praeclaresquamosus* Freem.	)	III	V					
Leucoagaricus serenus (Fr.) Bon & Boiffard	)	III	IV					
Hygrocybe konradii R. Haller	) A	III	III					
Xerocomus rubellus (Krbh.) Mos.	)	II	II					
Hydnocystis piligera Tul.	)	II	I					
Flammulaster carpophiloides (Kühn.) Watl.	)	I	V					
Volvariella murinella (Quél.) Mos.	)	I	II					
Hebeloma hiemale Bres.		I		III				
Mycena capillaripes Peck		III			I			
Xerocomus subtomentosus (L.: Fr.) Quél.		I				I		
Ripartites tricholoma (A.-S.: Fr.) Karst.		II					II	
Crucibulum laeve (Bull. ex DC) Kambly	)	IV						II
Citocybe font-queri Heim	)	II						IV
Ramaria flaccida (Fr.) Bourd.	)	II						III
Mycena leptocephala (Pers.: Fr.) Gill.	) B	I						V
Hemimycena gracilis (Quél.) Sing.	)	I						V
Dacrymyces stillatus Nees: Fr. ss. str.	)	I						IV
Bisporella citrina (Batsch.: Fr.) Korf & Carpenter	)	I						III
Krombholziella lepida (Bouch. ex Ess.) Alessio	)		V	V				
Hygrophorus russula (Schaeff.: Fr.) Quél.	) C		V	IV				
Hygrophorus dichrous Kühn. & Romagn. var. *fuscovinosus* Bon	)		IV	IV				
Boletus queletii Schulz.	)		IV	II				

TABLE 1 continued

		JS	EOWc	EOWh	HT	hCC	smCC	FW
Average altitude (m)		8	119	333	421	498	785	810
Total number of species		83	309	177	141	194	183	184
Lepista flaccida (Sow.: Fr.) Kumm.	)		IV	I				
Russula foetens Fr.	)		III	I				
Boletus satanas Lenz.	)		III	I				
Inocybe lacera (Fr.) Kumm.	)		II	I				
Cortinarius multiformis (Fr.) Fr.	)		II	II				
Lycophyllum immundum (Bk.) Kühn.	)		I	IV				
Hygrophorus arbustivus Fr.	)		I	III				
Sarcosphaera eximia (Durien & Lex) R.Mre.	) C		I	II				
Hygrocybe acutoconica (Clemens.) Sing.	)		I	II				
Cortinarius bulliardi (Fr.) Fr.	)		I	II				
Boletus impolitus Fr.	)		I	I				
Dermocybe cinnamomeolutea (Orton) Mos.	)		I	I				
Hygrophorus penarius Fr.	)		I	I				
Lyophyllum capniocephalum (Bull.: Fr.) Kühn.	))		I	I				
Lyophyllum transforme (Britz.) Sing.	)		I	I				
Tremella mesenterica Retz.: Fr.			V		III			
Lactarius violascens (Otto) Fr.			III		III			
Cantharellus cinereus Pers.: Fr.			II		II			
Cortinarius decipiens Fr.			II		I			
Cortinarius purpurascens Fr.			I		V			
Lyophyllum semitale (Fr.) Kühn.			I		III			
Amanita spissa (Fr.) Kumm.			I		I			
Geastrum fimbriatum Fr.			I		I			
Gymnopilus spectabilis (Fr.) Sing.			I		I			
Hygrocybe coccinea (Schaeff.: Fr.) Kumm.			I		I			
Russula alutacea (Pers.: Fr.) Fr.			IV			III		
Pseudocraterellus sinuosus (Fr.) Corner ex Heinem.		III			II			
Omphalotus olearius (DC: Fr.) Sing.			III			II		
Coprinus picaceus Bull.: Fr.			III			I		
Cortinarius croceo-coeruleus (Pers.: Fr.) Fr.			I			IV		
Lactarius controversus Pers.: Fr.			I			III		
Helvella crispa Scop.: Fr.			I			I		
Ramaria decurrens (Pers.) Pet.			V				IV	
Russula amoenicolor Romagn.			V				II	
Amanita crocea (Quél.) Sing.			II				III	
Inocybe cervicolor (Pers. ex Pers.) Quél.			II				II	
Pulveroboletus cramesinus (Secr.) Sing.			II				II	
Russula minutula Vel.			I				II	
Lepiota castanea Quél.			IV					II
Clavulinopsis corniculata (Schaeff.: Fr.) Corner			IV					II
Mycena stylobates (Pers.: Fr.) Kumm.			III					IV
Inocybe geophylla (Fr.: Fr.) Kumm. var. *lilacina* (Peck) Gill.			III					II
Byssomerulius corium (Fr.) Parm.			II					II
Cortinarius integerrimus Kühn.			I					IV
Schizopora paradoxa (Schrad.: Fr.) Donk			I					III
Hygrophorus discoideus (Pers.: Fr.) Fr.			I					II
Mycena haematopoda (Pers.: Fr.) Kumm.			I					II

TABLE 1 continued

		JS	EOWc	EOWh	HT	hCC	smCC	FW
Average altitude (m)		8	119	333	421	498	785	810
Total number of species		83	309	177	141	194	183	184
Inocybe similis Bres.	)			IV	III			
Ramaria aurea ss. lat.	)			II	III			
Lactarius deliciosus (L.: Fr.) S.F.Gray	)			II	II			
Suillus granulatus (L.: Fr.) Kuntze	)			II	II			
Mycena vulgaris (Pers.: Fr.) Quél.	)			II	I			
Suillus boudieri (Quél.) Watl.	)			II	I			
Cantharellus lutescens Pers.: Fr.	) D			I	V			
Mycena seynii Quél.	)			I	V			
Scleroderma citrinum Pers.	)			I	V			
Hydnum repandum L.: Fr. var. *albidum* (Quél.) Rea	)		I	III				
Tricholoma aurantium (Schaeff.: Fr.) Rick	)			I	II			
Clitocybe dealbata (Sow.: Fr.) Kumm.	)			I	I			
Cortinarius crystallinus Fr.				II		III		
Russula emetica Fr. var. *silvestris* Sing.				II		I		
Boletus appendiculatus Schaeff.: Fr.				I		II		
Cortinarius odoratus (Joguet ex Mos.) Mos.				I		I		
Cortinarius delibutus Fr.				I		I		
Entoloma incanum (Fr.) Hesler				I		I		
Trametes hirsuta (Wulf.: Fr.) Pilat				I		I		
Cortinarius dionysae R.Hry.				I				II
Entoloma sericellum (Bull.: Fr.) Kumm.					I	I		
Tremella foliacea (Pers. ex S.F.Gray) Pers.					I	I		
Schizophyllum commune Fr.: Fr.					III		III	
Paxillus involutus (Batsch: Fr.) Fr.					II		II	
Pseudoclitocybe obbata (Fr.) Sing.					I		II	
Amanita umbrinolutea Secr.					I		II	
Tricholomopsis rutilans (Schaeff.: Fr.) Sing.	) E				III			IV
Mycena inclinata (Fr.) Quél.	)					V	V	
Fistulina hepatica Schaeff.: Fr.	)					V	IV	
Trametes pubescens (Schum.: Fr.) Pilat	)					IV	III	
Marasmius cohaerens (Pers.: Fr.) Fr.	)					III	IV	
Russula laurocerasi Melz. var. *fragrans* Rom.	)					III	IV	
Clitocybe costata Kühn. & Romagn.	)					II	III	
Rutstroemia firma (Pers.) Karst.	)					II	II	
Hypholoma sublateritium (Fr.) Quél.	)					I	V	
Boletus erythropus (Fr.: Fr.) Pers.	) F					I	II	
Cantharellus friesii Quél.	)					I	II	
Collybia fusipes (Bull.: Fr.) Quél.	)					I	II	
Cortinarius pumilus (Fr.) Lge.	)					I	II	
Mycena galopoda (Pers.: Fr.) Kumm. var. *nigra* (Fl. Dan.)	)					I	II	
Rutstroemia echinophila (Bull. ex Mer.) Höhn	)					I	II	
Stropharia coronilla (Bull.: Fr.) Quél.	)					I	II	
Clitocybe brumalis (Fr.: Fr.) Kumm.						I		III
Oudemansiella radicata (Relhan.: Fr.) Sing.						I		III
Russula nigricans (Bull.) Fr.						I		III
Marasmius wynnei Bk. & Br.							III	II

TABLE 1 continued

	JS	EOWc	EOWh	HT	hCC	smCC	FW
Average altitude (m)	8	119	333	421	498	785	810
Total number of species	83	309	177	141	194	183	184
Pluteus atricapillus (Secr.) Sing.						II	IV
Calyocbe ionides (Bull.: Fr.) Donk.						II	II
Cantharellus amethysteus Quél.						II	II
Cortinarius paleaceus Fr.						II	II
Exidia glandulosa (Bull. ex St.-Am.) Fr. ss. Donk.						II	II
Lycoperdon umbrinum Pers. ex Pers.						II	II
Pseudoclitocybe cyathiformis (Bull.: Fr.) Sing.						II	II
Mycena meliigena (Berk. & Curt. in Cke.) Sacc.	V						
Geastrum saccatum Fr.	V						
Hemimycena lactea (Pers.: Fr.) Sing.	IV						
Pithya cupressina (Fr.) Fuckel	IV						
Melanoleuca stridula (Fr.) Métr.	IV						
Geastrum pseudolimbatum Hóllos	III						
Stereum rugosum (Pers.: Fr.) Fr.		V					
Lactarius decipiens Quél.		V					
Russula maculata Quél. & Roz.		V					
Russula pectinatoides Peck.		V					
Hygrocybe obrussea (Fr.) Wünsche		V					
Hygrocybe reai R. Mre.		V					
Gyroporus castaneus (Bull.: Fr.) Quél.		V					
Clavaria vermicularis Sw.: Fr.		IV					
Clavulinopsis pulchra Peck.		IV					
Camarophyllus pratensis (Pers.: Fr.) Kumm.		IV					
Lactarius rugatus Kühn. & Romagn.		IV					
Russula vinosobrunnea (Bres.) Romagn.		IV					
Suillus bovinus (L.: Fr.) Kuntze				V			
Cortinarius spilomeus (Fr.: Fr.) Fr.				V			
Gymnopilus penetrans (Fr.: Fr.) Murr.				V			
Lyophyllum loricatum (Fr.) Kühn.				V			
Gomphidius roseus (L.) Fr.				V			
Russula sardonia Fr. em. Romagn.				III			
Laccaria proxima (Boud.) Pat.				III			
Cortinarius fulvescens Fr. (ss. Favre)				III			
Amanita citrina Schaeff.: Fr. var. *alba* Price				III			
Cortinarius hoeftii Weinm. ap. Fr.				III			
Ramaria fumigata (Pk.) Corner					III		
Psilocybe crobula (Fr.) M.Lge. ex Sing.					III		
Stropharia aeruginosa (Kurt.: Fr.) Quél.						III	
Flammulina velutipes (Curt.: Fr.) Sing.						III	
Cortinarius alboviolaceus (Pers.: Fr.) Fr.						III	
Inocybe fuscidula Vel. var. *fuscidula*							V
Mycena flavoalba (Fr.) Quél.							V
Mycena zephirus (Fr.: Fr.) Kumm.							V
Mycena flavescens Vel.							V
Mycena erubescens Höhn.							V
Cortinarius flexipes Fr.							V
Collybia confluens (Pers.: Fr.) Kumm.							V
Pseudohydnum gelatinosum (Scop.: Fr.) Karst.							V
Galerina stylifera (Atk.) Smith & Sing.							V
Lactarius salmonicolor Heim & Lecl.							V
Russula viscida Kudr.							V
Trichaptum abietinum (Dicks.: Fr.) Ryv.							V

TABLE 1 continued

	JS	EOWc	EOWh	HT	hCC	smCC	FW
Average altitude (m)	8	119	333	421	498	785	810
Total number of species	83	309	177	141	194	183	184
Mycena polyadlpha (Lasch.) Kühn.							IV
Hygrophorus pudorinus (Fr.) Fr.							IV
Oudemansiella melanotricha (Dörfelt) Mos.							IV
Russula puellaris Fr.							IV
Calocera viscosa (Pers.: Fr.) Fr.							III
Rhodocybe nitellina (Fr.) Sing. (ss. Kühn.)							III
Lachnellula subtilissima (Cke.) Dennis							III
Mycena amicta (Fr.) Quél.							III
Russula cavipes Britz. (ss. Heim)							III
Russula violeipes Quél. f. *citrina* Quél.							III
Russula laurocerasi Melz.							III
Clitocybe vibecina (Fr.) Quél.							III
Tephrocybe coracina (Fr.) Mos.							III
Agaricus dulcidulus Schulz.							III
Amylostereum chailletii (Pers.) Boudin							III
Podofomes corrugis (Fr.) Pouz.							III
Sporadic species	43	78	39	29	18	20	57

Lado, C. (1993). Myxomycetes of mediterranean woodlands. In D.N. Pegler, L. Boddy, B. Ing & P.M. Kirk (Editors). Fungi of Europe: Investigation, Recording and Conservation, pp. 93–114. Royal Botanic Gardens, Kew.

8. MYXOMYCETES OF MEDITERRANEAN WOODLANDS

C. LADO

Real Jardín Botánico, CSIC, Plaza de Murillo, 2. 28014 Madrid, Spain

Abstract

Despite the abundance and worldwide occurrence of the Myxomycetes, floristic studies in the Mediterranean region are lacking and there are still some totally unexplored countries. The region includes a wide range of climatic and ecological conditions, which support one of the richest Myxomycete floras in the world. According to our database, 413 species, which represent approximately 70% of the world species, are reported from that area. This richness is examined in relation to altitudinal vegetation zones. The more characteristic woodlands with their specific myxoflora, e.g. *Physarum brunneolum*, or *Didymium laxifilum* linked with holm oak (*Quercus ilex*), and some of the more characteristic types of Mediterranean vegetation, e.g. holm oak woods, cork trees, and Mediterranean coniferous forests with *Pinus halepensis*, *P. pinaster* or *Juniperus thurifera*, are highlighted. The species phenology, which is noticeably different from that in the rest of Europe, is discussed.

Introduction

Myxomycetes (plasmodial slime moulds) are common inhabitants of decaying plant material in forest ecosystems throughout the world. Despite their abundance and widespread occurrence, floristic and ecological studies in many regions of Europe are still lacking. Such is the case for the Mediterranean basin, where the information we have from some countries is meagre or non-existent, and where there are still large unexplored areas of great ecological value, such as National Parks and nature reserves. Nevertheless, the past few years have seen much progress in the knowledge of the Mediterranean Myxomycetes, especially those towards the west (Pirola, 1968; Cochet, 1977; Cochet & Bozonnet, 1980, 1984; Orsino & Traverso, 1983, 1987; Credaro & Pirola, 1984, 1986; Almeida, 1987, Lado, 1991). This enables us to make a coherent estimate of the floristic and ecological significance of Myxomycetes in the Mediterranean region.

Geographical limits and General Features of the Mediterranean

The Mediterranean region, north of the tropics, is formed by similar lands which belong to three continents, Europe, Africa and Asia, and almost enclose a large sea (Fig. 1). It extends between ca. 30° and 45° north, with a surface area of approximately 2,300,000 km² (Quézel, 1985).

It is not a homogeneous geographical unit, having rugged massifs often fragmented by deep narrow valleys, highlands and enormous sedimentary basins, jagged and winding coastlines with numerous islands. The very wide range of climatic and ecological conditions, which lead to the highest level of European biodiversity, is also noticeable. The annual rainfall ranges between 100 mm in some subdesert zones and 3,000 mm on

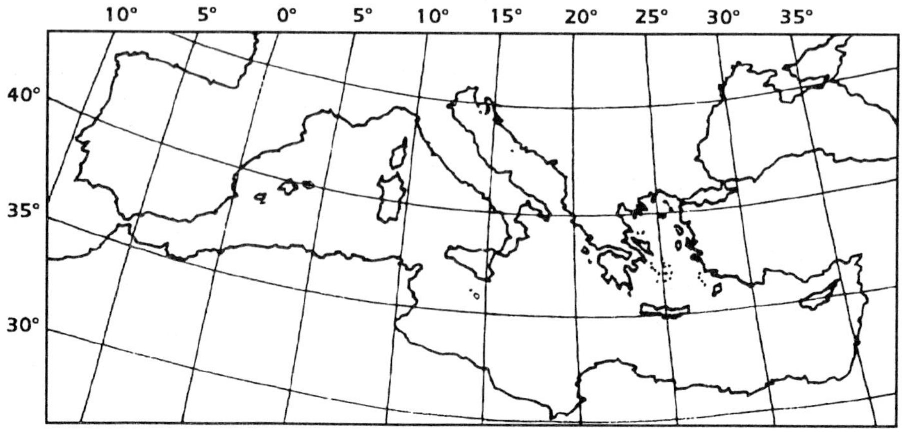

FIG. 1. Geographic situation of the Mediterranean region.

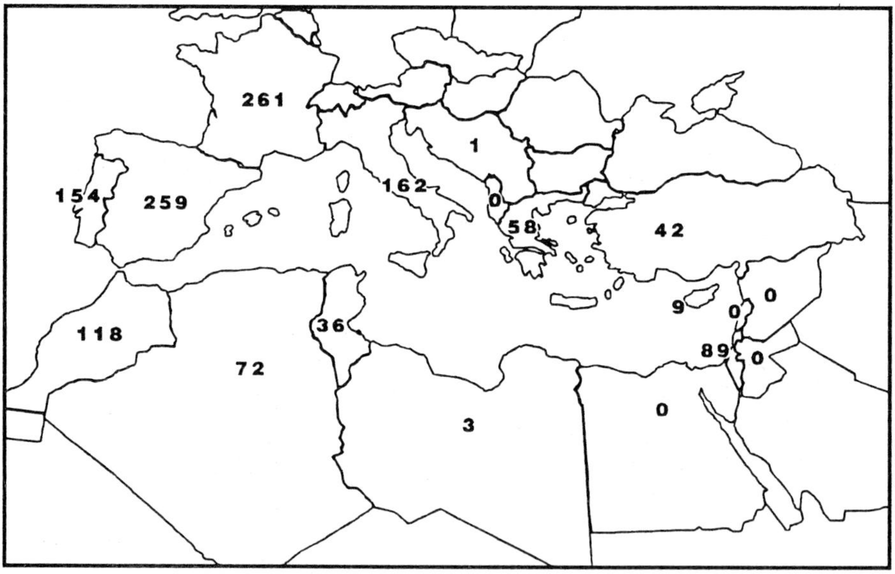

FIG. 2. Number of species of Myxomycetes reported for the circum-Mediterranean countries.

certain mountains. Average annual temperatures range from 5–18°C (Quézel, 1985). A common feature in all regions is a summer dry period of more than two months.

With regard to the vegetation, it is the richest and most varied of Europe (Unesco, 1969; Ozenda et al., 1979, Quézel, 1985) and one of the richest per square kilometre in the world.

Floristic Wealth and Inventory

Knowledge of the Mediterranean myxoflora is fragmentary as indicated by the variation in number of species recorded in circum-Mediterranean countries (Fig. 2). Some countries are relatively well studied, such as France, Italy, Morocco, Portugal and Spain (Buchet et al., 1912, 1920; Buchet, 1942; Faurel et al., 1964, Pirola, 1968; Pirola & Credaro, 1971; Rammeloo, 1973; Cochet, 1977; Cochet & Bozonnet, 1980, 1984; Orsino & Traverso, 1983, 1987; Credaro & Pirola, 1984, 1986; Lado, 1991). There are little studied countries, including Algeria, Greece, Israel, Tunisia and Turkey, partially explored by Maire et al., 1926; Alexopoulos, 1959; Faurel et al., 1964; Ramon, 1969; Härkönen & Uotila, 1983; Mitchell & Kylin, 1984; Binyamini, 1986, 1987, 1991; and Härkönen, 1988. Other countries have been almost unexplored, including Albania, Cyprus, Egypt, Jordan, Lebanon, Libya, Malta, Syria and former Yugoslavia (El-Buni & Rattan, 1981; Ing, 1987).

In spite of their irregular study 413 species have been catalogued for the whole region (excluding varieties) which represents approximately 70% of the species known worldwide (Nannenga-Bremekamp, 1991).

Knowledge of the ecology of the Myxomycetes is very meagre, although recently some intensive studies (Maimoni-Rodella & Gottsberger, 1980; Eliasson, 1981; Stephenson & Studlar, 1985; Stephenson, 1988, 1989) have attempted to correlate distribution patterns with a number of environmental factors. Others have discussed relationships with kinds of wood, types of habitat, substrate pH, time of the year, types of plant communities, etc. (Emoto, 1939; Pirola & Credaro, 1975, 1988; Eliasson, 1977; Härkönen, 1977, 1978;

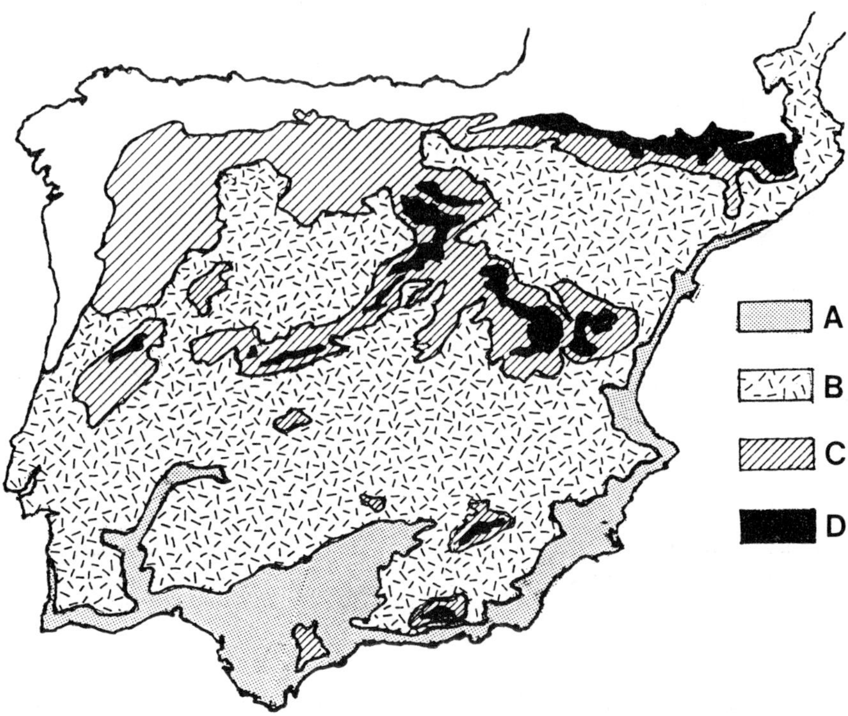

FIG. 3. Mediterranean altitudinal vegetation zones on the Iberian Peninsula. A. Thermo-Mediterranean. B. Meso-Mediterranean. C. Supra-Mediterranean. D. Oro-Mediterranean and Crioro-Mediterranean.

Stojanowska, 1980a, 1980b), but only Pirola & Credaro, (1975), provide information on some myxomycete communities in the Mediterranean basin.

Myxomycetes Associated with Mediterranean Woodlands

Knowledge of the distribution and ecology of Myxomycetes in Mediterranean woodlands is necessary for a more complete understanding of the relative importance and role of Myxomycetes in terrestrial ecosystems (Stephenson, 1988). Due to the dispersed and fragmentary nature of the information, only the numerous data (more than 2,000 records) on the Iberian Peninsula in which are found almost all types of Mediterranean woodlands are considered here. Altitudinal vegetation zones (Fig. 3) in the Mediterranean region (Ozenda et al., 1979; Quézel, 1985; Rivas-Martínez, 1987) are treated separately.

A. Infra-Mediterranean Level

The climate is subdesertic (Table 1). It occurs in far western Morocco, but is not represented in the Iberian Peninsula nor in Europe. Few tree species are present: *Argania spinosa*, *Acacia cyanophylla* and *A. gummifera*. Floristic data are limited to a few winter records (November to March) by Malençon & Bertault (1968, 1969) when there is minimal insolation and highest atmospheric humidity. Species present are tolerant of low water, e.g. *Arcyria cinerea* (Bull.) Pers., or use the dew and water accumulated in the tissues of fallen dead succulent plants, e.g. *Badhamia gracilis* (T.Macbr.) T.Macbr. and *Didymium vaccinum* (Durieu & Mont.) Buchet.

B. Thermo-Mediterranean Level

Climatic features are represented in Table 1. The area covered on the Iberian Peninsula is a narrow coastal strip (Fig. 3). The forests are primarily evergreen and formed by sclerophyllous stands of *Olea europaea* or *Ceratonia siliqua*, and conifers such as *Pinus halepensis*. There are remains of natural woodlands although most areas have been affected by human activity (farming, grazing, fires, forestry practices etc.).

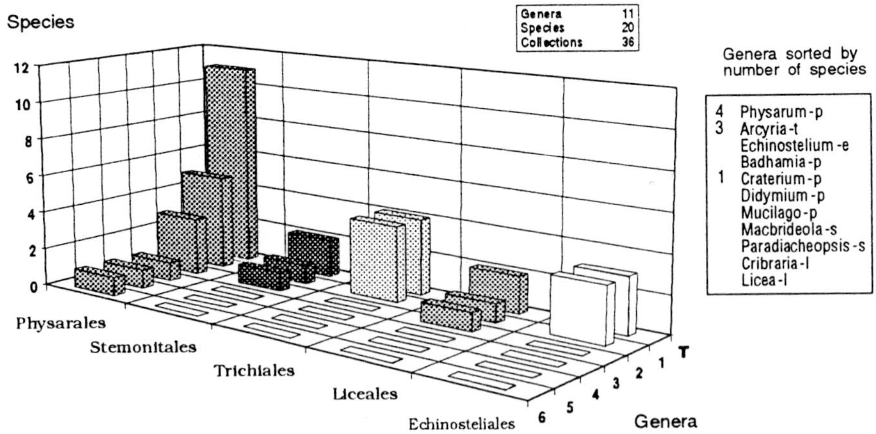

FIG. 4. Genera of Myxomycetes on *Olea europaea*. T. Total species. e. *Echinosteliales*. l. *Liceales* p. *Physarales*. s. *Stemonitales*. t. *Trichiales*.

TABLE 1. Climatic parameters of the altitudinal vegetation zones

	Infra-Med.	Thermo-Med.	Meso-Med.	Supra-Med.	Oro-Med.	Crioro-Med.
Annual precipitation	< 200 mm	200–400 mm	400–600 mm	600–800 mm	800–1200 mm	>1200 mm
Mean annual temperature	>19°C	17–19°C	13–17°C	8–13°C	4–8°C	<4°C
Average altitude	0–200 m	0–400 m	0–800 m	800–1600 m	1600–2200 m	2200–3200 m
Bioclimatic type	per-arid	arid	semi-arid	sub-humid	humid	per-humid

Olea europaea has been little studied. Available data (Fig. 4) refer to samples taken from scattered, old specimens and to bark cultures in moist chambers. There are barely 20 species in 11 genera. It is difficult to draw conclusions except to point out the presence of small-sized corticolous species of *Echinostelium* de Bary, *Macbrideola* H.C. Gilbert or *Paradiacheopsis* Hertel, such as *E. apitectum* K.D. Whitney, *E. colliculosum* K.D. Whitney & H.W. Keller, *M. decapillata* H.C. Gilbert or *P. solitaria* (Nann.-Bremek.). These species have short life cycles (often completing them in 24 h), and are probably adapted to xeric environments, as they can survive dry periods and make use of seasonal rainfall.

Data on *Pinus halepensis* (Fig. 5) contrast with those of *O. europaea*. The former woods are less widely disturbed, more capable of regeneration, more extensive and better studied. The Myxomycete diversity of these Mediterranean conifer woods is of note; 78 species in 28 genera have been reported, i.e. twice as many genera, and three times as many species as in *O. europaea*. Genera such as *Didymium* Schrad. and *Stemonitis* Roth, with seven taxa each, and *Physarum* Pers. and *Arcyria* F.H. Wigg. with 11, are the best represented, while *Hemitrichia* spp. are unrecorded. The more regularly found and frequent species (Table 2) are: *Comatricha nigra* (Pers. ex J.F. Gmel) J. Schröt. (47 records), *Lycogala epidendrum* (L.) Fr. (28), *Arcyria obvelata* (Oeder) Onsberg (22), *Physarum nutans* Pers. (20), *Didymium squamulosum* (Alb. & Schwein.) Fr. (15), *A. ferruginea* Saut. (14), *P. brunneolum* (W. Phillips) Massee (11), *Badhamia foliicola* Lister (11) and *Licea minima* Fr. (10). All these, except *P. brunneolum*, are ubiquitous and of worldwide distribution.

Some species have much narrower distributions: *Arcyria annulifera* Lister & Torrend, known only from Portugal (Torrend, 1909), whence it was described, and Spain (Gràcia & Llimona, 1980; Gràcia, 1987), and appears linked with the thermophile conifers; *Physarum listeri* T. Macbr., described by Lister & Lister (1904) as *P. luteoalbum* Lister & G. Lister from Ventimiglia, on the Ligurian coast of Italy, and other thermic localities; and *Macbrideola oblonga* Pando & Lado described recently (Pando & Lado, 1988) on bark of *Juniperus thurifera* and collected on *Pinus halepensis* (Carilla & Gràcia, 1991). Other species are rare worldwide but relatively frequent in these warm habitats: *Badhamiopsis ainoae* (Yamash.) T.E. Brooks & H.W. Keller, *Echinostelium colliculosum* K.D. Whitney & H.W.

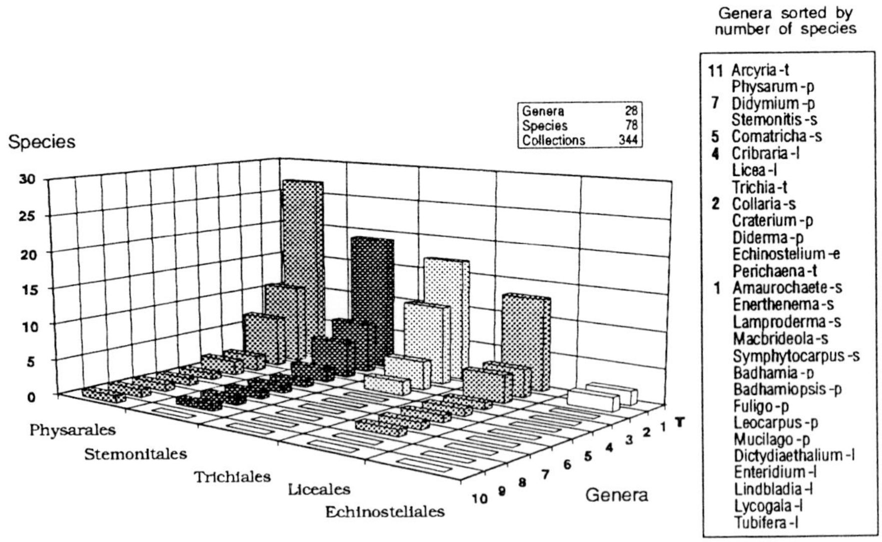

FIG. 5. Genera of Myxomycetes on *Pinus halepensis*.

TABLE 2. Species of Myxomycetes in the woodlands study

	Pinus halepensis	*Pinus pinaster*	*Pinus sylvestris*	*Juniperus thurifera*	*Olea europaea*	*Quercus ilex*	*Quercus suber*	*Quercus pyrenaica*	*TOTAL*
AMAUROCHAETE									
atra	2	3	3						8
ARCYRIA									
affinis		2	1				4		7
annulifera	3	1							4
cinerea	5	2	8		1	9	4	6	35
denudata	2		7				7	1	17
ferruginea	14	10	2					1	27
globosa							1		1
incarnata	9	14	26	2	1	12	2	4	70
insignis	1								1
major	1				1			1	3
minuta	1	5				2	3		11
obvelata	22	13	10			1	3	2	51
oerstedtii	3	3	3						9
pomiformis	7	15	5		1	8	2	1	39
versicolor			2						2
BADHAMIA									
affinis				7					7
capsulifera							1		1
foliicola	11	7	1	1	1	18	6	2	47
macrocarpa		1		1		1		1	4
obovata						7	4		11
panicea				2				7	9
utricularis		2		1	1	2	2	2	10
versicolor				18	5	5			28
BADHAMIOPSIS									
ainoae	2			2		1			5
CALOMYXA									
metallica		1		4		1			6
CLASTODERMA									
pachypus								2	2
COLLARIA									
arcyrionema	1				1				2
rubens	1	3				4			8
COLLODERMA									
macrotubulatum		1							1
COMATRICHA									
alta	1								1
elegans	2	1	3	9					15
ellae	1	4	5					10	
laxa	4			3		4		1	12
lurida		4		2		2		1	9
nigra	47	22	26			12	2	8	117
pulchella	4	3	2			1	1		11
tenerrima						1	1		2
CRATERIUM									
aureum								2	2
leucocephalum	6	1	2		2	26	1	8	46
minutum	1	1				17			19
CRIBRARIA									
argillacea	3	12	7			2			24
aurantiaca		3	1						4

TABLE 2 continued

	Pinus halepensis	*Pinus pinaster*	*Pinus sylvestris*	*Juniperus thurifera*	*Olea europaea*	*Quercus ilex*	*Quercus suber*	*Quercus pyrenaica*	TOTAL
cancellata	6	15	15						36
microcarpa		2	1						3
mirabilis			13					13	
persoonii			2						2
piriformis			2						2
rufa			5						5
violacea	1				3				4
vulgaris	1	11	21						33
DIACHEA									
leucopodia		1							1
DIACHEOPSIS									
nannengae			1						1
DIANEMA									
corticatum			1						1
harveyi				4		1			5
DICTYDIAETHALIUM									
plumbeum	1					2			3
DIDERMA									
asteroides	2	2			1	4		1	10
chondrioderma						2			2
cinereum								2	2
globosum						1			1
hemisphaericum				1					1
niveum			1						1
radiatum						1		1	2
sauteri							1		1
spumarioides	3					5		2	10
trevelyanii		4							4
umbilicatum		3	1	1		3		2	10
DIDYMIUM									
anellus								1	1
clavus						2	2	3	7
crustaceum						1			1
difforme	1		1	5		1			8
dubium						2			2
eximium	1	2				3		1	7
laxifilum		1				18	3	1	23
melanospermum	1	4	3			12		2	22
minus	1	2	1			4			8
muscorum		1				4		2	7
nigripes	2	1	1			27	19		50
ovoideum								1	1
quitense	2								2
serpula								1	1
squamulosum	15	9	3		8	29	3	7	74
sturgisii				27					27
trachysporum				5					5
ECHINOSTELIUM									
apitectum				1	1				2
brooksii						1			1
colliculosum	1			18	1	1	1		22
corynophorum				1					1
fragile				2		1			3

TABLE 2 continued

	Pinus halepensis	*Pinus pinaster*	*Pinus sylvestris*	*Juniperus thurifera*	*Olea europaea*	*Quercus ilex*	*Quercus suber*	*Quercus pyrenaica*	*TOTAL*
minutum	5		2	1	1	1	2	1	13
ENERTHENEMA									
papillatum	8	15	10			3	1	2	39
ENTERIDIUM									
liceoides		1	3						4
lycoperdon	8	1	5			4	9	1	28
olivaceum				4					4
splendens							1		1
FULIGO									
septica	3	9	20			2	3	5	42
HEMITRICHIA									
abietina		1	1						2
calyculata		1							1
clavata			5						5
leiotricha								1	1
LAMPRODERMA									
sauteri			12						12
scintillans	3	1				3		1	8
LEOCARPUS									
fragilis	2	10	5			36	1	1	55
LEPIDODERMA									
granuliferum				20					20
LEPTODERMA									
iridescens			2						2
LICEA									
castanea	1			41					42
kleistobolus	2	1	9		1				13
marginata	2						1		3
minima	10	8	11			3			32
nannengae				19					19
parasitica			1	16		2	2		21
perexigua			3						3
pusilla			2						2
testudinacea		6							6
variabilis		2	2						4
LINDBLADIA									
tubulina	2	1	4						7
LYCOGALA									
epidendrum	28	31	40			2	3	1	105
exiguum		1	1						2
flavofuscum							1		1
MACBRIDEOLA									
cornea				76		1			77
decapillata				20	1	1			22
macrospora							1	1	2
oblonga	4			2		4			10
synsporos						1			1
METATRICHIA									
vesparium			5				4		9
MUCILAGO									
crustacea	4				2	1			7
OLIGONEMA									
fulvum				1					1

TABLE 2 continued

	Pinus halepensis	*Pinus pinaster*	*Pinus sylvestris*	*Juniperus thurifera*	*Olea europaea*	*Quercus ilex*	*Quercus suber*	*Quercus pyrenaica*	*TOTAL*
PARADIACHEOPSIS									
fimbriata		1	1			1			3
solitaria					1		1		2
PERICHAENA									
chrysosperma		1							1
corticalis	8		1						9
depressa	1								1
vermicularis				17		2			19
PHYSARUM									
bethelii	1					2		1	4
bitectum	2	1				9		7	19
bivalve						5	1	4	10
bogoriense						2		5	7
brunneolum	11					9		4	24
cinereum						8	1	2	11
compressum					1	4	2		7
contextum	2							1	3
decipiens				1	3	3	1		8
diderma						1			1
flavicomum							1		1
leucophaeum	2	3	1			7	2	19	
leucopus						1			1
listeri	3								3
luteolum								1	1
melleum						3			3
newtonii	2					1			3
nudum								1	1
nutans	20	8	9		1	37	5	16	96
oblatum				1		1	1		3
penetrale		1							1
pusillum	1			1		2			4
robustum						1			1
serpula	3				2	1			6
vernum			2	3		2			7
viride	1	4	11		3	1		1	21
xanthinum						1			1
PROTOTRICHIA									
metallica		1	1						2
STEMONITIS									
axifera	1	22	1						24
emotoi				1					1
flavogenita	1					2			3
fusca	5	13	7			4		4	33
herbatica	3								3
nigrescens	1	2		1	1	1			6
smithii		2	1				1		4
splendens	1	5							6
virginiensis	2	6	3			2			13
STEMONITOPSIS									
amoena		5	3						8
hyperopta		3	1						4
typhina						1	2	1	4

TABLE 2 continued

	Pinus halepensis	*Pinus pinaster*	*Pinus sylvestris*	*Juniperus thurifera*	*Olea europaea*	*Quercus ilex*	*Quercus suber*	*Quercus pyrenaica*	TOTAL
SYMPHYTOCARPUS									
amaurochaetoides			1						1
flaccidus	3	7	2						12
implexus			1						1
TRICHIA									
affinis							1		1
botrytis	5	5	7			6		3	29
contorta	5	2		6		7	1	2	23
decipiens	1	1	9			12	3	3	29
favoginea		1	16			3	4		24
flavicoma								3	3
lutescens	4	4							8
persimilis							1		1
scabra							2		2
sordida			3						3
varia			1			2	9	5	17
TUBIFERA									
ferruginosa	2	4	10						16

Keller, *Licea marginata* Nann-Bremek., *Physarum newtonii* T. Macbr. and especially *P. brunneolum*.

The sporulation phenology on *Pinus halepensis* (Fig. 6) shows a maximum of fruiting in the spring (April, correlating with spring rains), and moderate but sustained values through autumn and winter, the latter contrasting with the near absence of myxomycete

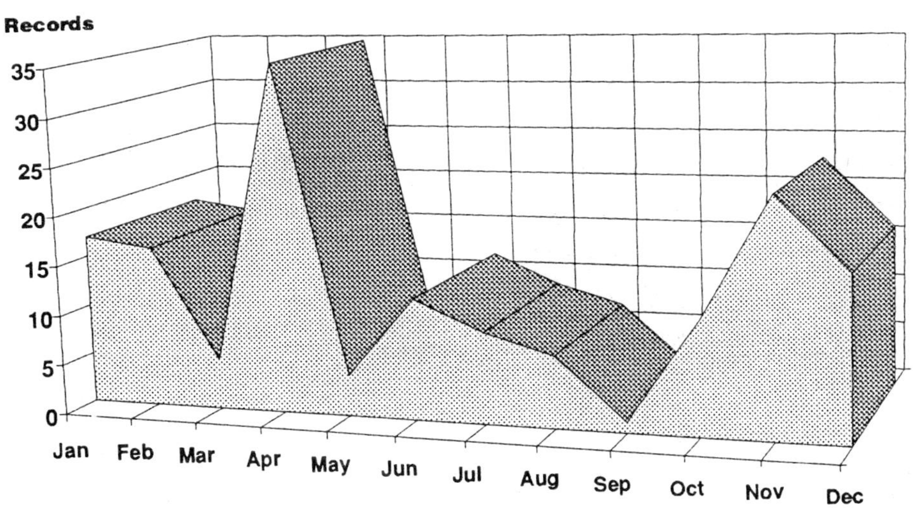

FIG. 6. Sporulation phenology of Myxomycetes on *Pinus halepensis*.

activity in non-Mediterranean Europe Bjørnekaer & Klinge, 1963). The higher ambient humidity of the winter season aided by lower insolation, lower evapo-transpiration, nocturnal dews, relatively mild daytime temperatures and absence of night frosts enables plasmodia to develop in the presence of more water. During the summer (six months from May to October) scarcely any activity occurs and some of the samples are old, effete fruitbodies.

C. Meso-Mediterranean Level

It is very well represented on the Iberian Peninsula (Fig. 3), extending over a broad area of medium altitude (North and South plateau) with mostly sclerophyllous forests of *Quercus ilex* s.l. (Barbero et al., 1992). The conifer woods are typified by *Pinus pinaster*.

The evergreen oak woods (*Q. ilex* s.l.) are the most characteristic of the Iberian vegetation. Species diversity is the highest (90 Myxomycete species in 28 genera; Fig. 7) of all vegetation zones. This coriaceous substrate, poor for water storage, is correlated with the high numbers of species having carbonates. Fifty six percent of the total species are *Physarales*: there are 21 taxa of *Physarum*, 11 of *Didymium*, and 6 of *Diderma* Pers. At the species level (Table 2) *Physarum nutans* (37 records), *Leocarpus fragilis* (Dicks.) Rostaf. (36), *Didymium squamulosum* (29), *D. nigripes* (Link) Fr. (27) and *Craterium leucocephalum* (Pers. ex J.F. Gmel) Ditmar (26) are the most frequent. They are all cosmopolitan, and found on twigs, leaves and plant debris.

Taxa such as *Physarum brunneolum* or *Didymium laxifilum* G. Lister & J. Ross are very rare or absent elsewhere in Europe, finding their ideal habitat in these sclerophyllous Mediterranean woods. Others, such as *P. bogoriense* Racib. or *P. melleum* (Berk. & Broome) Massee are reminiscent of more tropical environments (Martin & Alexopoulos, 1969; Eliasson, 1991).

Some lignicolous species such as *Badhamia obovata* (Peck) S.J. Sm., *Trichia botrytis* (J.F. Gmel.) Pers., *T. contorta* (Ditmar) Rostaf. and *T. decipiens* (Pers.) T. Macbr. are common, whereas species which are frequent elsewhere, e.g. *Fuligo septica* (L.) F.H. Wigg., *Mucilago crustacea* F.H. Wigg. or *Lycogala epidendrum* (L.) Fr., are very rare here. The absence of

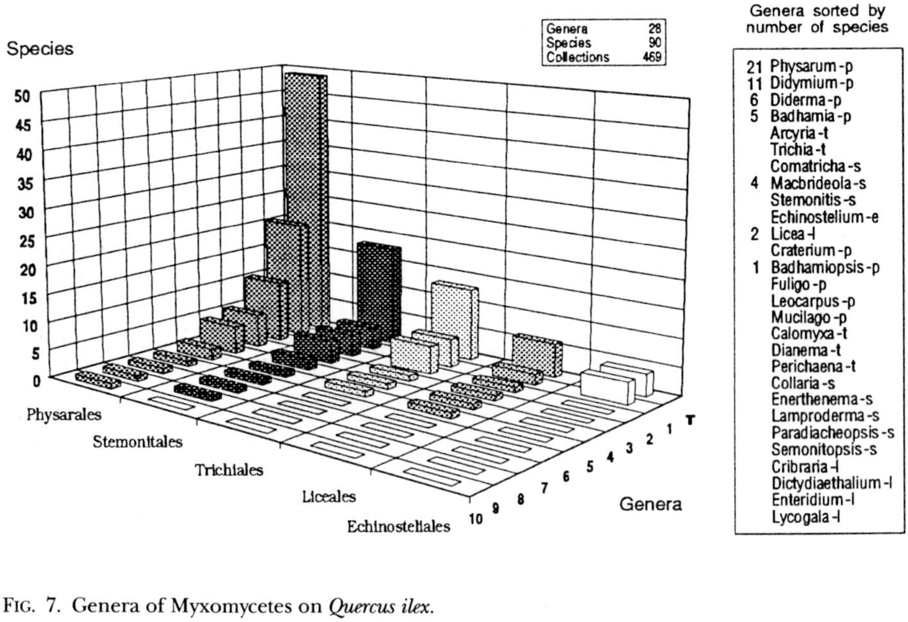

FIG. 7. Genera of Myxomycetes on *Quercus ilex*.

Hemitrichia species is striking. Corticolous species constitute an important part of the myxoflora. *Echinostelium* and *Macbrideola* species, e.g. *E. brooksii* K.D. Whitney, *E. fragile* Nann.-Bremek., *E. colliculosum*, *M. synsporos* (Alexop.) Alexop., *M. decapillata* and *M. oblonga* are found frequently.

The higher annual precipitation (Table 1); up to 600 mm with maxima in spring and autumn, conditions the occurrence of Myxomycete fruitbodies (Fig. 8). Patterns are similar to those found by Härkönen (1988) in Turkey: they remained constant from February to May, coinciding with spring rains; this is followed by an almost inactive summer correlating with high temperature and low water availability; then an increase in the autumn period, peaking in November, which is later than other European latitudes.

Cork oak (*Quercus suber*) forms a very similar Mediterranean woodland, but with a stronger Atlantic influence. Its Myxoflora is similar to that of the holm oak (*Q. ilex*), but the higher humidity is suitable for other species. e.g. *Metatrichia vesparium* (Batsch) Nann.-Bremek. ex G.W. Martin & Alexop. and *Stemonitopsis typhina* (F.H. Wigg.) Nann.-Bremek.

The conifer woods, represented by *Pinus pinaster*, have high species diversity: 77 species in 30 genera (Fig. 9). There are fewer foliicolous species than lignicolous species, e.g. *Arcyria* (9 species), *Stemonitis* (6 species), and *Cribraria* Pers. or *Lindbladia* Fr. species which are almost exclusively associated with conifer wood. A possible explanation is in the nature of the wood, which is more porous and has a higher water holding capacity, thus favouring the development of plasmodia. The genus *Physarum* (5 species) is poor compared with the 21 species on *Quercus ilex*.

The most abundant species (Table 2) are *Lycogala epidendrum* with 31 records, *Stemonitis axifera* (Bull.) T. Macbr. and *Comatricha nigra* with 22, *Arcyria pomiformis* (Leers) Rostaf., *Cribraria cancellata* (Batsch) Nann.-Bremek. and *Enerthenema papillatum* (Pers.) Rostaf. with 15, *A. incarnata* (Pers. ex J.F. Gmel) Pers. with 14 and *A. obvelata* and *S. fusca* with 13. All occur worldwide and are associated with conifer wood, but contrast markedly with previous vegetation zones.

Hemitrichia Rostaf. species such as *H. abietina* (Wigand) G. Lister and *H. calyculata* (Speg.) M.L. Farr are found here but not in other Mediterranean vegetal communities.

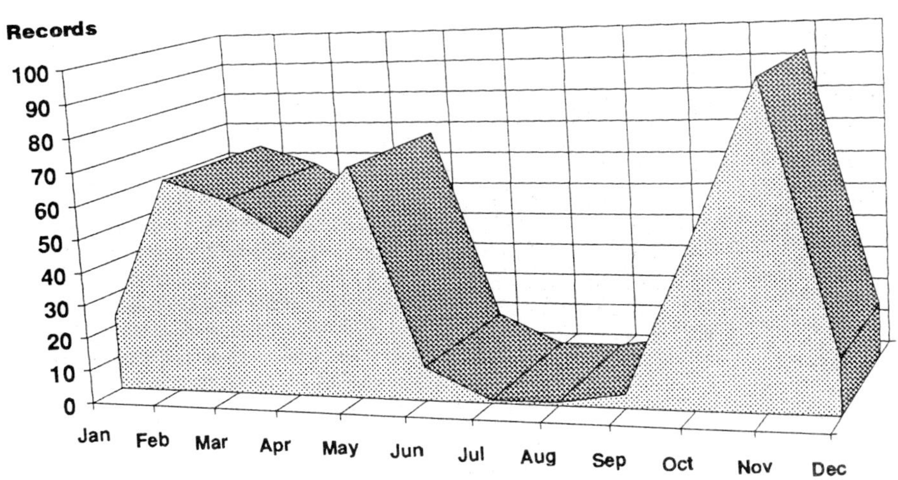

FIG. 8. Sporulation phenology of Myxomycetes on *Quercus ilex*.

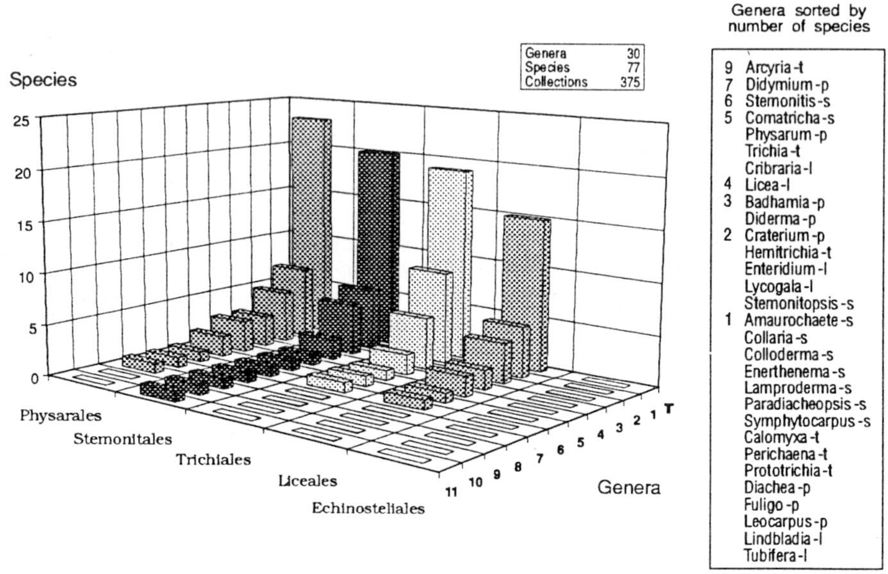

FIG. 9. Genera of Myxomycetes on *Pinus pinaster.*

Species such as *Amaurochaete atra* (Alb. & Schwein.) Rostaf., *Cribraria argillacea* (Pers. ex J.F. Gmel) Pers., *C. vulgaris* Schrad., *Licea minima* Fr., *L. testudinacea* Nann.-Bremek., *L. variabilis* Schrad., *Stemonitis splendens* Rostaf., *Symphytocarpus flaccidus* (Morgan) Ing & Nann-Bremek. ex Nann.-Bremek., *Trichia lutescens* (Lister) Lister and *Tubifera ferruginosa* (Batsch) J.F. Gmel. are not rare.

D. Supra-Mediterranean Level

The annual precipitation (600–800 mm) is high, but there are always summer droughts (Table 1). Geographically it comprises montane regions with high plateaus throughout (Fig. 3). Deciduous forests of *Quercus* (*Q. pyrenaica, Q. faginea*) are found in the main area. On colder and drier land they are replaced by conifers, e.g. *Juniperus thurifera.*

Q. pyrenaica are representative of deciduous woodlands, and contain 59 species in 20 genera (Fig. 10). Lignicolous Myxomycetes, e.g. *Arcyria* (7 species) and *Trichia* (5 species), are moderately represented while the *Stemonitales* are surprisingly few. Again, the most common species are those which have carbonate deposits, the more numerous being *Physarum* (12 species) and *Didymium* (9). *Physarum nutans* with 16 records (Table 2), *Comatricha nigra* and *Craterium leucocephalum* with 8, and *Badhamia panicea* (Fr.) Rostaf., *Didymium squamulosum* and *Physarum bitectum* G. Lister with 7 are the more abundant. Some species, e.g. *P. bivalve* Pers., *P. bogoriense* and *P. brunneolum* are frequently recorded. Species in *Echinostelium* and *Licea* are lacking, but this may reflect lack of study of corticolous habitats.

With regard to sporulation phenology (Fig. 11), there are again maxima in April and November, correlating with spring and autumn rains. The summer drought is milder due to the altitude and to summer storms.

In areas with low annual precipitation (ca. 700 mm) relatively high altitude (1,000 m) and poor soils, *Juniperus thurifera*, is often present, and is woodlands one of the richest for

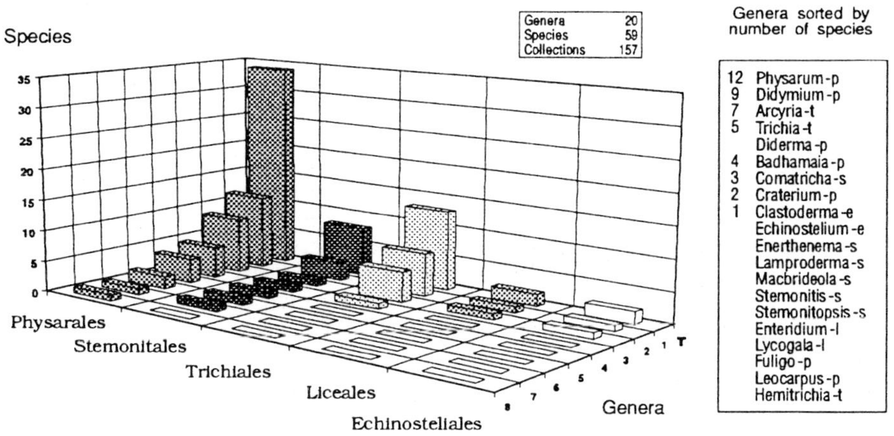

FIG. 10. Genera of Myxomycetes on *Quercus pyrenaica.*

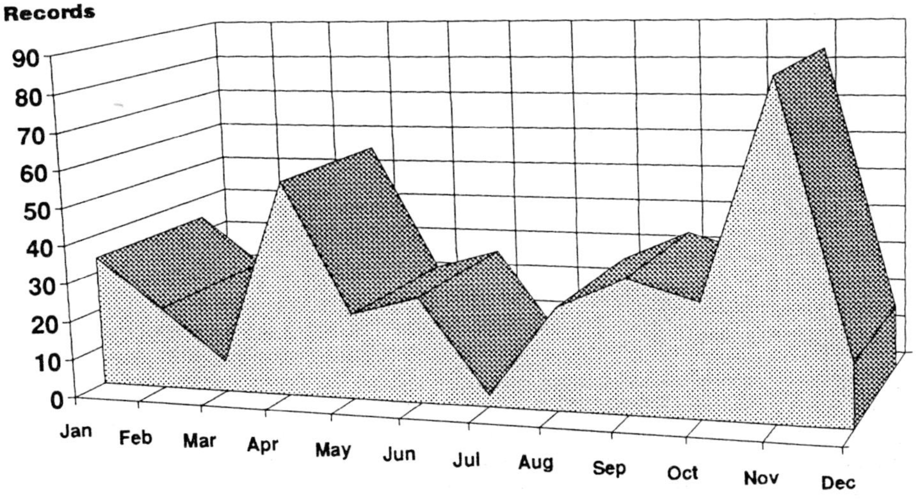

FIG. 11. Sporulation phenology of Myxomycetes on *Quercus pyrenaica.*

corticolous Myxomycetes (Pando & Lado, 1987a, 1987b, 1988, 1990). Three to five species per moist chamber culture are found, which is more than obtained in cultures from the boreal zone of Europe (Härkönen, 1977, 1978), tropical Africa (Härkönen, 1981) and other Mediterranean areas such as Turkey (Härkönen & Uotila, 1983).

Echinostelium (Fig. 12), including *E. apitectum, E. colliculosum, E. corynophorum* K.D. Whitney and *E. fragile, Licea* with *L. nannengae* Pando & Lado and *L. perexigua* T.E. Brooks

& H.W. Keller, and *Macbrideola* with *M. cornea*, *M. decapillata* and *M. oblonga*, are typically corticolous, and are representative of this habitat. Other species, e.g. *Badhamiopsis ainoae*, *Lepidoderma granuliferum* (W. Phillips) R.E.Fr. and *Didymium sturgisii* Hagelst. also appear regularly (Table 2). Interestingly, the species catalogue is similar to that obtained by Keller & Brooks (1976a, 1976b, 1977), Whitney (1980) and Whitney & Keller (1980) on American juniper species. Again, it is the short-lived corticolous species, completing their life cycles in a few hours, with less biomass and lower water requirements, which are common in more xeric environments.

Typically, lignicolous species, e.g. *Arcyria*, *Cribraria*, *Lycogala* or *Stemonitis*, which are frequent on other conifers, are absent from *Juniperus thurifera*. This may be due to the continental climate; the low rainfall often followed by strong, dry winds keeps wood dry. It is the less demanding corticolous species, favoured by water retained by bryophytes, algae and lichens sharing a similar ecological niche, which survive.

Also of interest is the strategy involving sclerotization by some corticolous species, e.g. *Lepidoderma granuliferum* and *Didymium sturgisii*, which are larger and have a higher water demand for fruiting. Sclerotia have been observed repeatedly covering 3–4 m of bark along the trunk under such extreme weather (e.g. in January or July). These sclerotia produce plasmodia and fruit bodies in moist chamber cultures in as little as 24 h. This also occurs in desert species (Blackwell & Gilbertson, 1980, 1984).

E. Oro-Mediterranean Level

This level is in the high mountains of the Centre, North and South of the Iberian Peninsula (Serra da Estrela, Sierra de Gredos, Sierra de Guadarrama, Picos de Urbión, Serranía de Cuenca, Sierra de Cazorla, Sierra Nevada and south slopes of the Picos de Europea and Pyrenees among others) (Fig. 3), but always above 1600 m altitude. Precipitation is high (Table 1) but largely as snow, and highland coniferous stands of *Pinus sylvestris* and *P. nigra* prevail.

Pinus sylvestris woods are possibly the most widespread and doubtless the best studied in the Iberian Peninsula. Seventy six species in 20 genera have been recorded (Fig. 13), and

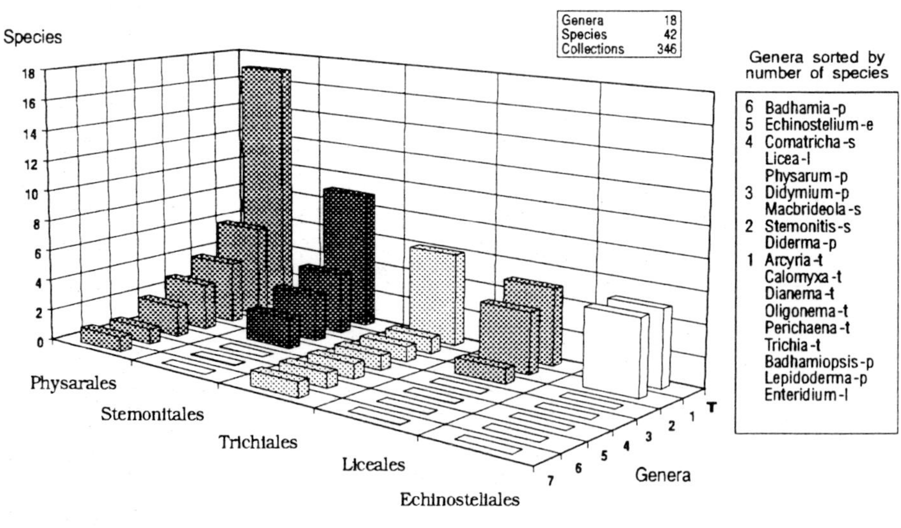

FIG. 12. Genera of Myxomycetes on *Juniperus thurifera*.

there is a high species diversity. The *Physarales* are least represented, while there are more genera and species of *Liceales*, *Trichiales* and *Stemonitales*. *Arcyria* (9 species) and *Cribraria* (9 species) are the best represented and are typically associated with conifer wood.

The more abundant species (Table 2) are: *Lycogala epidendrum* (40 records), *Arcyria incarnata*, *Comatricha nigra*, *Cribraria vulgaris*, *Fuligo septica* (>20 records) and *A. obvelata*, *Cribraria cancellata*, *Enerthenema papillatum*, *Lamproderma sauteri* Rostaf., *Licea minima*, *Physarum viride*, *Trichia favoginea* (Batsch) Pers. and *Tubifera ferruginosa* (>10). Lignicolous species are clearly predominant, and more hydrophilic genera producing large fruit bodies, e.g. *Amaurochaete atra* (Alb. & Schwein.) Rostaf., *Enteridium lycoperdon*, *Fuligo septica*, *Lycogala epidendrum* and *Symphytocarpus* spp., are also present.

Others such as *Arcyria oerstedtii* Rostaf., *Comatricha ellae*, *Dianema corticatum* Lister, *Enteridium liceoides* (Lister) G. Lister, *Hemitrichia abietina*, *Licea variabilis*, *Lindbladia tubulina* Fr. and *Prototrichia metallica* (Berk.) Massee are less frequent but nearly all are typically associated with conifer wood (Lister, 1925; Martin & Alexopoulos, 1969).

Due to the altitude (1,600–2,000 m) and snow, species characteristic of high mountains, e.g. *Arcyria versicolor* W. Phillips and other nivicolous species, e.g. *Diderma niveum* (Rostaf.) T. Macbr. or *Enerthenema melanospermum* T. Macbr. & G.W. Martin, are encountered.

Sporulation phenology (Fig. 14) approximates that of the rest of Europe (cf. Bjørnekaer & Klinge, 1963). The peak in fruiting occurs from May to November, July being the most favourable as it coincides with the end of thawing of snow. Values drop in August and September, but water supply through frequent summer storms enables the development of numerous fruit bodies. The November peak is attributed to the autumn rains. The activity is very sparse in winter due to low temperature and snowfall.

F. Crioro-Mediterranean Level

Low mean annual temperature (less than 4°C; Table 1), which results in a very short period of biological activity, and high annual precipitation in the form of snow are characteristic of this level. Geographically (Fig. 3) it is only found in the peaks on the

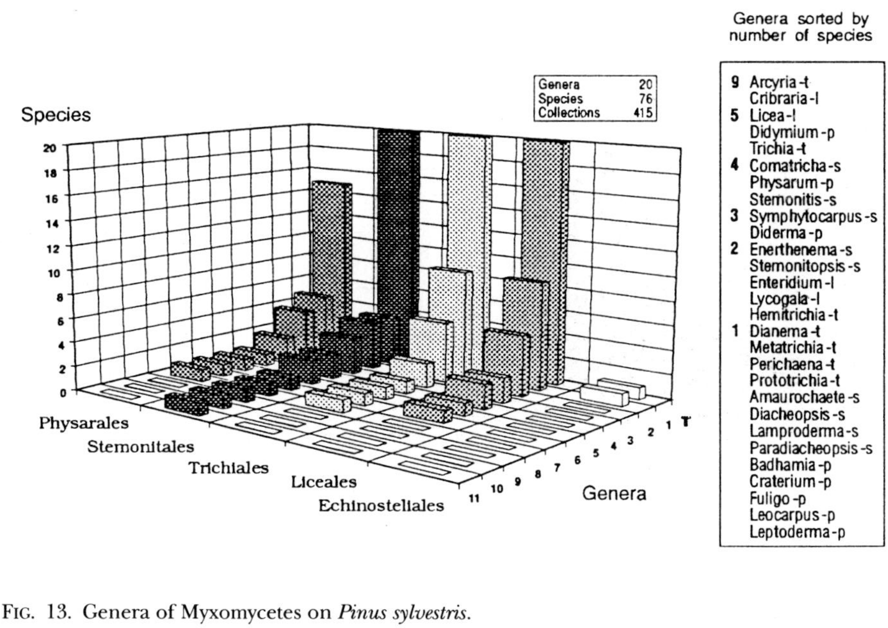

FIG. 13. Genera of Myxomycetes on *Pinus sylvestris*.

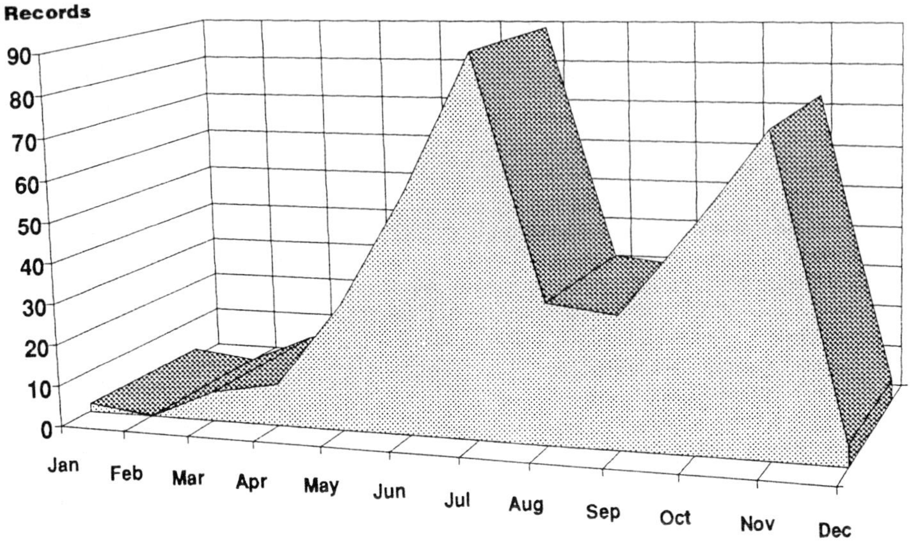

FIG. 14. Sporulation phenology of Myxomycetes on *Pinus sylvestris*.

highest mountain ranges of the Iberian Peninsula. Woodlands are lacking, there being only shrubs and psychrophytic meadows.

The Myxomycete data are incomplete but there appears to be a highly specialized group, i.e. the nivicolous species, which complete their life cycles near melting snow. So far, and according to personal unpublished data (Fig. 15), 21 species in 12 genera have been identified. *Lamproderma* and *Diderma* are the best represented, and species such as

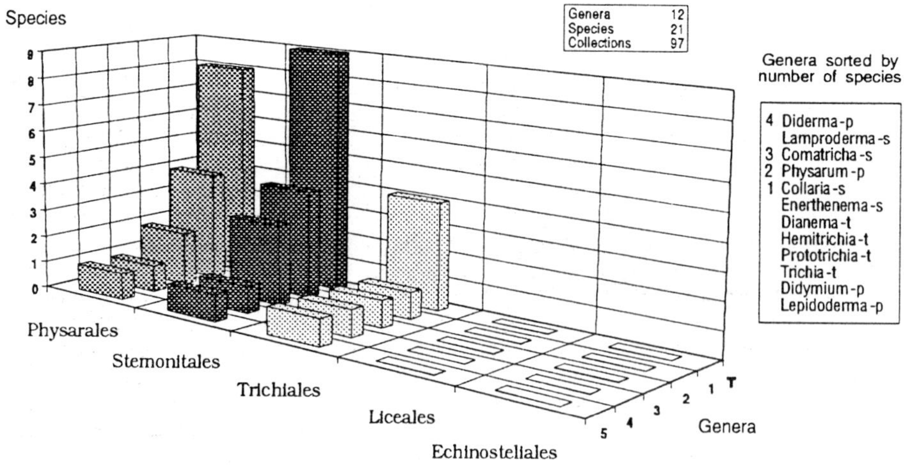

FIG. 15. Myxomycete genera near melting snow.

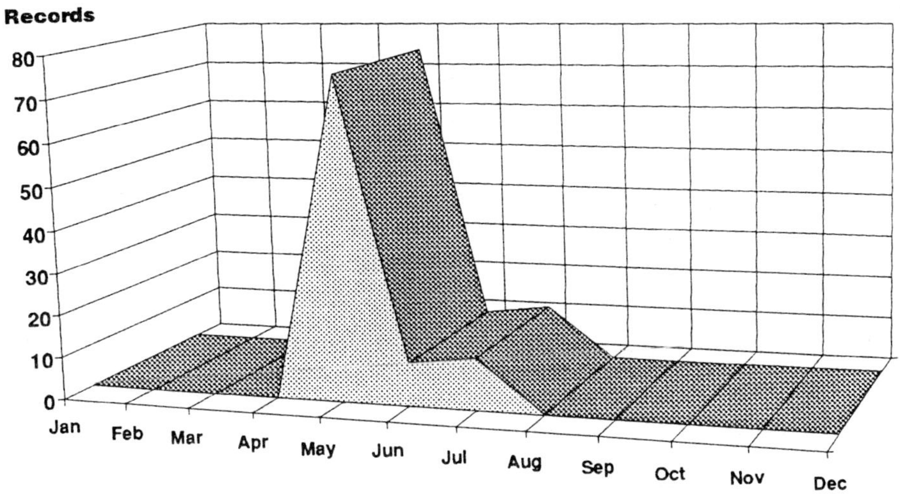

FIG. 16. Sporulation phenology of nivicolous Myxomycetes.

Collaria chionophila Lado, described recently from these territories; (Lado, 1992), *Comatricha suksdorfii* Ellis & Everh., *Enerthenema melanospermum*, *Lamproderma echinosporum* Meyl., *L. maculatum* Kowalski, *L. sauteri*, *Lepidoderma carestianum* (Rabenh.) Rostaf., *Physarum albescens* Ellis ex T. Macbr. or *Prototrichia metallica* occur regularly.

The sporulation phenology (Fig. 16) is reduced to the snow thawing period usually from April to July. There are a few differences between this curious myxoflora and that known from other mountain regions of Europe such as the Alps or Jura (Bozonnet *et al.*, 1991). Due to the lower latitude these species occur at higher altitudes and slightly earlier, with a maximum in May.

Discussion

It is axiomatic that the known distribution of many groups of organisms coincides with the distribution and/or interests of the pertinent field biologists (Gray & Alexopoulos, 1968). This is clearly the case with the Myxomycetes. There are still vast unexplored areas, and, in Europe, which is possibly the most studied, information is scarce in a substantial portion of the Mediterranean basin. To propose patterns of distribution is therefore risky. Nonetheless comments, correlations, predictions and establishment of working hypotheses, such as some of the above, can be made, for example, the high presence of species with carbonates on substrates subject to more xeric conditions, the significant correlation between North American and Spanish species found on *Juniperus* species, and the obvious correlation between the climate of the area and species phenology, which results in European taxa appearing at very different times of the year depending on latitude, should be noted.

It has been stated that the Mediterranean region, due to biogeographic, orographic, and climatological factors, has one of the richest myxofloras of the world. It is at present risky to advance a figure for the total flora of the Mediterranean region, because many zones have not been studied. However, between 410–450 species is probably a reasonable approximation. This species diversity, which represents 70% of the world flora, should justify the implementation of conservation practices aimed at preserving this natural heritage.

Finally, the lack of information on numerous areas, the floristic diversity mentioned above, its phenology which differs from the rest of Europe, its relative proximity to many study centres, and the accessibility to scientific institutions providing valuable logistic support, should render the Mediterranean woodlands among the most appealing for future myxomycete studies.

Acknowledgements

I thank Dr E. Descals (CSIC, Mallorca) for help in the English translation and for a critical reading of the manuscript, Mr A. García for his technical assistance with figures, and the Organising Committee of the CEM XI for their kind invitation and support. This contribution reports on a project within the "Flora Micológica Ibérica" (PR87–0370) and was supported by the DGICYT and CSIC.

References

Alexopoulos, C.J. (1959). Myxomycetes from Greece. *Brittonia* **10**(1): 25–40.

Almeida, M.A. (1987). Sinopse dos Myxomycetes de Portugal. Micologia, Departamento de Biologia Vegetal, Faculdade de Ciências de Lisboa, Lisboa.

Barbero, M., Loisel, R. and Quézel, P. (1992). Biogeography, ecology and history of Mediterranean *Quercus ilex* ecosystems. *Vegetatio* **99–100**: 19–34.

Binyamini, N. (1986). Myxomycetes from Israel: I. *Nova Hedwigia* **42**(2–4): 379–386.

Binyamini, N. (1987). Myxomycetes from Israel: II. *Nova Hedwigia* **44**(3–4): 351–364.

Binyamini, N. (1991). Myxomycetes from Israel: III. *Trans. Mycol. Soc. Japan* **32**(4): 485–493.

Bjørnekaer, K. and Klinge, A.B. (1963). Die dänischen Schleimpilze. *Friesia* **7**(2): 149–280.

Blackwell, M. and Gilbertson, R.L. (1980). Sonoran desert Myxomycetes. *Mycotaxon* **11**(1): 139–149.

Blackwell, M. and Gilbertson, R.L. (1984). Distribution and sporulation phenology of Myxomycete in the Sonoran Desert of Arizona. *Microbial Ecol.* **10**: 369–377.

Bozonnet, J., Meyer, M. and Poulain, M. (1991). Especes nivales de Myxomycetes. *Soc. Hist. Nat. Pays Montbéliard* **1991**: 51–72.

Buchet, S. (1942). Nouveaux matériaux pour la flore française des Myxomycètes. *Bull. Soc. Mycol. France* **57**: 56–103.

Buchet, S., Chermezon, H. and Evrard, F. (1912). Matériaux pour la flore française des Myxomycètes. *Bull. Soc. Mycol. France* **28**: 299–325.

Buchet, S., Chermezon, H. and Evrard, F. (1920). Matériaux pour la flore française des Myxomycètes (II). *Bull. Soc. Mycol. France* **36**: 106–121.

Carilla, J. and Gràcia, E. (1991). Mixomicetes corticícolas de Aragón. I. *Revista Iberoamer. Micol.* **8**(1): 3–7.

Cochet, S. (1977). Les Myxomycètes de France. *Bull. Soc. Mycol. France* **93**(3): 159–200.

Cochet, S. and Bozonnet, J. (1980). Les Myxomycètes de France. Nouveautés et compléments. *Bull. Soc. Mycol. France* **96**(4): 115–120.

Cochet, S. and Bozonnet, J. (1984). Les Myxomycètes de France. Nouveautés et compléments (II). *Bull. Soc. Mycol. France* **100**(3): 39–64.

Credaro, V. and Pirola, A. (1984). Aggiunte alla flora mixomicetologica italiana. *Atti. Ist. Bot. Lab. Crittog. Univ. Pavia Ser.* 7 **3**: 103–108.

Credaro, V. and Pirola, A. (1986). Nuove acquisizioni mixomicetologiche per l'Italia settentrionale. *Atti Ist. Bot. Lab. Crittog. Univ. Pavia, Ser.* 7 **5**: 111–125.

El-Buni, A.M. and Rattan, S.S. (1981). Check list of Libyan Fungi. Al Faateh Univ., Fac. Sci., Tripoli.

Eliasson, U. (1977). Ecological notes on *Amaurochaete* Rost. (Myxomycetes). *Bot. Not.* **129**: 419–425.

Eliasson, U. (1981). Patterns of occurrence of myxomycetes in a spruce forest in south Sweden. *Holarc. Ecol.* **4**: 20–31.
Eliasson, U. (1991). The myxomycete biota of the Hawaiian Islands. *Mycol. Res.* **95**(3): 257–267.
Emoto, Y. (1939). Untersuchungen über die Entwicklung der Myxomyceten auf faulenden Hölzern. *J. Jap. Bot.* **9**: 253–257.
Faurel, L., Feldmann, J. and Schotter, G. (1964). Catalogue des Myxomycètes de l'Afrique du Nord. *Bull. Soc. Hist. Nat. Afrique N.* **55**: 7–39.
Gràcia, E. (1987). Una nueva cita de *Arcyria annulifera* Torrend. In. Libro de resúmenes VII Simp. Nac. Bot. Criptog., Madrid p. 126.
Gràcia, E. and Llimona, X. (1980). Contribución al conocimiento de la flora y distribución de los mixomicetes en el Mediterráneo occidental. III. Sureste de España: Murcia. *Anales Univ. Murcia, Ci.* **34**(1, 2, 3, 4): 3–21.
Gray, W.D. and Alexopoulos, C.J. (1968). Biology of the Myxomycetes. Ronald Press, New York.
Härkönen, M. (1977). Corticolous Myxomycetes in three different habitats in southern Finland. *Karstenia* **17**: 19–32.
Härkönen, M. (1978). On corticolous Myxomycetes in northern Finland and Norway. *Ann. Bot. Fenn.* **15**: 32–37.
Härkönen, M. (1981). Gambian Myxomycetes developed in moist chamber cultures. *Karstenia* **21**: 21–25.
Härkönen, M. (1988). Some additions to the knowledge of Turkish Myxomycetes. *Karstenia* **27**: 1–7.
Härkönen, M. and Uotila, P. (1983). Turkish Myxomycetes developed in moist chamber. *Karstenia* **23**: 1–9.
Ing, B. (1987). Corticolous Myxomycetes from Cyprus. *Mycotaxon* **30**: 195.
Keller, H.W. and Brooks, T.E. (1976a). Corticolous Myxomycetes: IV. *Badhamiopsis*, a new genus for *Badhamia ainoae*. *Mycologia* **68**(4): 834–841.
Keller, H.W. and Brooks, T.E. (1976b). Corticolous Myxomycetes: V. Observations on the genus *Echinostelium*. *Mycologia* **68**(6): 1204–1220.
Lado, C. (1991). Catálogo comentado y síntesis corológica de los Myxomycetes de la Península Ibérica e Islas Baleares (1788–1990). *Ruizia* **9**: 1–144.
Lado, C. (1992). *Collaria chionophila*, a new Myxomycete from Spain. *Anales Jard. Bot. Madrid* **50**(1): 9–13.
Lister, A. (1925). A monograph of the Mycetozoa. Ed. 3. British Museum (Natural History), London.
Lister, A. and Lister, G. (1904). Notes on Mycetozoa. *J. Bot.* **42**: 129–140.
Maimoni-Rodella, R.C.S. and Gottsberger, G. (1980). Myxomycetes from the forest and the Cerrado vegetation in Botucatu, Brazil: A comparative ecological study. *Nova Hedwigia* **34**: 207–245.
Maire, R., Patouillard, N. and Pinoy, E. (1926). Myxomycètes de l'Afrique du Nord. *Bull. Soc. Hist. Nat. Afrique N.* **17**: 38–43.
Malençon, G. and Bertault, R. (1968). Champignons du Maroc. *Bull. Soc. Sci. Nat. Maroc.* **47**(3–4): 237–282.
Malençon, G. and Bertault, R. (1969). Champignons du Maroc II. *Bull. Soc. Sci. Nat. Maroc.* **49**: 69–80.
Martin, G.W. and Alexopoulos, C.J. (1969). The Myxomycetes. University of Iowa Press, Iowa.
Mitchell, D.W. and Kylin, H. (1984). Some Tunisian Myxomycetes. *Bull. Brit. Mycol. Soc.* **18**(1): 64–65.
Nannenga-Bremekamp, N.E. (1991). A Guide to Temperate Myxomycetes. Biopress Limited, Bristol.
Orsino, F. and Traverso, M. (1983). Mixomiceti della Liguria. *Rivista Micol. Ital.* **12**(3): 23–28.
Orsino, F. and Traverso, M. (1987). Nuovi Mixomiceti della Liguria. *Rivista Micol. Ital.* **16**(2): 3–7.
Ozenda, P., Noirfalise, A., Tomaselli, R. and Trautmann, W. (1979). Carte de la végétation des Etats membres du Conseil de l'Europe. Échelle: 1/3 000 000. E. Conseil de l'Europe, Strasbourg.

Pando, F. and Lado, C. (1987a). Myxomycetes cortícícolas ibéricos, I: Especies sobre *Juniperus thurifera*. *Bol. Soc. Micol. Madrid* **11**(2): 203–212.
Pando, F. and Lado, C. (1987b). Fragmenta chorologica occidentalia (Fungi), 812–830. *Anales Jard. Bot. Madrid* **44**(1): 143–146.
Pando, F. and Lado, C. (1988). Two new species of corticolous Myxomycetes from Spain. *Mycotaxon* **31**(2): 299–303.
Pando, F. and Lado, C. (1990). A survey of the corticolous Myxomycetes in Peninsular Spain and Balearic Islands. *Nova Hedwigia* **50**(1–2): 127–137.
Pirola, A. (1968). Una lista de Mixomiceti italiani. *Giorn. Bot. Ital.* **102**: 21–32.
Pirola, A. and Credaro, V. (1971). Contributo alla flora mixomicetologica italiana. *Giorn. Bot. Ital.* **105**(4): 157–165.
Pirola, A. and Credaro, V. (1975). Tentativo di inquadramento sinecologico di mixomiceti italiani. *Not. Fitosociol.* **10**: 111–130.
Pirola, A. and Credaro, V. (1988). Obsservazioni sulla distribuzione e sull'ecologia dei Mixomiceti in Italia. *Braun-Blanquetia* **2**: 203–222.
Quézel, P. (1985). Definition of the Mediterranean region and the origin of its flora. In C. Gómez-Campo (editor), Plant conservation in the Mediterranean area, pp. 9–24. Dr W. Junk Publishers, Dordrecht, Boston, Lancaster.
Rammeloo, J. (1973). Contribution à la connaissance des Myxomycetes du Maroc (1er note). *Bull. Soc. Sci. Nat. Maroc* **53**: 31–35.
Ramon, E. (1969). Myxomycetes of Israel. *Israel J. Bot.* **17**: 207–211 (1968 publ. 1969).
Rivas-Martínez, S. (1987). Memoria del mapa de series de vegetación de España. Ed. Icona, Madrid.
Stephenson, S.L. (1988). Distribution and ecology of Myxomycetes in temperate forests. I. Patterns of occurrence in the upland forests of southwestern Virgina. *Canad. J. Bot.* **66**: 2187–2207.
Stephenson, S.L. (1989). Distribution and ecology of Myxomycetes in temperate forests. II. Patterns of occurrence on bark surface of living trees, leaf litter, and dung. *Mycologia* **81**: 608–621.
Stephenson, S.L. and Studlar, S.M. (1985). Myxomycetes fruiting upon bryophytes: coincidence or preference? *J. Bryol.* **13**: 537–548.
Stojanowska, W. (1980a). The seasonal variation of Myxomycetes Flora in the Muszowicki Las Bukowy Reserve (Lower Silesia). *Fragm. Florist. Geobot.* **26**(1): 103–113.
Stojanowska, W. (1980b). Comparison of Myxomycetes of the forest in Skarszyn and of the beech reserve in Muszkowice. *Acta Mycol.* **16**(2): 221–230.
Torrend, C. (1909). Sur une nouvelle espèce de Myxomycète: *Arcyria annulifera* Lister & Torrend. *Bol. Soc. Portug. Ci. Nat.* **2**(3): 212–213.
Unesco (1969). Carte de la végétation de la région méditerranéenne. Notice explicative. Ed. Unesco, Paris.
Whitney, K.D. (1980). The Myxomycete genus *Echinostelium*. *Mycologia* **72**(5): 950–987.
Whitney, K.D. and Keller, H.W. (1980). A new species of *Echinostelium*. *Mycologia* **72**(3): 640–643.

Bujakiewicz, M. (1993). Fungi of the Alpine and Subalpine Zones of the Babia Góra Massif. In D.N. Pegler, L. Boddy, B. Ing & P.M. Kirk (Editors). Fungi of Europe: Investigation, Recording and Conservation, pp. 115–120. Royal Botanic Gardens, Kew.

9. FUNGI OF THE ALPINE AND SUBALPINE ZONES OF THE BABIA GÓRA MASSIF

M. BUJAKIEWICZ

Department of Plant Ecology and Environment Protection, Adam Mickiewicz University, Al. Niepodleglości 14, 61–713 Poznán, Poland

Abstract

A preliminary survey of the macrofungi occurring in the alpine and subalpine community types of Babia Góra and in the transition zone between spruce forest and subalpine vegetation is presented. The communities studied were *Trifido-Supinetum*, snowbeds with *Anthelia juratzkana, Luzuletum spadiceae, Pinetum mughi carpaticum silicicolum, Pinetum mughi carpaticum calcicolum, Vaccinietum myrtillii, Calamagrostietum villosae, Hieracio-Nardetum, Saxifrago-Festucetum versicoloris, Athyrietum alpestris, Petasitetum kablikiani, Empetro-Vaccinietum, Adenostyletum alliariae, Rumicetum alpini, Athyrio-Sorbetum, Plagiothecio-Piceetum* (upper forest limit). General remarks are made on ecology of selected macrofungi growing in various habitats and on slopes of different aspect. Differences and similarities in the fungal species composition of different plant communities are discussed, and the indicative value of macrofungi is demonstrated.

Introduction

Babia Góra (49°30′N, 19°25′E) is located in the Western Carpathians and forms an isolated massif north-west of the Tatra Mountains. It runs along the parallel of latitude and has its north side sloping down more steeply with glacial relief and rich, diversified vegetation. The highest peak is 1725 m above sea level (a.s.l.).

The first mycocoenological studies on Babia Góra were done in the montane forest zones, below the upper forest limit line, on north and south slopes (Bujakiewicz 1979, 1981, 1982a, b). The average altitude of the upper forest limit line on Babia Góra is approximately 1390m a.s.l. Mycocoenological research above the upper forest limit line was started in 1984, and is continuing on 50 permanent plots, laid out in 17 syntaxa of vegetation (Table 1). Further details of the fungi of these regions are provided by Moser (1982) and Senn-Irlet (1988).

Preliminary Results

Data on selected species of macrofungi that seem to be indicative of syntaxonomical units of vegetation are presented (Table 2). The data are based on the presence of basidiomycete carpophores.

A. There is a significant concentration and complexity of fungus taxa and diversity of forms in an ecotone zone at the upper forest limit adjoining the dwarf pine zone (*Pinetum mughi carpaticum*) and the *Anthyrio-Sorbetum* association.

TABLE 1. Plant Associations and Observation Plots in Alpine Subalpine Zones on Babia Góra

Zone	Phytosociological Unit and Number of Plots	Total Plots
Alpine	*Trifido-supinetum* – 3	9
	Snowbeds with *Anthelia juratzkana* – 3	
	Luzuletum spadiceae – 2	
	Community with *Deschampsia caespitosa* – 1	
Subalpine	*Pinetum mughii carpaticum*: 12	37
	— *silicicolum* – 7	
	— *calicicolum* – 5	
	Vaccinietum myrtillii – 5	
	Calamagrostietum villosae – 4	
	Hieracio-Nardetum – 4	
	Saxifrago-Festucetum versicoloris – 3	
	Athyrietum alpestris – 2	
	Petasitetum kablikiani – 2	
	Empetro-Vaccinietum – 2	
	Adenostyletum alliariae – 2	
	Rumicetum alpini – 1	
Transition Zone	*Athyrio Sorbetum* – 2	4
	Plagiothecio-Piceetum (upper forest limit) — 2	
3 Zones	16 Plant associations and communities	50

B. There is a gradual, significant decrease in number of taxa of fungi as the altitude increases.

C. There are differences in mycofloristic composition between north and south slopes in some plant associations, e.g. in dwarf pine thickets (*Pinetum mughi carpaticum*). On northern slopes dwarf pine thickets are distinguished by the occurrence of (1) fungi associated with planted *Pinus cembra*, e.g. *Cortinarius miniatopus* s.l. and (2) various montane species, e.g. *Chroogomphus helveticus* (Singer) Moser. On southern slopes mycorrhizal symbionts seem to form more carpophores.

D. There is striking correspondence between macrofungi and syntaxonomical units of vegetation. Particularly significant is a group of fungi associated with the habitats of the *Vaccinio-Piceetea* class (*Plagiothecio-Piccetum, Pinetum mughi carpaticum, Vaccinietum myrtilli, Empetro-Vaccinietum*), e.g. *Lactarius rufus* (Scop.: Fr.), *Marasmius androsaceus* (L.: Fr.) Fr., *Pholiota scamba* (Fr.: Fr.) Moser, etc. Fungi associated with the habitats of the *Betulo-Adenostyletea* class (*Athyrio-Sorbetum, Athyrietum alpestris, Petasitetum kablikianii, Adenostyletum alliariae, Calamagrostietum villosae*) also form a significant group, e.g. *Mycena pterigena* (Fr.: Fr.) Kummer, *Typhula sclerotioides* (Pers.) Fr., *Melastiza flavorubens* (Rehm) Pfister & Korf.

E. There are differences in habitat within a plant association, resulting in differences in number of taxa and carpophores. For example, with regard to mycorrhizal symbionts, *Pinetum mughi carpaticum silicicolum* has fewer taxa (11) but more carpophores (73) than

TABLE 2. Range of Selected Species of Fungi in the Alpine and Subalpine Vegetation Zones of the Babia Góra Massif

Phytosociological unit	PP	Pmc	AS	Ata	PK	Aal	Cv	SFv	Vm	EV	HN	Ls	TS	Aj	Ts	Ls	HN	EV	Vm	SFv	Cv	Aal	PK	Ata	AS	Pmc	Ra	PP
Altitude in m	1310–1385	1370–1670	1350–1398	1350–1455	1455	1505–1510	1450	1340–1520	1410–1565	1505–1640	1580–1625	1705–1720	1705–1720	1710–1720	1705–1720	—	1520	—	1600–1625	—	1450–1515	—	—	—	—	1390	1400	—
Number of plots	2	7	2	2	2	2	2	3	3	2	3	2	3	3	3	—	1	—	2	—	2	—	—	—	—	5	1	—
Number of observations	12	12	11	12	13	12	11	12	10	11	11	13	14	12	14	—	11	—	10	—	11	—	—	—	—	12	10	—
Average No. of species per plot	55	24	73	45	37	25	34	24	19	17	5	6	10	4	10	—	10	—	7	—	13	—	—	—	—	26	28	—
Omphalina trigonospora														—														
Hygrocybe reai													▬		▬													
Hygrocybe miniata													—		—													
Brunnipila calycioides													▬		▬													
Omphaliaster asterosporus										—				—	—													
Omphalina luteovitellina									▬	—	—	—	▬	▬	▬	—	—		▬									
Clitopilus hobsonii							▬					—	—															
Dasyscypha rosea									▬	▬	▬						▬				▬							
Entoloma cetratum									▬										▬									
Omphalina ericetorum								—	▬	▬	—						—		▬									
Omphalina obatra								▬											▬									
Rickenella fibula								▬																				
Arrhenia acerosa								—																				
Entoloma asprellum								—																				
Helvella arctica								—																				
Laccaria altaica								—																				
Omphalina hudsoniana								—																				
Sclerotina curreyana							▬																					
Galerina laevis																												
Marasmius limosus					—	▬	▬														▬							
Peziza cfr. howsei					▬																							
Flagelloscypha kavinae					▬																							
Scutellinia macrospora					▬																							
Hebeloma stenocystis					▬																							
Hemimycena pseudocrispula					▬		—																					
Nochascypha filicina				▬	▬																							
Psilocybe crobula			▬	▬		—																						

TABLE 2. continued

Phytosociological unit	PP	Pmc	AS	Ata	PK	Aal	Cv	SFv	Vm	EV	HN	Ls	TS	Aj	Ts	Ls	HN	EV	Vm	SFv	Cv	Aal	PK	Ata	AS	Pmc	Ra	PP
Mycena longiseta		—	█	█																								
Mycena pterigena	—	▬	█	█																								
Typhula sclerotioides			—	—	█	█																						
Typhula athyrii			▬	▬		▬																						
Mycena cyanipes			▬	▬		▬																						
Melastiza flavorubens						▬																						
Cortinarius miniatopus s.l.		█																										
Dermocybe bataillei		█																										
Guepiniopsis alpina		█																										
Lachnellula suecica		█																								█		
Suillus variegatus		█																								▬		
Trichoscyphella fuckelii		█																										
Trichoscyphella fuscosanguinea		█																										
Amanita submembranacea		▬																										
Collybia succinea		▬																								▬		
Hygrophorus hypothejus		▬																								▬		
Russula decolorans		▬																								▬		
Chroogomphus helveticus	▬	█																								▬		
Hygrophorus olivaceoalbus	▬	█																										
Pholiota scamba	▬	█																										
Russula emetica	▬																											
Russula vinosa																										▬		
Cudoniella clavus																										▬		
Calyptella capula																											█	
Ombrophila violacea																											█	
Climacocystis borealis	▬																										▬	
Clitocybe ditopa	▬																											
Hygrophorus pustulatus	▬																											
Lactarius lignyotus	▬																											

TABLE 2. continued

Phytosociological unit	PP	Pmc	AS	Ata	PK	Aal	Cv	SFv	Vm	EV	HN	Ls	TS	Aj	Ts	Ls	HN	EV	Vm	SFv	Cv	Aal	PK	Ata	AS	Pmc	Ra	PP
Lyophyllum gibberosum	▬																											
Phellinus viticola	▬																											
Russula ochroleuca	▬																											

Explanations: — r rare: ▬ n numerous: ■ a abundant:
Aal = *Adenostyletum alliariae*: Aj = snowbeds with *Anthelia juratzkana*: AS = *Athyrio-Sorbetum*: Ata = *Athyrietum alpestris*: Cv = *Calamagrostietum villosae*: EV = *Empetro-Vaccinietum*: HN = *Hieracio-Nardetum*: Ls = *Luzuletum spadiceae*: PK = *Petasitetum Kabilkiani*: Pmc = *Pinetum mughi carpaticum*: PP = *Plagiothecio-Piceetum*: Ra = *Rumicetum alpini*: SFv = *Saxifrago-Festucetum versicoloris*: TS = *Trifido-Supinetum*: Vm = *Vaccinietum myrtillii*

Pinetum mughi carpaticum calcicolum (14 taxa, 45 carpophores). Only 30% of species are common to the two subassociations. *Amanita submembranacea* (Bon) Gröger is more abundant in poorer sites and prefers southern slopes.

F. The *Saxifrago-Festucetum versicoloris* association, endemic in the massif, forming 'grassmats' on abrupt sandstone rock-steps, has a distinctive mycological character. Such species as *Omphalina obatra* (Favre) P.D. Orton, *Arrhenia acerosa* (Fr.) Kühner, *Entoloma* cfr. *asprellum* (Fr.) Moser and *Helvella arctica* Nannf. have been found only in this specific habitat.

G. A rather small alpine zone, only 75 m of absolute elevation (1650–1725 m), has small patches of snow-beds with *Anthelia juratzkana* where, among other fungi, *Omphalina trigonospora* Lam., a rare and interesting species has been recorded. Patches of the *Luzuletum spadiceae* association are the poorest (frequency and taxa) in fungi. In the patches of the *Trifido-Supinetum* association, that dominates in the alpine zone, *Brunnipila calycioides* is a constant species, growing on dead blades of *Juncus trifidus*. It has been recently recorded from Babia Góra by Chlebicki (1990). *Hygrocybe reai* Maire and *H. miniata* (Fr.) Kummer are often noted, but *Omphalina luteovitellina* (Pilát & Nannf.) M. Lange (*Phytoconis viridis*) and *O. ericetorum* (Fr.: Fr.) M. Lange are the most common. They are associated with algae forming loose Basidiolichen relationships.

References

Bujakiewicz, A. (1979). Grzyby Babiej Góry. I. Mikoflora lasów. (Fungi of Mt. Babia Góra. I. Mycoflora of forests.) *Acta Mycol.* **15** (2): 213–294.

Bujakiewicz, A. (1981). Grzyby Babiej Góry. II. Wartość wskáznikowa macromycetes w zespolach lésnych. (Fungi of Mt. Babia Góra. II. Indicative value of macromycetes in the forest associations.) *Acta Mycol.* **17** (1–2): 63–125.

Bujakiewicz, A. (1982a). Grzyby Babiej Góry. III. Wartość wskaźnikowa macromycetes w zespolach lésnych (Fungi of Mt. Babia Góra. III. Indicative value of macromycetes in the forest associations.) *Acta Mycol.* **18**(1): 3–44.

Bujakiewicz, A. (1982b). Macromycetes as an element of forest structure on Babia Góra massif. Structur und Dynamik von Wäldern (Rinteln 13–16.4.1981). *Ber. Int. Symp. Int. Vereinigung Vegetationsk.*: 645–657.

Chlebicki, A. (1990). *Brunnipila calycioides* (Rehm.) Baral an alpine fungus new to Poland. *Acta Mycol.* **25**(2): 33–36.

Moser, M. (1982). Mycoflora of the transitional zone from subalpine forests to alpine tundra. In G.A. Laursen, et al (editors). Artic and alpine mycology: 1, pp 371–389. University of Washington Press, Seattle.

Senn-Irlet, B. (1988). Macromycetes in alpine snowbed communities: mycoenological investigations *Acta Bot. Neerl.* **37**(2): 251–263.

Rotheroe, M (1993). The Macrofungi of British Sand Dunes. In D.N. Pegler, L. Boddy, B. Ing & P.M. Kirk (Editors). Fungi of Europe: Investigation, Recording and Conservation, pp. 121–137. Royal Botanic Gardens, Kew.

10. THE MACROFUNGI OF BRITISH SAND DUNES

M. ROTHEROE

Cambrian Institute of Mycology, North Falcondale Drive, Lampeter, Dyfed SA48 7RX, U.K

Abstract

The sand-dune ecosystem as a macrofungal habitat is reviewed. Comparisons are made between different types of dune system in the British Isles. The unusual and distinct nature of the fungal flora of sand dunes is highlighted. Distribution, phenology and ecological strategy of some characteristic fungal components are discussed, with reference to both biotic and abiotic factors. The relationships between macrofungi and higher plant communities are considered, with particular reference to the potential value of use of the National Vegetation Classification (NVC).

Introduction

Sand dunes occupy approximately 9% of the coast length of England and Wales (some 400 km) and slightly more of Scotland (Fig. 1). They represent an important natural or semi-natural resource, both in terms of the proportion of land mass they cover and in the distinctive nature of the form and content of their varied habitats. Coastal dunes are of outstanding value for nature conservation as well as being important as sea defences and for recreation. Vegetation plays a large part in their formation and development, their dynamics and their continuing survival. The fungi have a rôle in facilitating colonisation by higher plants and in sustaining coastal dune systems, both by the physical action of fungal hyphae in accretion of sand particles and as mycorrhizal symbionts of the major dune-building grass, i.e. marram (*Ammophila arenaria* (L.) Link), and of dune shrubs, e.g. *Salix* species, later in the succession. Recent studies of macrofungi on British sand dunes (Rotheroe, Hedger & Savidge, 1987; Watling & Rotheroe, 1989) have highlighted both the richness of this ecosystem and the fact that, historically, it has been neglected mycologically. European studies (Arnolds, 1983; Bon, 1970, 1972 & 1975; Courtecuisse, 1984 & 1986; Høiland, 1975, 1977 & 1978) present a similar picture on a somewhat larger scale.

The following account is based largely on personal studies of Welsh sand dunes, since 1984, including an intensive fortnightly recording survey at Ynyslas National Nature Reserve, Dyfed, in 1985 and a similar survey in 1991/92 at Whiteford Burrows National Nature Reserve, West Glamorgan. The latter included a preliminary assessment of the potential value of the National Vegetation Classification (NVC) in describing the distribution of macrofungi in relation to plant communities.

The Dune Environment and Dune System Characteristics

Five main types of dune system have been identified (Ranwell & Boar, 1986). One of the commonest types is the spit dune system, forming a promontory at the mouth of an

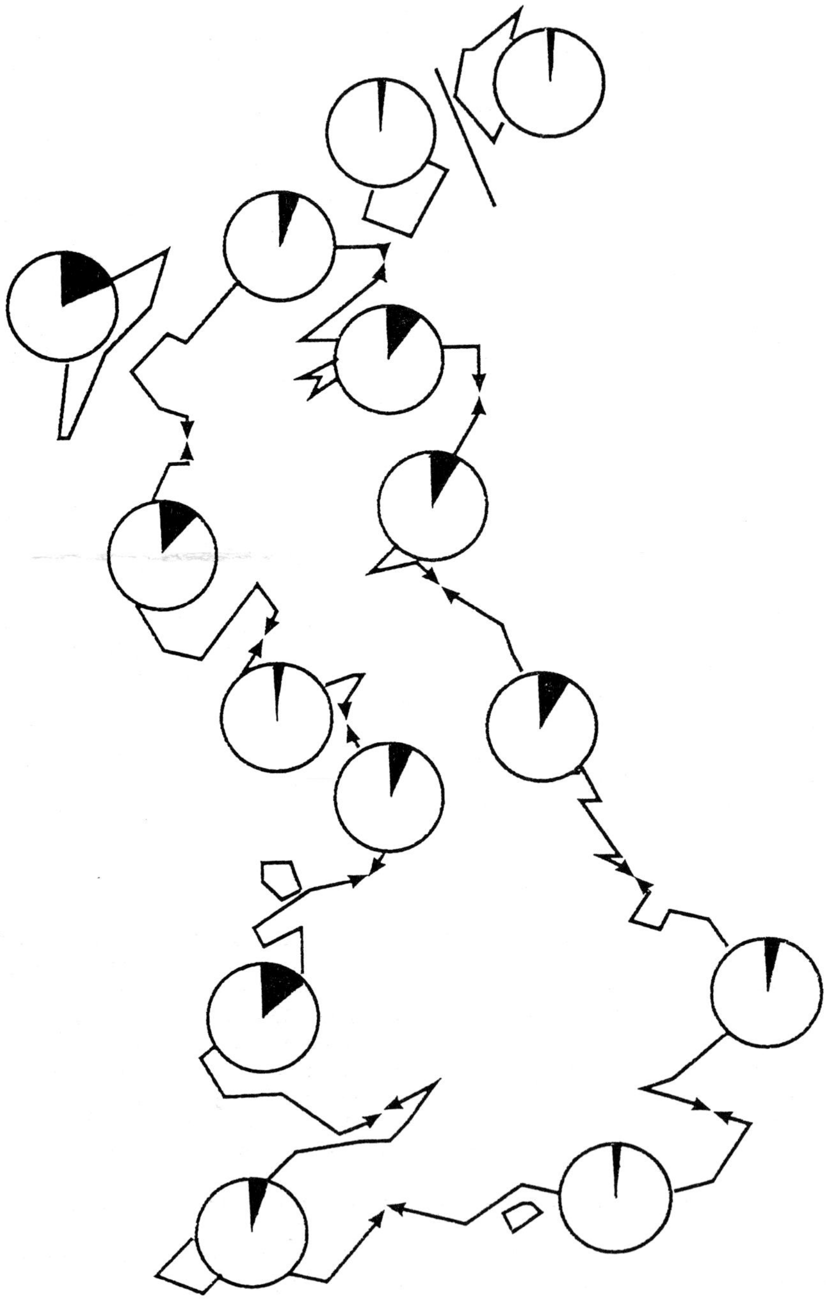

FIG. 1. Diagrammatic map showing extent of sand dunes in Britain (After Ranwell & Boar, 1986).

estuary. Bay dunes, where the system typically forms a crescent shape, are the most frequently encountered type in Britain. Because headland shelter may limit the availability of sand, bay dunes are often restricted to a narrow band along the coastline. Hindshore dune systems are formed where the prevailing wind is also the dominant wind and, in contrast to bay dunes, they can extend for several kilometres inland, e.g. Braunton Burrows in Devon, Culbin Sands in Scotland and Newborough Warren on Anglesey.

On most dune systems, but particularly on hindshore dunes, a simple line transect taken inland at right angles to the sea displays a vegetational succession which has been well documented. Indeed, the term "succession" as a basic concept in plant ecology dates back to a classical sand-dune study in North America at the end of the last century (Cowles, 1899). The zonation (Fig. 2) begins with strandline vegetation, followed by foredune or embryo dune and then mobile yellow dunes, where marram grass accelerates the accumulation of sand and the exposed uncolonised areas impart a characteristic colour. As vegetation cover increases the dunes become more consolidated in the form of semi-fixed or grey dunes, so-called because of the colour imparted by the lichens which colonise the sand. The development which follows depends on the amount of calcium carbonate from sea-shell fragments present in the system. If this is low, as on dunes built up from silica sand or old dunes on higher ground where leaching has occurred, then dune heath is formed, vegetated by "acid-loving" plants. If it is high (e.g. Whiteford Burrows, Fig. 2), a species-rich grassland develops which, with adequate grazing, is similar to that of chalk or limestone grassland. Interspersing dune ridges of the xerosere in this type of dune are calcareous dune slacks, damp or wet hollows close to the water table and often subject to winter flooding. These have been described as a partial or truncated hydrosere (Ranwell, 1959) and represent one of the few undisturbed damp lowland habitats which still survive in Britain. Dune development continues with the formation of scrub. In a natural system scrub vegetation would progress to climax woodland but this seral stage is virtually non-existent in Britain. It has been replaced in many locations by conifer plantation. Along the schematic transect line changes in several biotic and abiotic factors occur. Young dunes are unstable and deficient in water and plant nutrients. The transect inland follows gradients of nutrient availability, organic matter accumulation, pH levels and water status. Vegetation is also affected by aspect, degree of exposure and levels of grazing by rabbits or domestic animals.

The Fungal Component

Both saprotrophic and biotrophic nutritional modes are represented in the fungal component of sand dunes. The former can be seen near the start of the transect in the yellow dunes. Characteristic dune macrofungi at this stage include *Peziza ammophila* Dur. & Mont. (Pezizales), *Psathyrella ammophila* (Dur. & Lév.) Dennis & Pearson (Agaricales), *Phallus hadriani* Vent. ("Gasteromycetes") and three recently-described Agaricales, *Coprinus ammophilae* Court., *Hohenbuehelia culmicola* Bon and *Melanoleuca cinereifolia* Bon. All six species appear to be confined to sand dunes, particularly the yellow dune zone, and are presumably highly specialized. Saprotrophic sand-dune endemics also occur farther inland. Here, however, they are joined by a number of characteristic species which also grow elsewhere. They appear to have broader habitat preferences but, while they may be found most often in one particular zone (see E–H below), they can occur in the grey dunes, dune grassland or dune slacks.

The most commonly encountered agarics include the following (an asterisk indicates a species confined to sand dunes): *Agaricus devoniensis* Orton*; *A. semotus* Fr.; *Agrocybe subpediades* (Murr.) Watling*; *Calocybe carnea*; *Clitocybe barbularum*; *C. dealbata*; *Conocybe dunensis* T.J. Wallace apud Orton*; *Eccilia sericeonitida* Orton; *Fayodia pseudoclusilis*; *Hygrocybe acutoconica* (Clem.) Sing.; *H. aurantiolutescens* Orton; *H. conicoides* (Orton) Orton & Watling*; *Lepiota laevigata* Lange; *Lepista irina* (Fr.) Bigelow; *L. nuda* (Bull.: Fr.) Cooke;

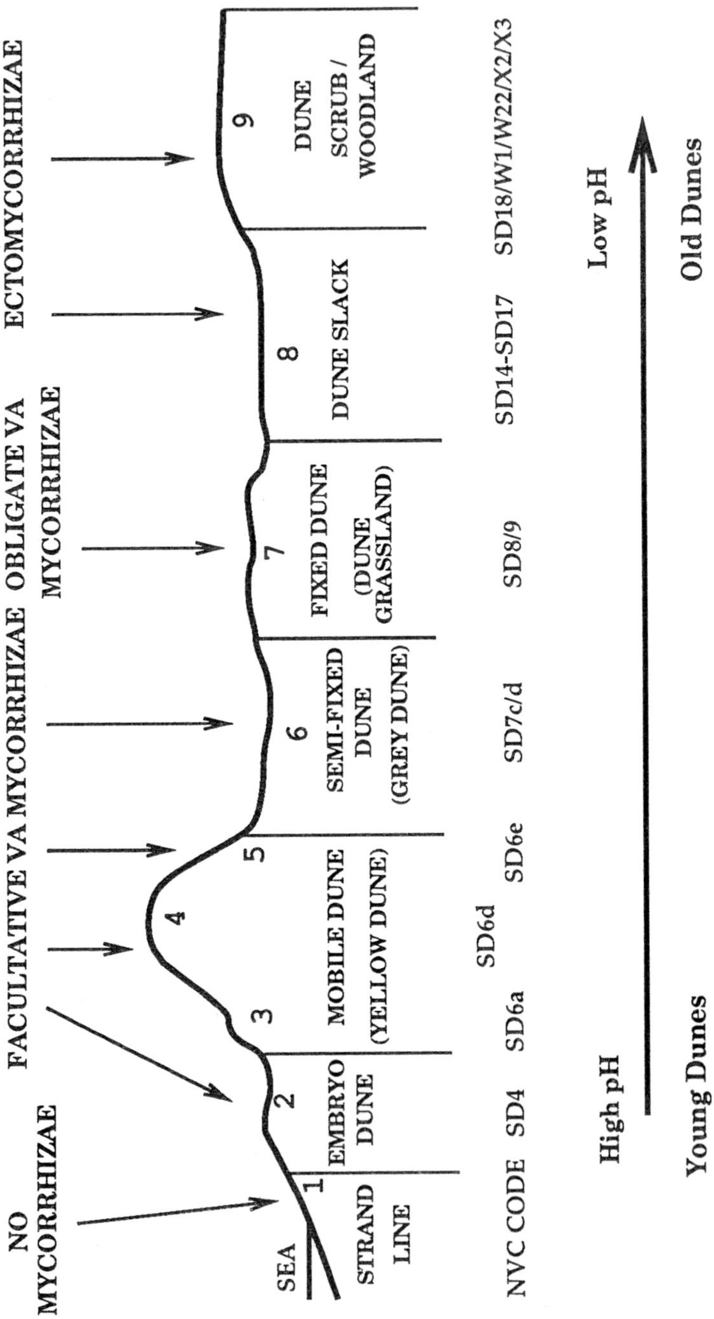

FIG. 2. Stylised diagram of a dune transect at Whiteford Burrows (with acknowledgements to Courtecuisse, 1984; Ranwell & Boar, 1986; and Read, 1989).

L. sordida (Fr.) Sing.; *Leucoagaricus naucinus* (Fr.) Sing.; *Macrolepiota konradii* (Huijsm.: Orton) Moser; *Marasmius litoralis* Quél.; *M. oreades* (Bolt.: Fr.) Fr.; *Melanoleuca exscissa* (Fr.) Sing.; *M. melaleuca* (Pers.: Fr.) Murr.; *Mycena flavo-alba* (Fr.) Quél.; *M. pura* (Pers.: Fr.) Kummer; *Omphalina pyxidata* (Bull.: Fr.) Quél.; *Rhodocybe popinalis* (Fr.) Sing.; *Stropharia caerulea* Kreisel and *S. coronilla* (Bull.: Fr.) Quél. Common gasteromycetes include: *Bovista plumbea* Pers.; *Geastrum nanum* Pers.; *Lycoperdon lividum* Pers.; *Tulostoma brumale* Pers.: Pers.; and *Vascellum pratense* (Pers.) Kreisel. The most frequently encountered ascomycetes are *Geopora arenosa* (Fuckel) Ahmad; *Geoglossum cookeianum* Nannfeldt and *Trichoglossum hirsutum* (Pers.: Fr.) Boud.

Among the biotrophs, mycorrhizal associations are undoubtedly the most important. The relationship between mycorrhizal infection and the various successional changes has been well characterised by Read (1989). Facultative vescicular-arbuscular (VA) mycorrhizae in the embryo, mobile and semi-fixed dunes give way to obligate VA mycorrhizal fungi in the fixed dune grasslands. The first appearance of biotrophic macrofungi is in the slacks where the more specialised ectomycorrhizal infections dominate the mycoflora. The main symbionts are creeping willow (*Salix repens*) with members of the Cortinariaceae. Two species of Russulaceae also occur here, *Lactarius controversus* (Fr.: Fr.) Fr. and *Russula persicina* var. nov. (see Rayner, 1985), and also characteristic are *Paxillus involutus* (Fr.) Fr. (Paxillaceae) and *Tricholoma cingulatum* (Almfelt: Fr.) Jacobasch (Tricholomataceae), which may be ectomycorrhizal. As scrub develops in the form of arborescent willows (*Salix* spp.), invading birch (*Betula*), hawthorn (*Crataegus*), blackthorn (*Prunus spinosa*) and other shrubs, increasing numbers of ectomycorrhizal species appear.

Macrofungi and National Vegetation Classification (NVC) Communities

Descriptions of community types (and NVC codes) for sand dunes are given by Rodwell (1989), but he omits the five dune-slack communities SD13, SD14, SD15, SD16 and SD17 and their subdivisions. The latter are described by Dargie (1990), who noted that a wholesale revision of slack types had taken place during the course of his fieldwork to the extent that field mapping could not in most cases be assigned to the new units. He concluded that all mapping of slack communities should therefore be regarded as very approximate. This is unfortunate since dune slacks are the richest of all sand-dune habitats for macrofungi. The following is a summary of the occurrences of macrofungi within the NVC groupings listed for Whiteford Burrows.

A. Strandline

(SD2/SD3; referred to at 1 on Fig. 2). Poorly developed and atypical at Whiteford Burrows with no characteristic macrofungal flora.

B. Embryo Dune

(SD4 *Elymus farctus* foredune community; 2). Depauperate mycoflora with occasional occurrence of *Psathyrella ammophila* and ruderals in the Bolbitiaceae.

C. Mobile (Yellow) Dune

(SD6a *Ammophila arenaria* mobile dune community, *Elymus farctus* sub-community; 3). As SD4.

D. Mobile (Yellow) Dune

(SD6e *Ammophila arenaria* mobile dune community, typical sub-community/*Festuca rubra* sub-community; 4 & 5). These comprise the classic yellow-dune zone characterised by the species listed at in the section on 'The Fungal Component'.

E. Semi-Fixed (Grey) Dune

(SD7c/d *Ammophila arenaria*–Festuca rubra semi-fixed dune community, *Ononis repens* sub-community/*Festuca rubra* sub-community; 6). Increased diversity of fungal species. Most common Agaricales are *Agaricus devoniensis, Agrocybe subpediades, Omphalina* spp. and *Rhodocybe popinalis*, together with *Tulostoma brumale* (Lycoperdales).

F. Fixed Dune (Dune Grassland)

(SD8/SD9 *Festuca rubra–Galium verum* fixed dune community/*Ammophila arenaria*–Arrhenatherum elatius dune grassland; 7). This grouping presents a considerable problem of interpretation. At Whiteford Burrows only a small area at the southern end of the reserve conforms to the SD8 and SD9 codes, with no clear sub-communities being described. Most of the area has a cover of rank, ungrazed vegetation which is not conducive to fungal fruiting. The areas of grazed turf, which occur throughout the site around the dune slacks, appear not to be included in this coding but are classified (probably as semi-fixed dune communities) in very broad mapping units, which in some cases lump together as many as seven codings of sub-community. In these circumstances it is not possible to ascribe fungal distribution to NVC types.

G. Dune Slack

(SD13 *Salix repens–Bryum pseudotriquetrum* dune-slack community). This community, which represents a relatively young and wet type, is apparently absent from Whiteford Burrows. At other locations it can be identified as the most likely habitat for *Agrocybe paludosa* (J. Lange) Kühn. & Romagn., *A. praecox* (Pers.: Fr.) Fayod and a number of *Cortinarius* species in the Telamonia section.

H. Dune Slack

(SD14/SD15/SD16/SD17 dune-slack communities and sub-communities; 8). As already mentioned, these codings are only approximate at Whiteford, and groups have been combined for mapping. Thus it is not possible to assign fungal species in NVC dune-slack communities with any degree of confidence.

I. Dune Scrub/Woodland/Plantation

(SD18 *Hippophaë rhamnoides* dune scrub/W22 *Prunus spinosa–Rubus fruticosus* agg. scrub/W1 *Salix cinerea–Galium palustre* woodland/X2 *Alnus glutinosa* slack woodland/X3 *Pinus nigra/sylvestris* woodland; 9). Intensive management has virtually eradicated *Hippophaë* from Whiteford Burrows, but this shrub has no known agaric associates. With regard to scrub and woodland, fungal distribution is similar to that in other ecosystems in relation to ectomycorrhizal, parasitic and saprotrophic modes.

It must be emphasised that systematic sampling was not undertaken during this preliminary assessment of fungal distribution relative to NVC communities. A realistic sampling scheme would, in fact, be hard to design since most habitats have elements of several communities or sub-communities. Comments are therefore based on a series of subjective observations at Whiteford Burrows. Many habitats can only be described as transition zones between one community and another. Further, it is well known that many fungi tend to fruit on ecotones between habitats. Thus, it is difficult to draw meaningful conclusions about the fungal distribution within NVC higher plant community codings.

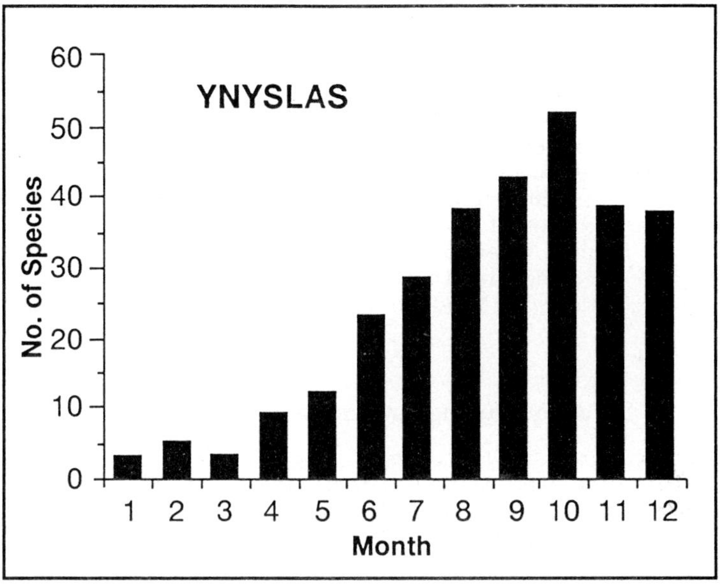

FIG. 3. Phenology of agaric fruiting (all species): Ynyslas.

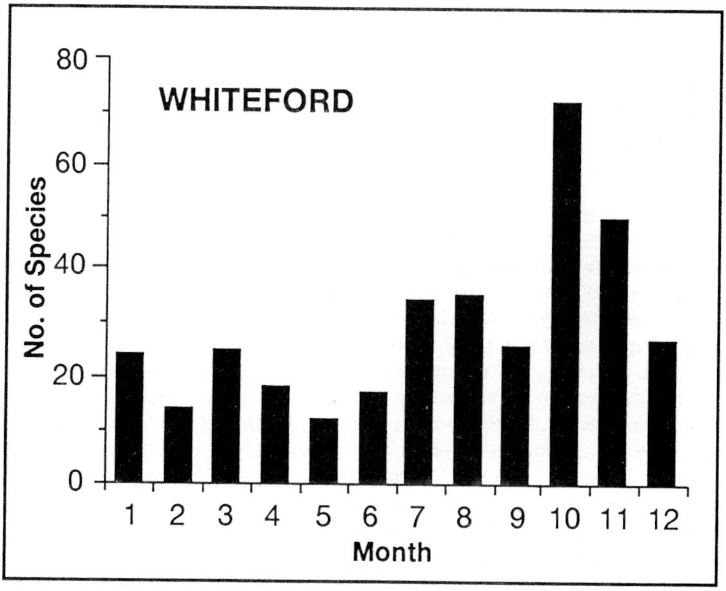

FIG. 4. Phenology of agaric fruiting (all species): Whiteford Burrows.

Phenology of Fungal Fruiting

The pattern of macrofungal fruiting on sand dunes differs considerably from that inland, and especially from inland woodlands. The coastal climate combined with low altitude greatly extends the potential fruiting period, which is particularly apparent where high rainfall is combined with milder winters, i.e. on the western coasts of England, Wales and Ireland and, to a lesser extent, of Scotland. The open sand dune habitats receive more precipitation than woodland soil, because less is intercepted by the canopy. Thus, fungal fruiting can be prolific throughout the year, being limited mainly by periods of low rainfall (Figs. 3 and 4). Plots of rainfall during the 10 days prior to surveys mirror the pattern of agaric fruiting. As previously noted (Watling & Rotheroe, 1989) monitoring throughout the year must be carried out to obtain the full picture of the dune mycoflora: 35% of the known mycoflora at Ynyslas fruits outside the traditional spring and autumn foray times.

Family Fruiting Patterns in the Agaricales

Eight families dominate the agaric flora of sand dunes in Britain (following the classification of Singer, 1986). These are the Agaricaceae, Bolbitiaceae, Coprinaceae, Cortinariaceae, Entolomataceae, Hygrophoraceae, Strophariaceae and Tricholomataceae. The Crepidotaceae, Paxillaceae and Pluteaceae are poorly represented, while the Amanitaceae, Boletaceae, Gomphidiaceae, Polyporaceae and Russulaceae make a substantial appearance only when tree plantations are present. Fruiting in the eight main agaric families (Figs. 6 & 7) is phenologically independent, with flushes sometimes occurring at perhaps unexpected times of the year, associated with specific environmental conditions. The very extended fruiting season of a number of the families is marked. That in the Coprinaceae, for example, is due in part to the fact that *Psathyrella ammophila* has been found fruiting continuously for ten months. Although the data are few, these patterns suggest that the family group in the agaricales may have some value as an ecological unit.

Comparisons Between Sand-Dune Locations

Agaric families can be used to characterise the profile of an individual sand-dune location, enabling broad comparisons to be made quickly and serving as an indication of the relative completeness of the mycoflora survey of the site (Fig. 7). There is a clear delineation of Ynyslas and Dyffryn, which do not have plantations, from the other four locations which do and are therefore well represented by members of the Russulaceae and Boletaceae. Aside from this, there appear to be few major differences between the mycofloras of sand dune systems in Britain. Members of the Agaricaceae are apparently more abundant in the south but few other generalisations can be made at present. The more favourable climatic environment in the west suggests that there is likely to be greater diversity there than on the east coast of Britain. A number of rare and unusual taxa have been recorded only from western dunes, including *Cyathus stercoreus* (Schw.) de Toni (Nidulariaceae), known only from Wales (four sites) and *Bovista limosa* Rostrup (Lycoperdaceae), recorded from four sites in Wales and Lancashire (Fig. 8). Three recently-described agaric species, *Coprinus ammophilae* (Fig. 9), *Hohenbuehelia culmicola* (Fig. 10) and *Melanoleuca cinereifolia* (Fig. 11), appear from general records to be rare. However, at locations which are visited regularly, the latter two taxa are such constant yellow-dune species that they have almost certainly been overlooked in the past and are probably widespread. *C. ammophilae* has been recorded from a number of sites in northern France (Courtecuisse, 1988) as well as from the Vendée (Courtecuisse, pers.

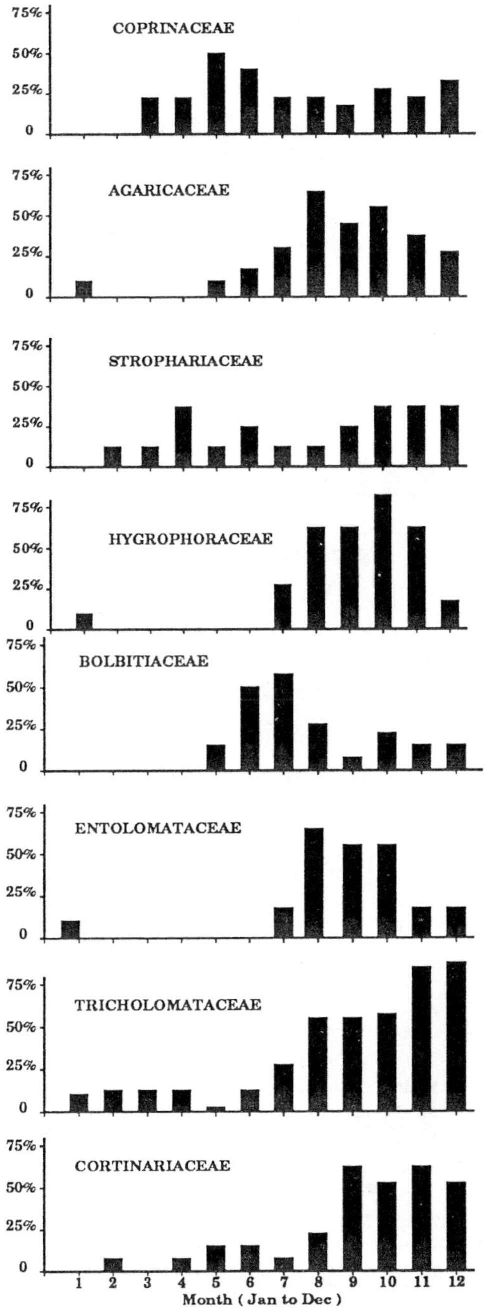

FIG. 5. Phenology of fruiting of 8 agaric families: Ynyslas.

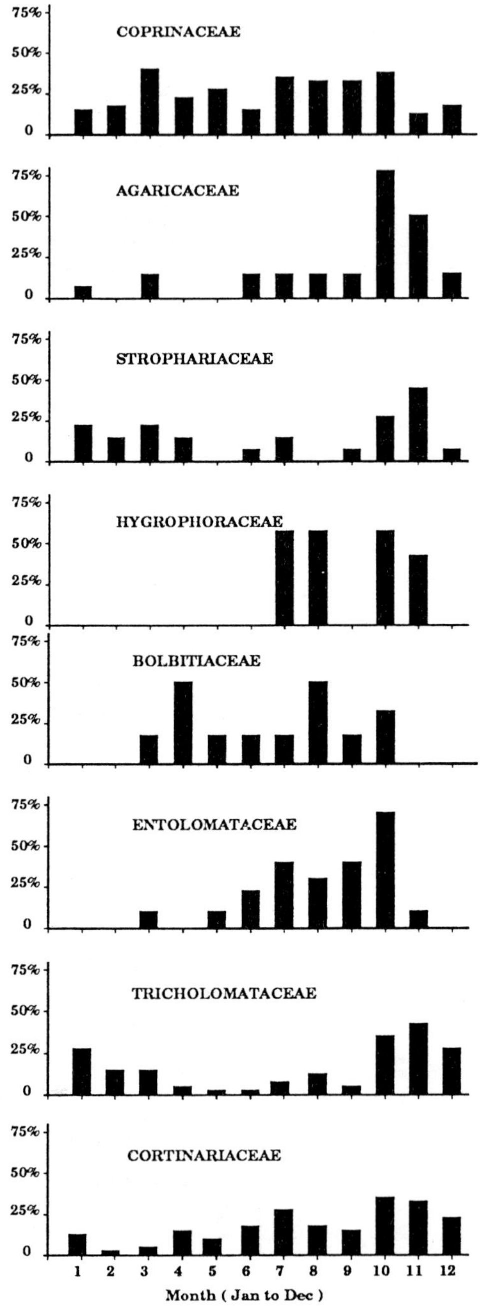

FIG. 6. Phenology of fruiting of 8 agaric families: Whiteford Burrows.

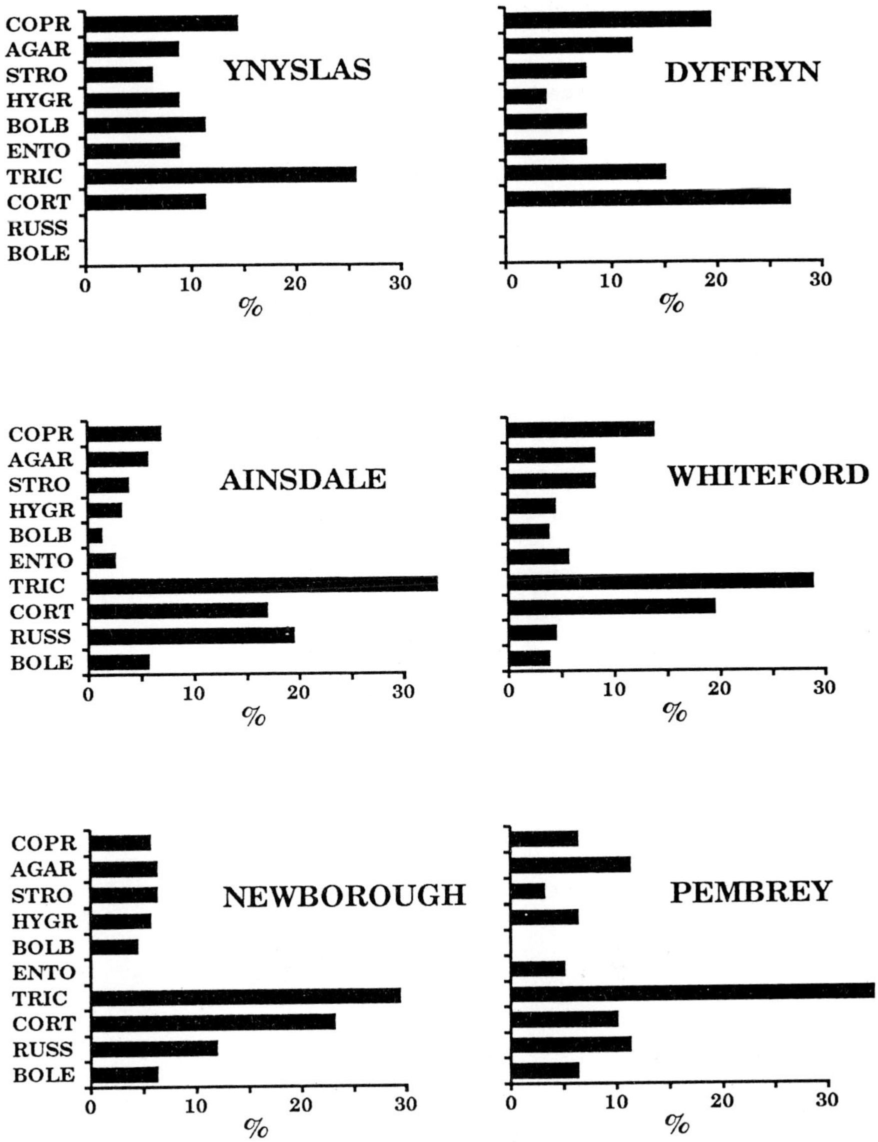

FIG. 7. Profiles of 10 agaric families in the mycofloras at 6 sand-dune sites. Family membership is expressed as % total number of all agaric species from each site.

comm.), also suggesting a widespread distribution for this taxon. One member of the Aphyllophorales which appears to be absent from the west coast of Britain is *Phellinus hippophaëcola* Jahn (Hymenochaetaceae), which grows on sea buckthorn (*Hippophaë rhamnoides*), but appears to be relatively common on east-coast dunes. This may be because, although widely planted and well established on western dunes, sea buckthorn is native only on the east coast between East Lothian and Sussex.

Sand-Dune Adaptations

The phenomenon of sand-dune ecotypes is well known amongst the phanerogams and a number of fungus species produce distinctly smaller carpophores when fruiting on

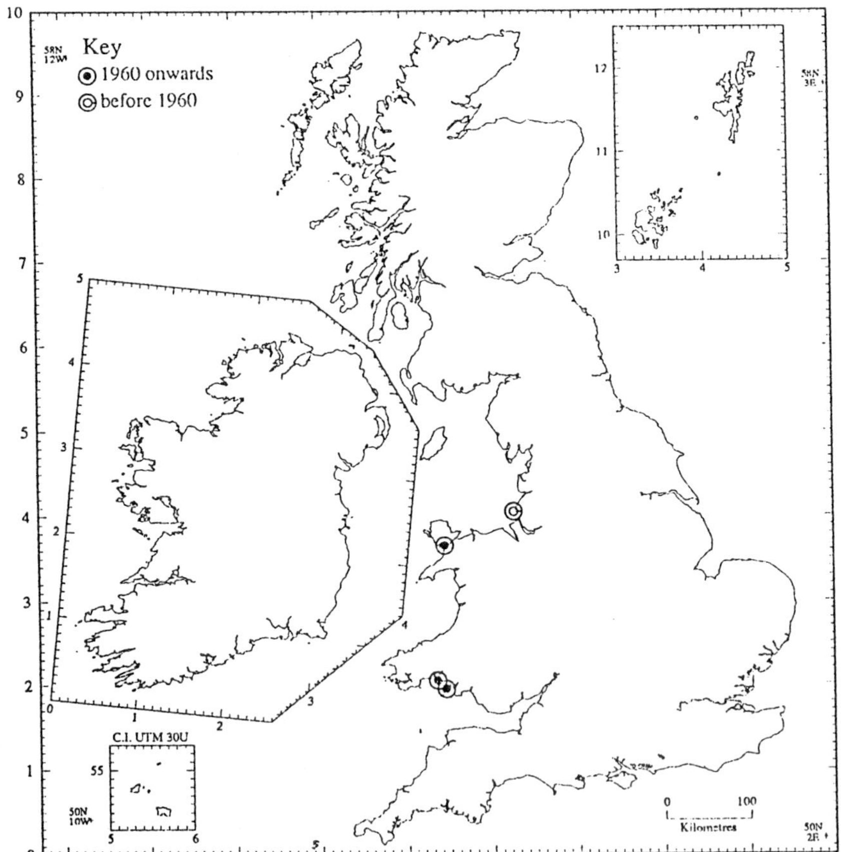

FIG. 8. Distribution map of *Bovista limosa*.

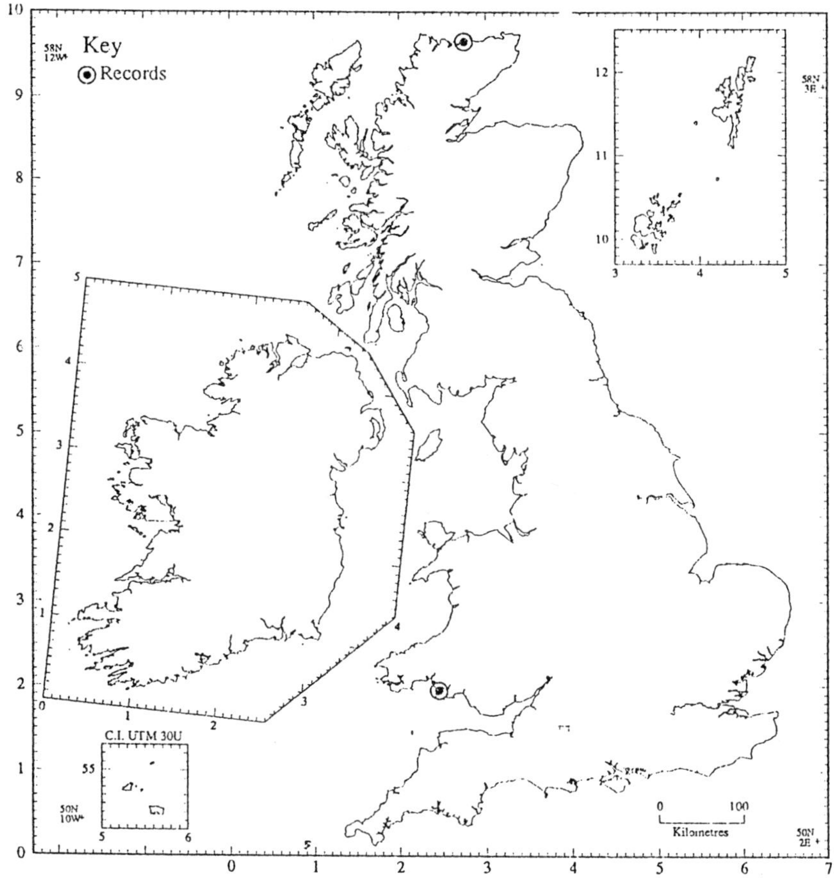

FIG. 9. Distribution map of *Coprinus ammophilae*.

sand dunes. Whether this has a genetic or merely an environmental basis is not known. *Paxillus involutus*, one of Europe's commonest agarics, has a wide ectomycorrhizal host range and is usually found producing relatively large fruit bodies (80 to 100 mm across the pileus), occurring either singly or in small groups. In dune slacks with *Salix repens*, however, the fungus produces massive flushes of several hundred fruit bodies which are invariably small (< 40 mm). Likewise, *Geoglossum cookeianum* and *Tricholoma cingulatum* fruit infrequently and produce smaller fruit bodies in very large numbers on dunes.

In the hostile environment of the yellow dunes, the habit of a large number of macrofungi of producing sturdy subterranean aggregations of mycelium mixed with sand, at the base of the stipe or pseudo stipe, has previously been noted (Watling & Rotheroe, 1989). These serve as a resource reservoir, where moisture and nutrients are

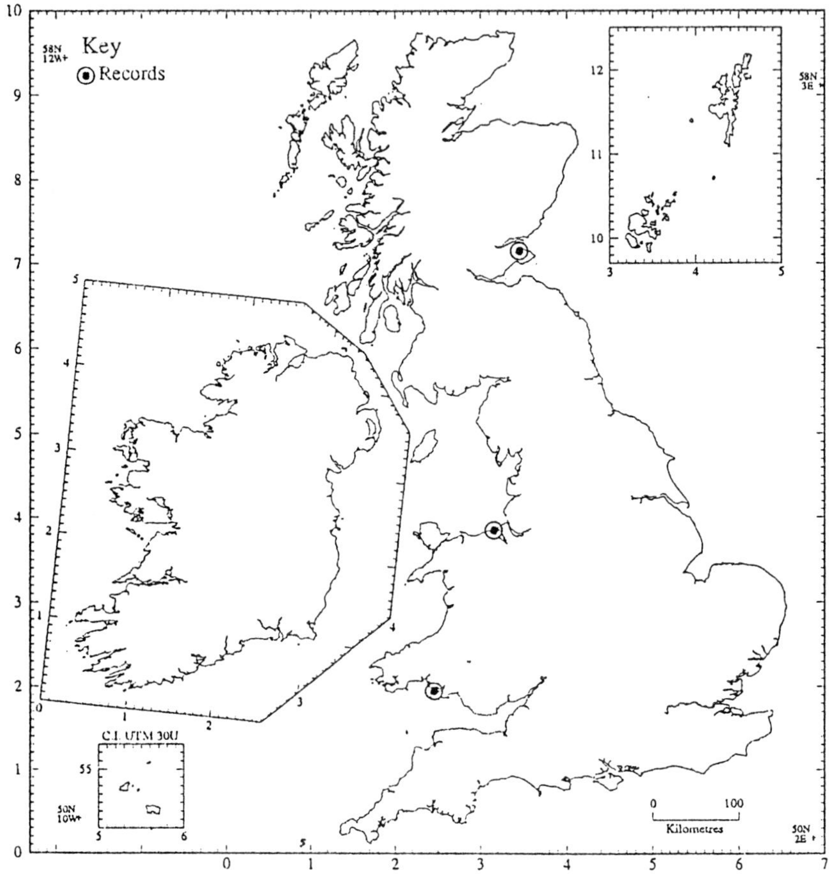

FIG. 10. Distribution map of *Hohenbuehelia culmicola*.

retained, as well as acting as an anchor in the shifting loose sand. Another common fruiting strategy of several yellow-dune fungi is the use of the bases of *Ammophila* stems for support. For example, the bird's nest fungus *Cyathus stercoreus*, a coprophilous fungus known in Britain only from the few Welsh locations, is invariably found fruiting attached at ground level to marram stems, although mycelial strands can be traced from the sporocarps to rotting rabbit pellets buried in the sand. In laboratory culture *C. stercoreus* sporocarps fruited in flasks of liquid medium formed on or close to sintered glass rods standing in the liquid (Brodie, 1975). This phenomenon suggested that some physical attribute of the rough glass rods in contact with the mycelium stimulated the formation of sporocarps, and that the physical nature of the environment might be a stronger determinant of fruiting than the chemical nature of the substrate.

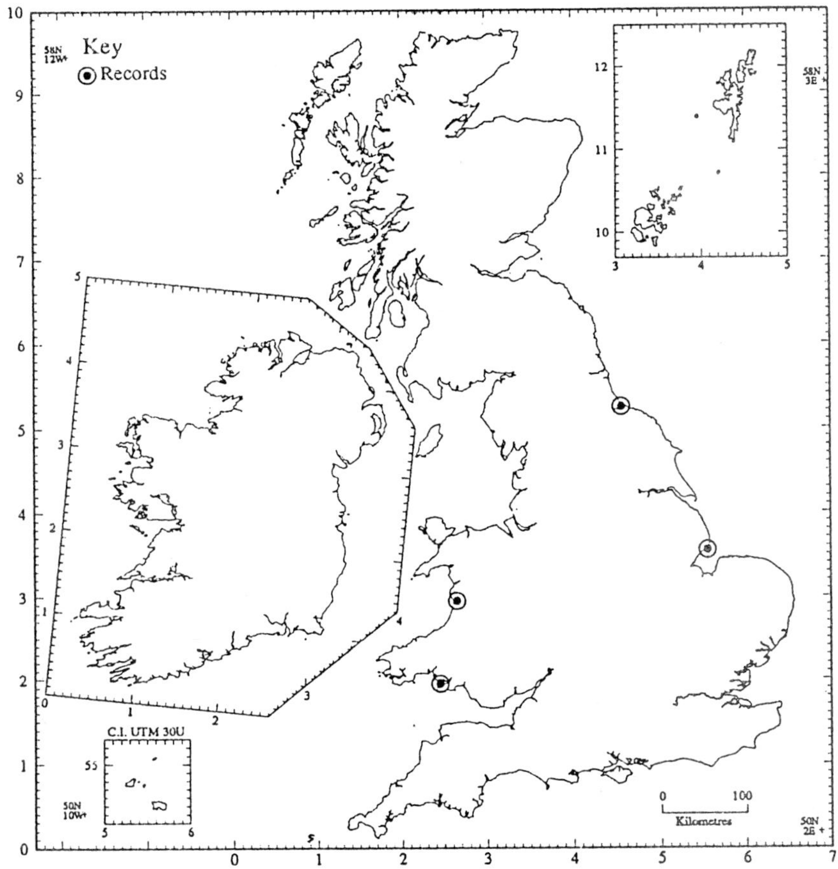

FIG. 11. Distribution map of *Melanoleuca cinereifolia*.

Conservation

Sand dunes are under considerable threat in Britain, as elsewhere in Europe, from coastal development, pollution, military activity, human trampling and increasing demands for recreational use. Fortunately, many dune systems are already established as nature reserves; others enjoy a degree of protection as country parks or Sites of Special Scientific Interest (SSSI). Yet there has been little or no mycological input when formulating the criteria for conservation and management of these sites, or for designation of SSSIs. However, the publication of a Provisional Red Data List of British Fungi (Ing, 1992) contributes important new evidence for making management decisions: for example, 14 macromycetes recorded from Whiteford Burrows and 9 from Ynyslas feature in the Red Data List.

Conclusions

Coastal sand dunes in Britain represent an important and relatively unspoiled ecosystem where the macromycete flora is more than a sub-set of the national mycoflora because it contains elements which are not found elsewhere. British sand-dune habitats are a refuge for rare and unusual species, which makes them eminently deserving of further study. In elucidating the rôle of fungi in relation to higher plant communities on sand dunes it seems unlikely, from casual observations, that the National Vegetation Classification will prove to be of great practical use. Frequent surveys throughout the year are necessary to evaluate the mycoflora of dunes because of likely phenological differences in fungal fruiting times. Growing awareness of the fungal potential of this ecosystem may lead mycologists to take a more active part in the conservation and management of sand-dune sites, particularly in view of the fact that many taxa found on sand dunes feature in the Red Data List of Fungi.

Acknowledgements

The author is grateful to the Countryside Council for Wales and the Joint Nature Conservation Committee for grants towards projects on Welsh and English dune systems and to Dr D.N. Pegler, Royal Botanic Gardens, Kew, for identification of critical species. Thanks are also due to Dr J.A. Cooper, Academic Computing Service, The Open University, for production of distribution maps.

References

Arnolds, E. (1983). Macrofungi 4.3. In Flora and vegetation of the Wadden Sea islands and coastal area. *Comm. Biol. Station Kampsweg P.D.* **257**: 61–74, 393–399.

Bon, M. (1970). Flore héliophiles des macromycètes de la zone maritime picarde. *Bull. Trimest. Soc. Mycol. France* **86**: 79–213.

Bon, M. (1972). Macromycètes du littoral boulonnais. *Doc. Mycol.* **3**: 9–46.

Bon, M. (1975). Agaricales de la Côte atlantique française. *Doc. Mycol.* **5**(17): 1–40.

Brodie, H.J. (1975). The Bird's Nest Fungi. University of Toronto Press. Toronto & Buffalo.

Courtecuisse, R. (1984). Transect mycologique dunaire sur la Côte d'Opale (France): I. Les groupements héliophiles et arbustifs de la xerosère. *Doc. Mycol.* **15**(57–58): 1–115.

Courtecuisse, R. (1986). Transect mycologique dunaire sur la Côte d'Opale (France): II. Les groupements de l'hygrosère. *Doc. Mycol.* **17**(66): 1–70.

Courtecuisse, R. (1988). Macromycètes interéssants, rares ou nouveaux: V. Coprinaceae. *Doc. Mycol.* **18**(72): 73–84.

Cowles, H.C. (1899). The ecological relations of the vegetation on sand dune of Lake Michigan. *Bot. Gaz.* **27**: 95–117, 167–202, 281–308, 361–391.

Dargie, T.C. (1990). National Sand Dune Vegetation Survey, South Wales Dune Survey, Site Report No. 51, Whiteford Burrows. Nature Conservancy Council, Peterborough.

Høiland, K. (1975). De obligate storsoppene p sanddyner i Norge med saerlig vekt pa forekomstene p Lista, Vest-Ågder. *Blyttia* **33**: 127–140.

Høiland, K. (1977). Storsopper i etablert sanddyne-vegetasjon p Lista, Vest-Ågder 1, Progressive systemer. *Blyttia* **35**: 139–155.

Høiland, K. (1978). Storsopper i etablert sanddyne-vegetasjon p Lista, Vest-Ågder 2, Eroderende systemer. *Blyttia* **36**: 69–86.

Ing, B. (1992). A provisional Red Data List of British fungi. *Mycologist* **6**: 124–128.

Ranwell, D.S. (1959). Newborough Warren, Anglesey: I. The dune system and dune slack habitat. *J. Ecol.* **47**: 571–601.
Ranwell, D.S. and Boar, R. (1986). Coast Dune Management Guide. Institute of Terrestrial Ecology, Huntingdon.
Rayner, R.W. (1985). Keys to the British Species of *Russula* (3rd Edn.). British Mycological Society, London.
Read, D.J. (1989). Mycorrhizas and nutrient cycling in sand dune ecosystems. In C.H. Gimmingham et al. (editors), Coastal Sand Dunes. *Proc. Roy. Soc. Edinburgh, B* **96**: 89–110.
Rodwell, J. (1989). National Vegetation Classification, Sand-dune, Strandline & Shingle Communities. Nature Conservancy Council, Peterborough.
Rotheroe, M., Hedger, J.N. and Savidge, J.P. (1987). The fungi of Ynyslas sand dunes. A preliminary survey. *Mycologist* **1**: 15–17.
Singer, R. (1986). Agaricales in Modern Taxonomy. J. Cramer, Weinheim.
Watling, R. and Rotheroe, M. (1989). Macrofungi of sand dunes. In C.H. Gimmingham et al. (editors), Coastal Sand Dunes. *Proc. Roy. Soc. Edinburgh, B* **96**: 111–126.

Mukhin, V.A.(1993). Rare Species of Wood-decaying Fungi from the West Siberian Plain. In D.N. Pegler, L. Boddy, B. Ing & P.M. Kirk (Editors). Fungi of Europe: Investigation, Recording and Conservation, pp. 139–145. Royal Botanic Gardens, Kew.

11. RARE SPECIES OF WOOD-DECAYING FUNGI FROM THE WEST SIBERIAN PLAIN

V.A. MUKHIN

Institute Plant & Animal Ecology, Ekaterinburg, Russia

Abstract

The ligniculous Basidiomycetes of the West Siberian Plain are investigated, with respect to their origin, distribution and frequency. 345 species have been recorded, a number are considered to be rare.

General Characteristics of the West Siberian Plain

The West Siberian Plain is a vast region of Eurasia. It is situated between the Urals mountains and the Enisey river (Fig. 1). The climate is continental, and the natural conditions of northern Eurasia are experienced. The main part of the West Siberian plain is occupied by forests: forest-tundra, northern, middle, southern taiga, subtaiga and forest-steppe. Studies on the biota of wood-decaying fungi of the West Siberian plain started in 1976.

FIG. 1. Location of West Siberia.

Systematic and Geographic Structure of the Basidiomycetes of the West Siberian Plain

The present West Siberian xylotrophic basidiomycetes biota is composed of 345 species in 5 orders and 23 families. About 95% are Aphyllophorales and Agaricales, 5% are Tremellales. Predominant families include Corticiaceae, Poriaceae, Hymenochaetaceae, Polyporaceae s.str., Tremellaceae, Tricholomataceae, Strophariaceae, Stereaceae, Crepidotaceae, Steccherinaceae, which include 91% of all species of the West Siberian mycobiota, similar to the boreal zone.

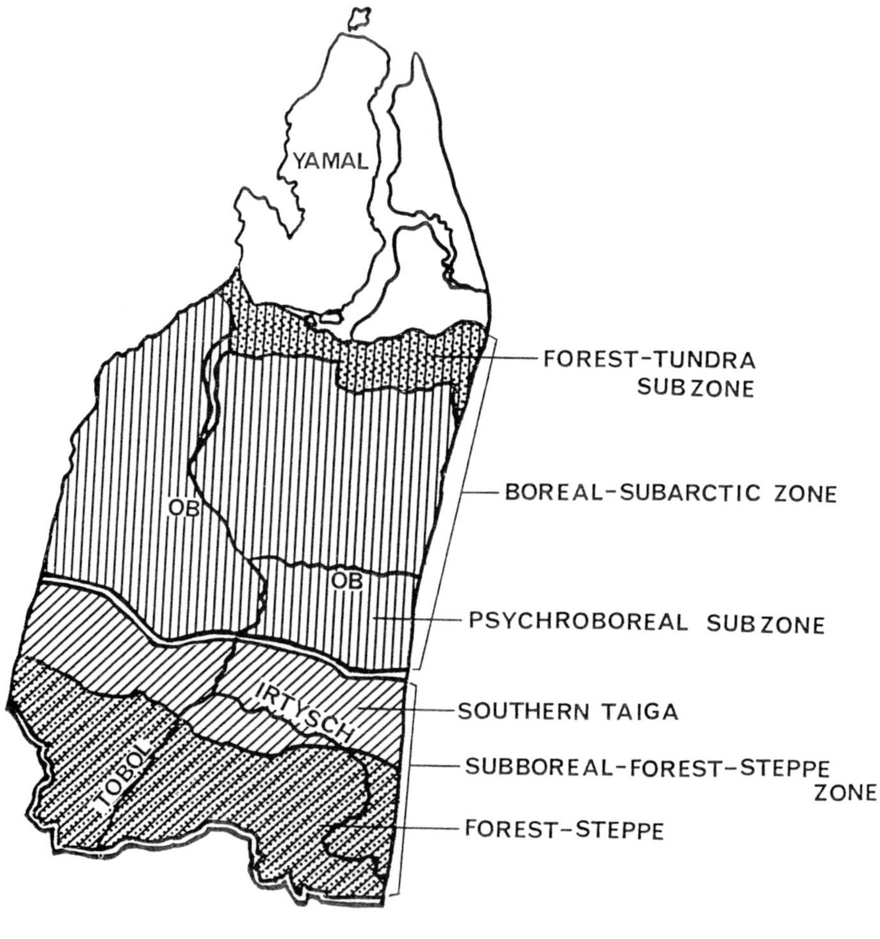

FIG. 2. Zonation of the xylotrophic basidiomycetes in the West Siberian Plain.

The majority of the West Siberian biota are pan-holarctic and eury-regional species. Regional endemics are almost entirely absent. In polypore composition it is similar to regional Eurasian biotas, especially those of the Russian Plain, Fennoscandia, and the Urals. Located in the zone of contact between Siberian and European biotas of xylotrophic basidiomycetes, it is remarkable in containing both "western" and "eastern" elements.

Age of the West-Siberian Mycobiota

The West-Siberian mycobiota was not continuous in its development since there were periods when it was totally or mostly deforested. In particular, 23,000–25,000 years ago, during the main thermal minimum of Pleistocene tundra landscapes predominated in the plain. The present biota of xylotrophic basidiomycetes was initiated in the early Holocene when forest vegetation, once again spread over the plain territory. Therefore, the West-Siberian mycobiota is a young landscape (Holocene), as determined from the time of its continuous development on a particular territory in relatively stable landscape-climatic conditions. The biota is a natural system formed by forest formation mycobiotas. They are of different ages exceeding the landscape age of the West Siberian xylotrophic

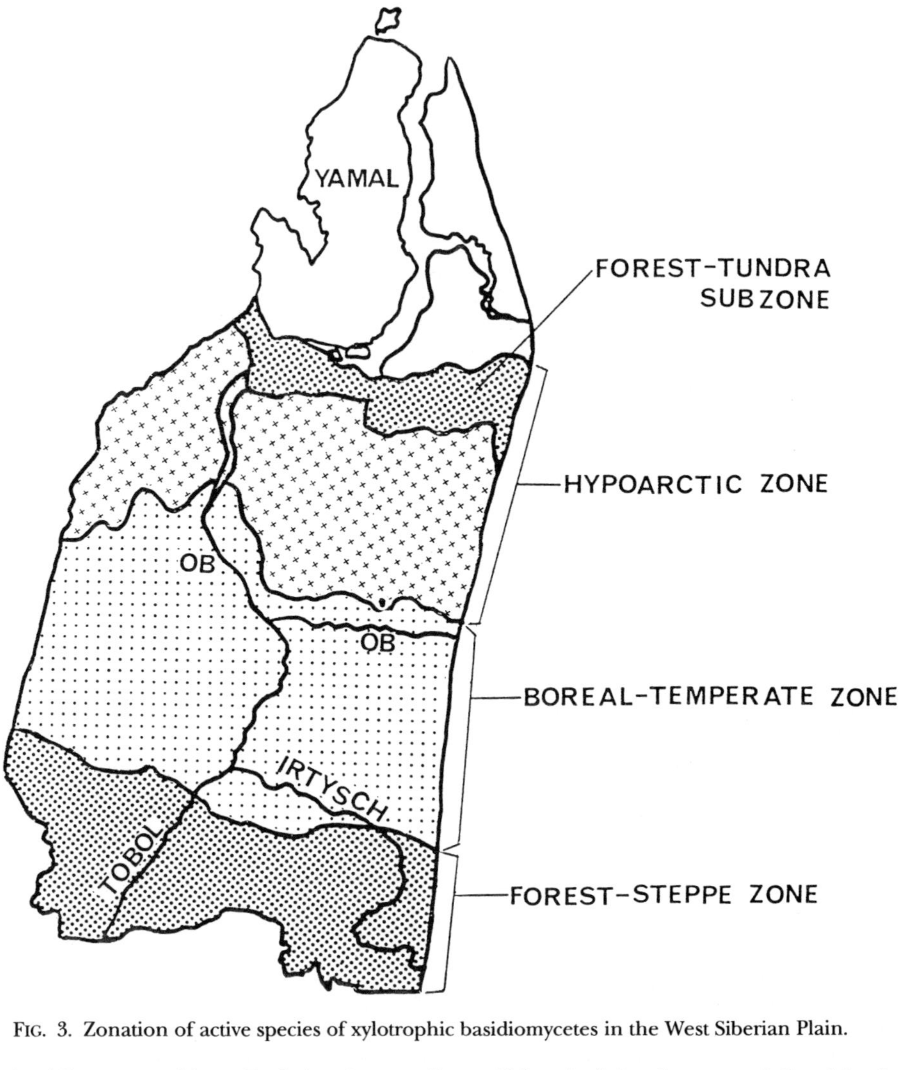

FIG. 3. Zonation of active species of xylotrophic basidiomycetes in the West Siberian Plain.

basidiomycetes biota. Each has its own line of historical development defined by the forests and formed outside of the plain. Therefore, we can say that the West Siberian xylotrophic basidiomycetes biota is of migratory origin and formed by species adapted to the plain natural conditions of the plain.

Zonality of the West Siberian Mycobiota

Gradient changes in the formation of mycobiotas and latitudinal alterations in the forest cover structure cause the zonal differentiation of the whole West Siberian plain mycobiota and the development of its spatial structure (Fig. 2). The last includes several zonal elements. They differ both in the fungi composition, their taxonomic variety, and the mycobiotas proportions. This zonal element reflects latitudinal specificity of ecological regimes and age of zonal landscapes.

Active Species of the West-Siberian Mycobiota

From an ecological position it is possible to recognise common and active species from rare species. The West-Siberian mycobiota includes 32 active species which represent not more than ten per cent of all species of the mycobiota. The significant renewal of active species in the latitudinal direction is one of the natural criteria of zonal boundaries. On this basis, we prepared a scheme of zonality of the West Siberian mycobiota (Fig. 3). It reflects specificity of modern ecological regimes in different parts of the plain. The following active species are typical of the boreal-subarctic mycobiota: *Phellinus chrysoloma* (Fr.) Donk, *P. pini* (Fr.) Ames, *P. igniarius* (Fr.) Quél., *Inonotus obliquus* (Fr.) Pilàt, *Trichaptum laricinum* (Karst.) Ryv., *Gloeophyllum sepiarium* (Fr.) Karst., *Cerrena unicolor* (Fr.) Murr., *Gloeoporus dichrous* (Fr.) Bres., *Plicatura nivea* (Fr.) Karst., *Peniophora aurantiaca* (Bres.) Höhn. & Litsch., *Tectella patellaris* (Fr.) Murr. In the subboreal-forest-steppe

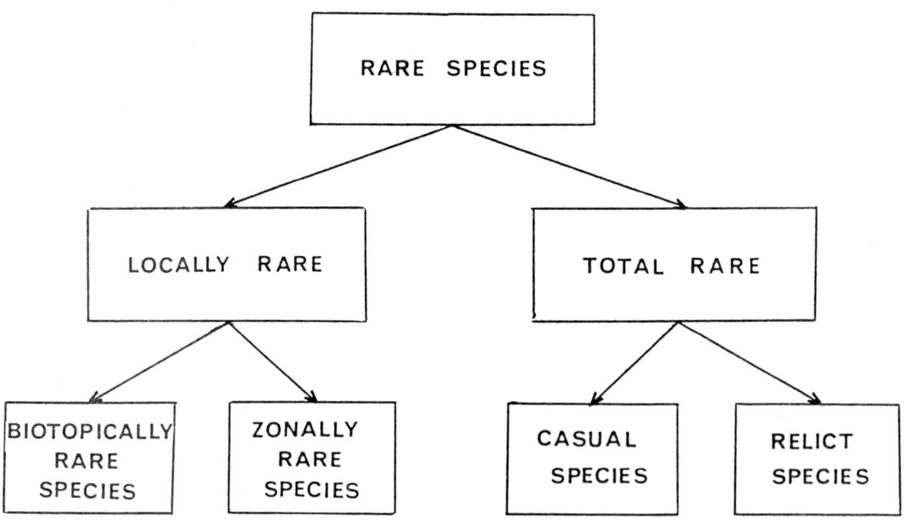

FIG. 4. Rare species groupings of wood-decaying fungi from the West Siberian Plain.

mycobiota the characteristic active species include *Stereum hirsutum* (Willd.: Fr.) Pers., *Daedaleopsis confragosa* (Bolton: Fr.) Schroeter, *Trichaptum biformis* (Fr.) Ryv., *Exidia glandulosa* Fr., *Pleurotus calyptratus* (Lindbl.) Sacc., *Hymenochaete mougeottii* (Fr.) Cooke,

Funalia trogii (Berk.) Bond & Singer, *Auriculariopsis ampla* (Lév.) Maire, *Fomitopsis rosea* (Alb. & Schwein.: Fr.) Karst., *Crepidotus mollis* (Schaeffer.: Fr.) Kummer. About 90% species of West Siberian mycobiota are rare.

Rare Species of the West Siberian Mycobiota

Rare species are represented by several groups: locally rare, biotopically rare, zonally rare and totally rare of 2 subgroups — casual and relict species (Fig. 4). Ecological-biological properties of the latter do not correspond to the present natural conditions of the plain. It is possible to select several groups of relicts on the base of the biogeographical analysis (Fig. 5). The first is mountain-taiga relicts (Fig. 6) which include such species as: *Leptoporus mollis* (Pers.: Fr.) Pilàt, *Ischnoderma resinosum* (Schrader: Fr.) Karst., *Phellinus nigrolimitatus* (Rom.) Boardot & Galzin, *P. ferrugineofuscus* (Karst.) Bourdot & Galzin, *Laetiporus sulphureus* (Bull.: Fr.) Murr., *Amylocystis lapponicus* (Rom.) Singer, *Pycnoporellus fulgens* (Fr.) Donk, *Fomitopsis cajanderi* (Karst.) Kotl. & Pouzar, *Antrodia heteromorpha* (Fr.) Donk, *Diplomitoporus crustulinus* (Bres.) Dom. Most of these are restricted to the dark-coniferous taiga and are most probably relicts of tertiary mountain-taiga dark-coniferous forests. *Laricifomes officinalis* (Vill.: Fr.) Kotl. & Pouzar, regarded by Murashkinsky (1939) as a mountain-taiga species, is a Pleistocene remnant. The second group is the nemoral relicts. Nemoral relicts in the West-Siberian biota of xylotrophic basidiomycetes is represented by *Aurantioporus fissilis* (Berk. & Curtis) Jahn, *Tyromyces kmetii* (Bres.) Band. & Singer, *Spongipellis spumeus* (Sow.: Fr.) Pat., *Aurantioporus alborubescens* (Bourdot & Galzin) Jahn, *Trametes foliaceo-dentata* (Nikol.) Dom., *T. cervina* (Schwein.) Bres., *T. ljubarskyi* Pilàt, *Phellinus conchatus* (Fr.) Quél., *Polyporus badius* (Pers.) Schwein., *P. squamosus* (Huds.) Fr. In modern distribution the species have certain connections with broad-leaved forests. They might be involved in the mycobiota: (a) in the tertiary period when coniferous-broadleaved forests were replaced by taiga vegetation; (b) in the Atlantic period of the Holocene during the last invasion of broad-leaved species southwards in the plain. Therefore, nemoral relicts are of different ages — tertiary, or pre-glacial (*Polyporus squamosus, P. badius, Spongipellis spumeus, Trametes ljubarskyi*) and holocenic (*Aurantioporus alborubescens, Trametes foliaceo-dentata, Tyromyces fissilis* and perhaps *Trametes cervina*). To the latter group of species West Siberia is eastern boundary of their ranges.

Rare species are the living history of the origin and development of mycobiota and they should be protected.

Protection of Rare Species of the West-Siberian Mycobiota

In accordance with the economic activity types historically formed on the plain protective measures are zonally specific. The status of hypoarctic forest (protective) is a sufficient condition for the conservation of xylotrophic basidiomycetes in the northern region of the plain. The mycobiota of forest-steppe and subtaiga zones of the plain are adapted to the long existing economic activity type. In both zones only the inventory of rare species and studies of their ecology and spread are required.

In the zone of the industrially-exploited forests (middle and southern taiga), the future of the mycobiota of dark coniferous formations is a cause for anxiety. It is composed of 178 species, 109 genera and 21 families. 45 species are endemic in this group of forest formation. The mycobiota of spruce forests includes 143 species, that of *Abies sibirica* forests — 99, that of *Pinus sibirica* forests — 63 species. In the first mycobiota, specific species (endemics of the formation) account for 12.6% in the second — 16.2% whilst in the third — only 4.8%.

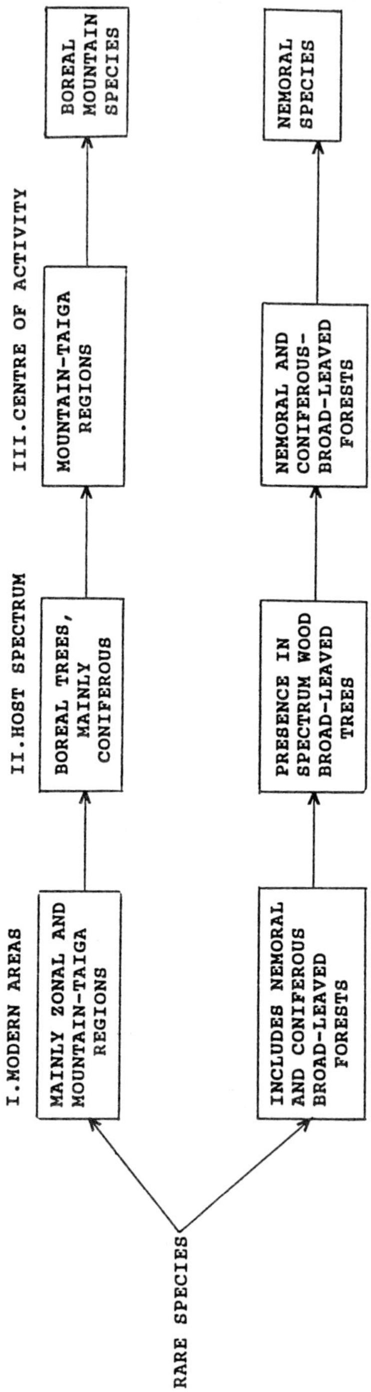

FIG. 5. Biogeographical status of rare species.

A developed specific, generic, and family endemism of the xylotrophic basidiomycetes biota of the dark coniferous formation is evidence of its relative antiquity. This is consistent with the idea that a dark coniferous taiga is the oldest and the most stable main body of the present taiga ecosystems. The characteristics and peculiarities of the contemporary West Siberian plain mycobiota are defined in many aspects, but basically by the mycobiotas of dark coniferous forests.

Human activity can change the formation structure of the plain forest cover, thus the very West Siberian xylotrophic basidiomycetes biota. Extremely undesirable is the decrease of their total taxonomic diversity due to the disappearance of rare relict species from area. To prevent this it is necessary to create a network of forest reservations especially in the middle and southern taiga zones, i.e. in the zone of industrially exploited forests. First, it is necessary to study the distribution and ecology of fungi.

I. **Moutain Taiga Relicts**

Amylocystis lapponicus
Diplomitoporus crustulinus
Ischnoderma resinosum
Leptoporus mollis
Phellinus ferrugineo-fuscus
Phellinus nigrolimitatus
Pycnoporellus fulgens (Tertiary)

II. **Nemoral Relicts**

(a) **Tertiary or Pre-Glacial**
Polyporus badius
Polyporus squamosus
Spongipellis spumeus
Trametes ljubarskyi

(b) **Holocenic**
Aurantioporus alborubescens
Trametes foliaceo-dentata
Tyromyces fissilis

III. **Pleistocene Relicts**

Laricifomes officinalis

FIG. 6. Relict fungi of the West Siberian Plain.

Benkert, D. (1993). Bryoparasitic Pezizales: Ecology and Systematics. In D.N. Pegler, L. Boddy, B. Ing & P.M. Kirk (Editors). Fungi of Europe: Investigation, Recording and Conservation, pp. 147–156. Royal Botanic Gardens, Kew.

12. BRYOPARASITIC PEZIZALES: ECOLOGY AND SYSTEMATICS

D. BENKERT

Institut für Spezielle Botanik und Botanische Sammlungen des Museums für Naturkunde an der Humboldt-Universität Berlin, Späthstrasse 80/81, D (O) — 1195 Berlin, Germany

Abstract

Parasitism is extremely rare within the order Pezizales, but some genera include obligate parasites of moss. Parasitism on mosses is a generic character of *Lamprospora* de Not., *Neottiella* (Cooke) Sacc., *Octospora* Hedw., and *Octosporella* Döbl.; it is correlated with morphological and anatomical characters. Consequently, these genera are clearly distinct from the similar but not parasitic genera *Byssonectria* Karsten, *Kotlabaea* Svrček, *Leucoscypha* Boud., *Moravecia* Benk., Caillet & Moyne, and *Ramsbottomia* Buckley. The neglect of host mosses in the past has prevented detailed knowledge of species differentiation within the associated fungal genera. Thus, over half of all bryophytic Pezizales species have been described in the last 20 years, i.e. since the close relationship between fungi and mosses has been realized. The paper considers those relationships, host mosses, the host mosses actually known, and their taxonomic and ecological position as well as the moss communities that include host mosses.

Introduction

Ecology should always be an important part of taxonomic work, though it is often rather disregarded, and is essential with the bryoparasitic Pezizales. On the one hand, systematics can be elucidated only by studying the ecological peculiarities of the taxa in question; on the other hand, clarified systematics enables more comprehensive studies of the position and role of the taxa in ecosystems.

The close relationship between some Pezizales and moss species was largely overlooked for a long time. In the early 1970s, I became aware that species of *Lamprospora* de Not., *Octospora* Hedw. and *Neottiella* (Cooke) Sacc. always grow in association with definite moss species, and that this was a generic character. This was also realized by Dennis & Itzerott (1973) who described several new species that were characterized by a combination of morphological pecularities with an association with definite moss species. Earlier, Hedwig (1789) had noted in his description of *O. leucoloma*, the type species of the genus *Octospora*, the association with the moss *Bryum argenteum*. In some cases the fungal species were named after an associated moss, e.g. *Peziza ricciae* Crouan; *Peziza orthotricha* Cooke & Ellis; *Lamprospora campylopodis* Buckley; *Neottiella catharinaea* McLennan & Halsey. Sometimes the presence of mosses was mentioned in descriptions as "on the ground amongst mosses" or "parmis les petites moùsses", more precisely. Boudier (1907) noted for *Humaria wrightii* (Berk. & Curtis) Boud. that it grows "presque toujours sur l'*Hypnum serpens*".

Where the apothecia are situated directly on the host, a relationship between fungus and moss is implicated. In many cases, however, the apothecia are formed on the soil amongst mosses and an association is not immediately apparent. Also, many non-parasitic species of Pezizales grow in moss covered areas. Thus, parasitism can only be detected with certainty by observation of appressoria on the moss.

The genera *Lamprospora, Neottiella, Octospora* and *Octosporella* are all bryoparasitic, whereas *Byssonectria, Kotlabaea, Moravecia* and *Ramsbottomia,* which are superficially similar are not bryoparasitic.

TABLE 1. Bryoparasitic species of Pyronemataceae described later than 1970.

Lamprospora

L. aneurae Benkert
L. cailletii Benkert
L.ditrichi Benkert
L. faroensis Benkert
L. gotlandica Benkert
L. hanffii Benkert
L. hispanica Benkert
L. kristiansenii Benkert
L. leptodictya Dissing
L. lubicensis ined.
L. moynei Benkert
L. norvegica Benkert, Aas & Kristiansen
L. paechnatzii Benkert
L. rehmii Benkert
L. retispora (Itzerott & Thate) Schumacher
L. rugensis Benkert
L. seaveri Benkert
L. tortulae-ruralis Benkert

Neottiella

N. aphanodictyon (Kobayasi) Dissing, Korf & Sivertsen

Octospora

O. bridei Caillet & Moyne
O. echinospora Caillet & Moyne
O. grimmiae Dennis & Itzerott
O. heterosculpturata Schumacher
O. hygrohypnophila Dissing & Sivertsen
O. longispora ined.
O. moravecii Khare
O. neglecta Dennis & Itzerott
O. roxheimii Dennis & Itzerott
O. subglobispora ined.

Octosporella

O. hemicrypta (Döbbeler) Döbbeler
O. ornithocephala Döbbeler
O. perforata (Döbbeler) Döbbeler
O. suboperculata Döbbeler & James in Döbbeler
O. urosperma Döbbeler

Host Mosses of Bryoparasitic Pyronemataceae: Significance and Problems

Before about 1970 a rather wide species concept existed in the Pyronemataceae, e.g. most species of *Lamprospora* with reticulate ascospores were united in *Lamprospora miniata* de Not. After detection of host specificity, however, it became possible to distinguish more species on the basis of the correlation between slight morphological differences and specific host mosses (Table 1). For example, in the *Lamprospora miniata* complex, the new species *L. rugensis* Benk., *L. tortulae-ruralis* Benk., and *L. hanffii* Benk. were discriminated, in the group of the fine-meshed species, *Lamprospora dictydiola* Boud. and *L. carbonicola* Boud. could be clearly distinguished on the basis of a slight but constant difference in spore size correlating with different host mosses (Benkert 1987). The two species had been previously discriminated by Boudier (1905) but became confused later on. Though the morphological difference is very slight, these species are ecologically very distinct in regard of the different ecological behaviour of the host mosses.

Though the discovery by Döbbeler (1979) of the obligate association of these fungi with mosses, and especially of their bryoparasitic behaviour, considerably improved our knowledge, there are still many gaps. There are several reasons for the latter. Firstly, few mycologists collect and study fungi with mostly rather hidden and inconspicuous apothecia as the Pyronemataceae. Related to this, there are few specimens deposited in herbaria; most of these belong to *Ramsbottomia* or other non-bryophilous species.

TABLE 2. Number of *Lamprospora* species known from individual countries up to now (out of a total of 31 species)

(based on specimens checked by the author, or credible information in the literature)

Germany	16
Norway	13
Czechoslovakia	12
France	12
U.S.A.	11
Great Britain	8
Sweden	6
Switzerland	6
Austria	6
Denmark	4
Australia	4 (+1)
India	(3)
Netherlands	3
Italy	2
Greenland	2

+ 11 countries with 1 species each

TABLE 3. Number of *Lamprospora* species known from individual continents up to now (out of a total of 31 species)

Europe	29
North America	12
Australasia	4 (+1)
Asia	1 (+3)
Africa	1
Antarctic	1

(in brackets: judging from literature)

Secondly, there is little information on the associated mosses. The mosses are neglected completely, or their presence is mentioned incidentally. Sometimes associated mosses are recorded but the host moss is not indicated. Thirdly, from many countries or even continents, there are no or only very few specimens available (Tables 2 and 3). Thus, our system of the bryoparasitic genera, including our knowledge of the mosses attacked, is almost exclusively based on the temperate areas of Europe and North America.

Taxonomic Position and Ecological Characteristics of the Host Mosses

Less liverwort genera (Table 4) are attacked than moss genera (Table 5). Within the mosses, the genera attacked by Pezizales nearly completely belong to the acrocarpous species (tufted mosses), few species of pleurocarpous (creeping) mosses being affected (Tables 5 and 6).

Many of the lost mosses are pioneers, colonizing the ground very rapidly, e.g. *Ceratodon purpureus*, *Bryum argentum*, *Barbula unguiculata*, and *Ephemerum serratum*. The pioneers take advantage of the short time before the more vigorous flowering plants begin to colonize the open habitat. They are r-strategists with short life cycles and rapid development. *Marchantia polymorpha* is a pioneer liverwort.

Other mosses avoid competition with flowering plants by colonizing habitats unfavourable to the latter. For example, mosses can dominate in swamps, bogs, fens etc.; on dry sandy soils; on rocks and walls; in shadowy gorges; in alpine localities; on tree bark; on pathways, road sides and similar localities; on burnt patches; on places treated with herbicides etc. The distribution of mosses will determine the possible distribution of the obligately parasitic Pyronemataceae.

TABLE 4. Bryophyta (Hepaticae)

(Survey on the system with host genera and number of bryoparasitic species(in brackets)

Sub-classes	Orders	Genera
Sphaerocarpidae		—
Marchantiidae	*Marchantiales*	*Marchantia*
		(*Lunularia*) (1)
	Ricciales	*Riccia* (1)
Jungermanniidae	*Calobryales*	—
	Metzgeriales	*Aneura* (1)
		Pellia (1)
	Jungermanniales	*Frulliana* (3)
		Lophocolea (1)
		Plagiochila (1)
		Porella (1)
		Radula (1)

TABLE 5. Bryophyta: Musci (sub-class Bryidae)

Orders and genera including host species of Pyronemataceae

Polytrichales:	*Atrichum* (1), *Oligotrichum* (2), *Pogonatum* (1), *Polytrichum* (4)
Dicranales:	*Campylopus* (2), *Ceratodon* (7), *Dicranella* (3), *Ditrichum* (1), *Pleuridium* (5)
Encalyptales:	*Encalypta* (1)
Pottiales:	*Aloina* (2), *Barbula* (4), *Phascum* (1), *Pottia* (2), *Pterygoneurum* (1), *Tortella* (3), *Tortula* (4)
Grimmiales:	*Desmatodon* (1), *Grimmia* (3), *Schistidium* (2)
Bartramiales:	*Philonotis* (1)
Funariales:	*Ephemerum* (4), *Funaria* (3), *Splachnum* (1), *Tetraplodon* (1)
Bryales:	*Bryum* (9), *Pohlia* (2)
Neckerales:	*Cryphaea* (1), *Orthotrichum* (1)
Hypnales:	*Amblystegium* (1), *Hygrohypnum* (1)

Sub-classes without host species: Sphagnidae, Andreaeidae

Orders without host species: Archidiales, Buxbaumiales, Fissidentales, Hookeriales, Schistostegales, Tetraphidales

(in brackets: number of bryoparasitic Pyronemataceae)

TABLE 6. Most frequently attacked families of Musci

Ditrichaceae:	13 species of bryoparasitic Pyronemataceae
Pottiaceae:	12
Bryaceae:	11
Funariaceae:	5
Grimmiaceae:	5
Polytrichaceae:	5
Ephemeraceae:	4

All other families of Musci with 0–2 species

Moss Coenology

Mosses are often major components of plant communities. For example, in woodland communities they form the so-called moss layer which often densely covers the ground, e.g. species of *Hypnum*, *Dicranum*, *Mnium* and *Hylocomium splendens*, *Pleurozium schreberi*, *Pseudoscleropodium purum*. They are often used to characterize associations and sub-associations, e.g. Dicrano-Pinetum, Polytricho-Carpinetum, Leucobryo-Pinetum. Within communities, mosses can also form relatively independent constituents on fallen trees, burnt patches etc. Furthermore, they often constitute independent moss communities (but sometimes associated with lichens and a few small flowering plants) in habitats unfavourable to flowering plants, as indicated above.

The bryoparasitic fungi therefore are often characteristic members of moss communities. Tables 7–9 give some examples where bryoparasitic Pezizales are significant components. The Funarietum hygrometricae (belonging to the class Funarietea hygrometricae; Table 7) community develops primarily in places that are disturbed by man's activities, e.g. burnt patches, on heaps of rubbish or fertilizers, on areas treated with herbicides, etc. The character species *Funaria hygrometrica* is known to be parasitized by three species of Pyronemataceae. *Marchantia polymorpha*, the character species of a sub-association on moister soils, is sometimes attacked by *Octospora ithacaensis* (Rehm) Khare. *Ceratodon purpureus* and *Bryum argenteum*, occurring in the Funarietum hygrometricae constantly and in large quantities, are both attacked by a wide range of *Lamprospora* and *Octospora* species. *Octospora rustica* (Velen.) Moravec shows remarkable behaviour in attacking *Ceratodon purpureus* almost only on burnt places, though the host moss is also very common in many other habitats.

TABLE 7. Moss commmunity Funarietum hygrometricae and associated host species of bryoparasitic Pyronemataceae

Character species: *Funaria hygrometrica*:

- **Lamprospora carbonicola*
- *Neottiella hetieri*
- *Octospora roxheimii*

Character species of a moister sub-association: *Marchantia polymorpha*:

- *Octospora ithacaensis*

Characteristic associated species:

Ceratodon purpureus

- **Lamprospora arvensis*
- *Lamprospora faroensis*
- **Lamprospora seaveri*
- *Octospora rubens*
- **Octospora rustica*
- *Octospora spec.*

Bryum argenteum

- **Octospora leucoloma*
- **Octospora tetraspora*
- *Octospora spec.*

(* = common and widely distributed species)

TABLE 8. Moss communities of the class Barbuletea unguiculatae and associated host species of bryoparasitic Pyronemataceae

Moss Species	Bryoparasitic Pyronemataceae
Aloina rigida	(*Lamprospora hispanica*)*
Barbula convoluta	*Lamprospora gotlandica* *Octospora crosslandii*
Barbula unguiculata	*Lamprospora miniata*
Bryum klinggraefii	*Octospora tetraspora*
Bryum microerythrocarpum	*Octospora melina*
Ephemerum serratum	*Octospora bridei* *Octospora echinospora* (*Octospora moynei*) *Lamprospora tuberculatella*
Phascum cuspidatum	*Octospora axillaris*
Pottia intermedia	*Lamprospora miniata* *Octospora crosslandii*

* no appressoria seen yet

The basiphilous and photophilous moss communities of the class *Barbuletea unguiculatae* preferably grow on fields, especially on stubble-fields in late autumn and winter. They consist mainly of species of the family Pottiaceae, and many of these are attacked by fungi (Table 8).

The *Grimmietea anadonti* comprises moss communities on limestone, and also occurs on walls, monuments, etc. (Table 9). Cushions of *Grimmia pulvinata* are particularly common and characteristic.

Life Style of Mosses and Phenology of Bryoparasitic Pyronemataceae

The bryoparasitic Pyronemataceae are adapted to the life style of their host mosses. Mosses are poikilo-hydric plants; they cannot regulate their water balance, hence they behave like a sponge with water contents reflecting that of the surroundings. Many mosses are extremely drought tolerant, surviving dry conditions in an inactive state. The bryoparasitic fungi grow on mosses only when the latter are in an active condition. Hence most bryoparasitic Pezizales develop in moist seasons, especially from October to March provided there is no frost or snow cover.

In Central Europe, apothecia of *Octospora* and allied genera are most common in late autumn (October to December) and early spring (March to May) although this group exhibits at least three different seasonal fruiting patterns (Fig. 1).

TABLE 9. Moss communities of the Grimmietea anodonti and associated host mosses of bryoparasitic Pyronemataceae

Moss Species	Bryoparasitic Pyronemataceae
Grimmia pulvinata	*Octospora grimmiae* *Octospora meslinii* *Octospora musci-muralis*
Schistidium apocarpum	*Octospora neglecta* *Octospora pseudoampezzana*
Tortella tortuosa	*Lamprospora cailletii* *Octospora crosslandii*
Tortula muralis	*Lamprospora dictydiola*
Tortula ruralis agg.	*Lamprospora retispora* *Lamprospora tortulae-ruralis*

With the first type, illustrated by *Lamprospora carbonicola* Boud. associated with *Funaria hygrometrica*, apothecia can be found in all seasons, but with peaks in spring and autumn and troughs in winter, due to frost, and in summer due to dryness. A second type, represented by *Octospora humosa* (Fr.) Dennis associated with Polytrichaceae, fruits from August to November. Thirdly, *Octospora musci-muralis* Graddon associated with *Grimmia pulvinata* only fruits from October to February with an optimum in January, at least in eastern Germany.

Conclusions

Much remains to be elucidated concerning taxonomy, distribution and ecology of the bryoparasitic Pezizales. Collections from all parts of the world are urgently required; specimen should include the moss substrate and as much apothecia as possible.

Some species, e.g. *Lamprospora miniata* de Not., *L. seaveri* Benk., and *Octospora crosslandii* (Dennis & Itzerott) Benk. seem to be complexes of several closely related species. A species complex is especially likely if a 'species' has several host moss species which are not taxonomically related.

In some species groups speciation is probably taking place. Whether several species can be distinguished on morphology can be clarified only by investigation of as many specimens as possible.

Acknowledgements

The author wishes to thank the directors and curators of numerous herbaria, many colleagues and friends for the loan of many valuable specimens, and finally the Organising Committee of the XIth Congress of European Mycologists and its chairman Dr. David N. Pegler for the invitation to the congress.

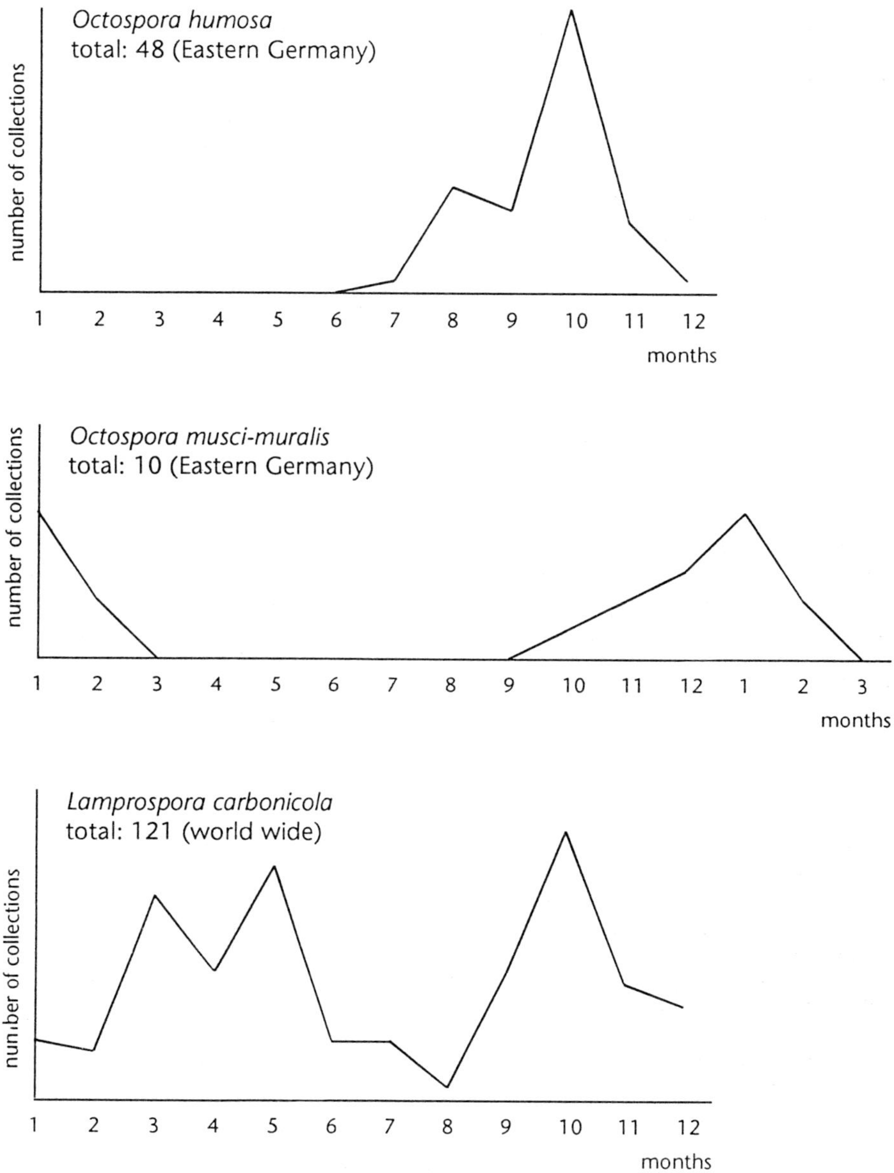

FIG. 1. Fruiting patterns of *Octospora* and allied genera in central Europe.

References

Aleev, Yu. G. (1986). Ecomorphology. 433 pp. Naukova dumka, Kiev.
Baltaeva, G.M. (1988). Bracket fungi of reservations of Uzbekistan. In: Izuchenie gribov v biocoenosakh. Abstr. of 4th All-Union conf. Sverdlovsk. p. 6.
Baltaeva, G.M. (1991). Some data on distribution of bracket fungi in Uzbekistan. *Mycol. & Phytopathol.* **25**(1): 15–21.
Benkert, D. (1987). Beiträge fur Taxonomie der Gattung *Lamprospora* (Pezizales). *Z. Mykol.* **53**(2): 195–271.
Boudier, J.L.E. (1905–1910). Icones mycologicae ou Iconographie de Champignons de France. Paris.
Boudier, J.L.E. (1907). Histoire et classification des Discomycètes d'Europe. Paris.
Braun-Blanquet, J. (1951). Pflanzensoziologie, Grundzüge der Vegetationskunde. 2 Aufl., 632 pp. Springer, Wien.
Dennis, R.W.G. and Itzerott, H. (1973). *Octospora* and *Inermisia* in Western Europe. *Kew Bull.* **28**: 5–23.
Döbbeler, P. (1979). Untersuchungen an moosparasitischen Pezizales aus der Verwandtschaft von *Octospora. Nova Hedwigia* **31**: 817–864.
Hedwig, J. (1789). Desciptio et adumbratio microscopico-analytica Muscorum frondosum. 2. *Lipsiae.*
Komarova, E.P. (1964). Key book of bracket fungi of BSSR. 343 pp. Nauka y tekhnika, Minsk.
Orlov, V.A. (1981). Floristic elements of aphyllophoraceous fungi of foreasian plateaux. *Izv. Akad. Nauk Turkmensk. SSR, ser. Biol. Nauk,* **3**: 33–39.
Parmasto, E. (1959). Bracket fungi of the Estonian SSR. *Trudy Bot. Inst. Akad. Nauk SSR, ser. 2, Sporov. Rast.,* **12**: 213–273.
Parmasto, E. (1965). Life forms of higher Basidiomycetes. In: Problems of Fungi and Lichens Investigation. pp. 64–68. Tartu.
Schwarzman, S.R. (1964). Heterobasidiomycetes (Auriculariales, Tremellales, Dacryomycetales) and Autobasidiomycetes (Exobasidiales, Aphyllophorales). Flora sporovykh rastenii Kazakhstana. Vol. 4, 715 pp. Ed. AN Kaz. SSR, Alma-Ata.
Serebriakov, I.G. (1962). Ecological morphology of plants. 378 pp. Vysscaya shkola, Moscow.
Tsvelev, N.N. (1977). On primary life form of Angiosperms. *Byull Moskovsk. Obshch. Isp. Prir., Otd. Biol.* **82**(1): 79–88.

Bondarceva, M.A. (1993). Life Forms of Higher Fungi in European Ecosystems. In D.N. Pegler, L. Boddy, B. Ing & P.M. Kirk (Editors). Fungi of Europe: Investigation, Recording and Conservation, pp. 157–170. Royal Botanic Gardens, Kew.

13. LIFE FORMS OF HIGHER FUNGI IN EUROPEAN ECOSYSTEMS

M.A. BONDARCEVA

V L Komarov Botanical Institute of the Russian Academy of Sciences, 2 Popov Street, St Petersburg 197376, Russia

Abstract

Life form systems for the whole of the fungi and a more detailed one for the order Aphyllophorales are presented. Distinguishing life forms of fungi is based on their adaptation to the substrate, on form of spore protection before maturity and mechanism of their discharge. Homological series in aphyllophoroid fungi life form systems are evident. Some examples illustrate that the life form concept is useful for comparing the aphyllophoroid fungi in different geographical zones.

Life Form Concepts for Fungi

The life form concept was originally suggested for higher plants, and much has been written on the subject (e.g. Raunkiaer, 1934; Braun-Blanquet, 1951; Serebriakov, 1962; Tsvelev, 1977; Allev, 1986). The idea received much critical appraisal and, life forms are now considered to be adaptations of plants to the environment. For a long time mycologists showed no interest in the life forms concept, and the few papers published in the USSR (Parmasto, 1965; Osipian, 1968, 1970; Bondarceva, 1971, 1972, 1974) were not noticed by the mycological community, neither was the general system of fungi life forms proposed by Riedl (1971). Since these authors worked independently their concepts are not connected to each other, making summary difficult.

Distinguishing life forms of fungi is based on the principles firstly worked out for higher plants, but having general biological significance. Life forms are a result of long adaptation of plants to local conditions, expressed in their appearance (Alekhin, 1950). Plants with the same life form are suggested as having the same adaptation to the environment (Shennikov, 1950). In this sense the term "life form" may be used for fungi, in connection with fruitbodies or sporiferous structures and vegetative organs, though the latter are not considered here.

As heterotrophic organisms, fungi interface with the climatic environment indirectly via the substrata on which they develop and whose physical and chemical properties change with time. Thus the first character in my system of fungal life forms is adaptation to the substratum. Another fundamental character is the form of spore protection before maturity, and the mechanism of their discharge. This character defines some physiognomical features of fruitbodies, as well as their role in distribution of the biosphere and for species protection, in accord with one of the definitions of Drude (1932). Genetically determined limits of variability in the process of adaptation to local environment was named "modus of adaptability" by Drude. Adaptation of plants to the dominating conditions of growth has its expression not only in physiognomical features, but also in biological properties. Raunkiaer (1934), for example, founded his system of life forms of higher plants mainly upon one character-complex — "the adaptation of plants to the surviving of the unfavourable season especially with regard to the protection of the survival buds or shoot-apices". For the fungi this principle may be expressed in the construction of the fruitbodies or sporiferous organs. In this case the principle of the physiognomicity is maintained and forms of biological protection of spores before maturity are also reflected.

The fungal life forms system founded on these two main characters was first published in Russian (Bondarceva, 1972) and is presented here in Table 1. The main groups are identified by environmental conditions in which spore discharge takes place, i.e. air, water, soil. The basic forms are also charactized by the substratum or some specific form of existence or distribution of fungi, for example, mycorrhizomycetes, planktomycetes. A group of multisubstratomycetes was also foreseen. To reflect forms of protection of the sporiferous organs, groups of gymno-, hemiangio- and angiosporous fungi were suggested. This system allows the estimation of the role of fungi in ecosystems, their domination of any forms under particular climatic and ecological conditions. Endogenous, budding and sclerotial forms are included for the registration of all forms of fungal presence in the ecosystem. The position of the basidial macromycetes in the general system of fungal life forms is presented in Table 2, and these fungi are clearly distributed between many life forms.

Although this system allows for the estimation of ecological function of the fungi it disregards the evolution of morphological structures in different taxa. In the more or less homogenous taxonomic groups of fungi, homologous series (sensu Zavarzin (1923) and Vavilov (1935)) of species with determined characters are seen. For example, an homologous series is seen in the Aphyllophorales, Agaricales, and another is seen in the Gasteromycetes.

When studying the distribution of aphyllophoroid fungi in the Leningrad region (Bondarceva, 1963 a, b), a connection between longevity, sporophore and hymenophore form and environmental conditions was noted, particularly with regard to temperature and moisture. Differences in temperature at different seasons affect the duration of sporophore production. This is more in relation to fungal life forms than to progressive evolution. It is illustraterd by data on the Aphyllophorales of the Leningrad region (Table 3). Homologous series are clear in the system of life forms of the Aphyllophorales (Bondarceva, 1974, Table 4).

Description by words is too long when trying to give the whole characteristic, e.g. "xylophilous annual resupinate with poroid hymenophore", so I have suggested the following abbreviation: first letters of terms of longevity (Annual, Perennial), of substrate (Xylophilous, Humicolous) and of the sporophore form (Resupinate, Lateral, Stipitate). Hymenophore type is determined by numbers (1 – smooth to reticulate, 2 – hydnoid, 3 – poroid). All these abbreviations are listed in Table 4).

Ecological and Geographical use of Life Forms Concept for Higher Fungi

The life forms concept may be used in all types of ecological and geographical descriptions, although comparative significance of the method is not always clear. The specific composition of the aphyllophoroid fungi in different forest types and other localities of the Leningrad region has been studied (Bondarceva, 1963 b) and life forms analysis is presented in Table 5. The life forms composition is typical of the boreal zone, over 50% of the species are annual resupinates, and very few are perennial resupinates (4.1%).

Comparison of the data in Table 5 is made for different types of sporophores independently of longevity. The lateral form group includes both effuso-reflexed and sessile species and under conditions discussed the species with effuso-reflexed sporophores predominate. This group on the whole (lateral annual plus lateral perrenial) is not so great as the resupinate one (resupinate annual plus resupinate perennial). Although the annual species (XAL–1, 2, 3) predominate here too, the difference between this group and the perennial lateral species (XPL, 2s, 3s, 3ol) is not as impressive (XAL – 1 + 2 + 3 = 19.6%; XPL – 2s + 3s + 3ol = 7.1%) as in the resupinate group (XAR – 1 + 2 + 3 = 50.2%; XPR – 1s + 3s = 4.1%). The stipitate species prevail on wood (xylophilous, XAS – 1, 2, 3) and on the ground (humicolous, HAS – 1, 2, 3) forming 18.7% overall (6.5 of XAS plus 12.2% of HAS).

Comparison of aphyllophoroid fungi in different types of fir (*Abies*), pine (*Pinus*) and mixed forests reveals no real difference of life forms, only of specific composition (Table 6). The highest number of species (82) are found in fir forests, second highest in mixed forests and next pine forests. Resupinate species predominate in the fir forests, but their number is less than in pine and mixed forests, and notably less than in flood lands and cleared spaces. There is no apparent difference between stipitate species on wood in different forest types. Perennial life forms have a similar frequency in most ecosystems, but there are notably less in pine forests. Humicolous species were infrequently found in mixed forests or flood lands, but prevailed in pine forests, and were common in fir forests and cleared spaces. Among stipitate species, in fir types clavarias (HAS-1) prevail. In pine forests hydnums (HAS-2), constitute the greatest number, among polyporoid species scutigers (HAS-3) prevail. In cleared spaces clavarias (HAS-1) predominate, polypores are scarce (HAS-3), and hydnums are absent.

The main difference between aphyllophoroid fungi in forest types lies in the composition of species with different sporophore forms (Table 6). These data are summarisized more concisely in Table 7. Resupinate forms predominate in fir forests —50.9%, less frequently in flood lands — 36.7%, and least in cleared spaces — 21.4%. Lateral forms are distributed more in cleared spaces, and stipitate humicolous forms in pine forests.

One of the necessary characteristics of life form is physiognomicity, difficult and subjective exponents, as the consistency of the context was not taken into consideration, nor was the hyphal system type, because this last feature cannot be determined in the field. The significance of these characters can be seen by examining polypores in different climatic zones in the former USSR (Table 8; Bondarceva et al., in press). The first three columns show data for two central Asian regions with dry climates and for Kazakhstan, a territory with diverse climatic conditions, but where polypores grow principally in mountains where conditions are similar to those of central Asian republics. The next three territories lie in boreal and partly frondose zones. The last one is the Urals, which stretches meridionally from polar to steppe zone.

There is a clear difference in the ratio of different sporophore groups in arid and boreal zones. The number of fragile, annual species is greater in the boreal zone than in the arid ones. In arid zones, the number of species with monomitic construction is less than 30%, whereas in boreal zones this is over 40%. There is no appreciable difference in the type of wood decay, since the mountain forests where polypores are principally distributed are composed predominantly of conifers, which are the main hosts for brown rot fungi.

The potential for using life form characteristics for comparison of remote geographical zones is illustrated in Table 9. Numbers of species with different life forms were determined from the information in Ryvarden & Johansen (1980) and Gilbertson & Ryvarden (1986, 1987), as well as my own unpublished data on Leningrad region and Cuba. Table 9 shows the similarity of life form composition in tropical countries: in East Africa and Cuba there is a predominance of lateral form of sporophore, and also perenniality which contrasts with boreal (Leningrad region) and mixed (North America) conditions.

Conclusion

Despite the absence of statistical treatment of data, clear difference in exponents are evident, suggesting that this method has potential. The life form concept may be used for all the types of ecological descriptions, though it is more useful for comparison of life forms of different types of localities, of mycobiotic composition in remote regions. There are some characteristic features of boreal and arid zones: in boreal zones annual resupinate forms with fragile sporophores predominate; in arid zones — lateral annual and wintering, with coriaceous and corky sporophores are common. In tropical countries species with lateral forms of sporophores also predominate, and the proportion of perennial species is higher in comparison with the boreal territories. Anthropogenic influence may also affect the life form composition. Besides the ecological value, the life form concept may aid understanding of evolutionary tendencies.

Both homological series of life forms and omissions from these series are helpful in demonstrating evolutionary limitations of the group.

Acknowledgements

The author is grateful to the Organizing Committee of the XI Congress of European Mycologists, particularly Dr Pegler, for making it possible for me to participate in the Congress. My special thanks are also to the patient translator for linguistic improvements. I also appreciate Mrs L Svishcz and Mrs V Lositskaya, from Komarov Botanical Institute, for technical assistance.

References

Alekhin, V.V. (1950). Geography of Plants, 3rd ed. 120 pp. Moscow.

Bondarceva, M.A. (1963a). On the anatomical criterion in the systematics of aphyllophoraceous fungi. *Bot. Zhurn.* **48**(3): 362–372.

Bondarceva, M.A. (1963b). Review of the Order Aphyllophorales of Leningrad region. Thesis on C.Sci. degree. Leningrad (unpublished).

Bondarceva, M.A. (1971). Life forms of aphyllophoraceous fungi. In Abstracts of the 6th Symposium of Mycologists and Lichenologists of Baltic Republics. Riga pp. 111–115.

Bondarceva, M.A. (1972). Principles of life forms distinguishing of the Fungi. *Ecology* **5**: 52–58.

Bondarceva, M.A. (1974). Life forms of basidial macromycetes. *Novosti Syst. Nizsh. Rast.* **11**: 29–40.

Bondarceva, M.A., Svishch, L.G. and Baltayeva, G.M. (in press). Some regularities of distribution of polyporaceous wood-destroying fungi. *Mycol. & Phytopathol.* **26**(6) (in press).

Drude, O. (1932). Pflanzengeographische Oecologie. In E. Abderhalden (editor), Handbuch der biologischen Arbeitsmetoden. Abt. 11, 5. pp. 1–56.

Gilbertson, R.L. and Ryvarden L. (1986). North American Polypores. Vol. 1. 433 pp. Fungiflora, Oslo.

Gilbertson, R.L. and Ryvarden, L. (1987). North American Polypores. Vol. 2. 434–885 pp. Fungiflora, Oslo.

Osipian, L.L. (1968). To the question on life forms of fungi. In Abstracts of Symposium of Mycologists & Lichenologists of Baltic Republics. Vilnius, pp. 124–128.

Osipian, L.L. (1970). To the question on life forms of imperfect fungi. *Mycol. & Phytopathol.* **4**(2): 169–173.

Raunkiaer, C. (1934). The Life Form of Plants and Statistical Plant Geography. 632 pp. Oxford.

Riedl, H. (1971). Die Lebensformen der Pilze. *Österr. Akad. Wiss. Math.-Naturwiss. Kl., Sitzungsber., Abt. 1, Biol.* **179** (7): 77–101.

Ryvarden, L. and Johansen, I. (1980). A preliminary polypore flora of East Africa. 636 pp. Fungiflora, Oslo.
Shennikov, A.P. (1950). Ecology of Plants. 376 pp. Moscow.
Vavilov, N.I. (1935). Low of Homological Series and Hereditary Variability. 2nd ed. 56 pp. Moscow, Leningrad.
Zavarzin, A.A. (1923). Parallelism of structures as a fundamental principle of morphology. *Izv. Biol. Nauchno-Issl. Inst. Permsk. Gosud. Univ.* **2**(4): 135–140.

TABLE 1. System of the Life-Forms of Fungi

AEROMYCETES	
Aeroaxylomycetes	Aeromycorrhizomycetes
Aerohemixylomycetes	Aerohumusomycetes
Aeroxylomycetes	Aerocopromycetes
Aerozoomycetes	Aeromultisubstratomycetes
Aeromycomycetes	
HYDROMYCETES	
Hydroaxylomycetes	Hydromycomycetes
Hydrohemixylomycetes	Hydroplanktomycetes
Hydroxylomycetes	Hydromultisubstratomycetes
Hydrozoomycetes	
GEOMYCETES	
Geoaxylomycetes	Geomycomycetes
Geohemixylomycetes	Geohumusomycetes
Geoxylomycetes	Geomultisubstratomycetes
Geozoomycetes	

TABLE 2. Life-Forms of Basidial Macromycetes

AEROAXYLOMYCETES

On fallen leaves and dry stems of grasses

Gymnosporous unprotected: Clavariaceae
Gymnosporous protected: Polyporaceae, Agaricaceae

AEROXYLOMYCETES

On living trees and dead wood

Gymnosporous unprotected: Tremellaceae, Corticiaceae and other families with the smooth, uneven, grandinioid, odontioid, raduloid or reticulate surface
Gymnosporous protected: Polyporaceae, Agaricales
Hemiangiosporous: Agaricales, *Cryptoporus volvatus* (Polyporaceae)

AEROMYCOMYCETES

On other fungi

Gymnosporous unprotected: Corticiaceae, some species of Hydnaceae s.lato
Gymnosporous protected: Polyporaceae, Agaricaceae

AEROMYCORRHIZOMYCETES

Mycorrhiza forming

Gymnosporous unprotected: Clavariaceae
Gymnosporous protected: Boletales, Agaricales, Scutigeraceae
Hemiangiosporous: Agaricales

AEROHUMUSOMYCETES

On soil

Gymnosporous unprotected: Lachnocladiaceae, Corticiaceae, Clavariaceae, Hydnaceae s.lato, *Thelephora terrestris* (Thelephoraceae)
Gymnosporous protected: Boletales, Agaricales, Cantharellaceae, Coltricia perennis (Hymenochaetaceae)
Hemiangiosporous: Agaricales, Gasteromycetes
Angiosporous: Gasteromycetes

AEROCOPROMYCETES

On animal excrement

Gymnosporous protected: Agaricales
Hemiangiosporous: Agaricales

GEOHUMUSOMYCETES

In soil

Angiosporous: Gasteromycetes

TABLE 3. Anatomical Construction and Longevity of Polyporous Sporophores from Different Ecological Groups

Ecological Group Perennial	Annual	Sporophores Wintering
Hygrophilous	(M) *Anomoporia bombycina*	
	(M) *Leptoporus mollis*	
	(M) *Oligoporus caesius*	
	(M) *O. rennyi*	
	(M) *Physisporinus sanguinolentus*	
	(M) *Pycnoporellus fulgens*	
	(M) *Tyromyces gloeocystidiatus*	
Mesophilous	(T) *Ganoderma lucidum*	(T) *Amauroderma* spp.
(T) *Ganoderma applanatum*	(MT) *Junghuhnia nitida*	(MT) *Bjerkandera adusta*
(D) *Fomitopsis pinicola*	(MT) *Inonotus radiatus*	(MT) *B. fumosa*
(D) *F. rosea*	(MT) *I. rheades*	(T) *Ganoderma* spp.
(D) *F. cajanderi*	(MT) *I. hispidus*	
(D) *F. feei*	(DB) *Laetiporus sulphureus*	
(T) *Fomes fomentarius*	(DB) *Piptoporus betulinus*	
(D) *Heterobasidion annosum*		
(D) *Phellinus* spp.		
Xerophilous		(D) *Antrodia xantha*
(T) *Daedalea quercina*		(T) *Cerenna unicolor*
(T) *Gloeophyllum sepiarium*	(MT) *Coltricia perennis*	
(T) *G. striatum*		(D) *Diplomitoporus lenis*
(T) *Microporus luteus*		(T) *Lenzites betulina*
		(T) *Fomes* spp.

Note: M = monomitic, MT = monomitic with thick-walled hyphae, D = dimitic with skeletal hyphae, DB = dimitic with binding hyphae, T = trimitic.

TABLE 4. System of Life-Forms of Aphyllophoraceous Fungi

AEROXYLOMYCETES

ANNUAL, WINTERING

Resupinate

1. XAR-1. Surface from smooth to reticulate. Examples: Corticium s. lato, Cytidia, Peniophora s. lato, Coniophora, Serpula
2. XAR-2. Surface hydnoid. Examples: Radulum, Schizopora, *Steccherinum dichroum*
3. XAR-3. Hymenophore poroid. Examples: resupinate polypores

Lateral

4. XAL-1. Hymenial surface from smooth to reticulate. Examples: Stereum, Merulius tremellosus
5. XAL-2. Hymenial surface hydnoid. Examples: *Steccherinum ochraceum, Cerrena unicolor*
6. XAL-3. Hymenophore poroid, daedaleoid, lamellate. Examples: Spongipellis, Fistulina, Cyclomyces, Daedaleopsis

Stipitate

7. XAS-1. Hymenial surface smooth. Examples: Thelephora, Clavicorona, Sparassis
8. XAS-2. Hymenial surface hydnoid. Examples: *Hericium coralloides*
9. XAS-3. Hymenophore poroid. Examples: Polyporus, Bondarzewia, Amauroderma

PERENNIAL

Resupinate

10. XPR-1s. Hymenium smooth, stratified. Examples: Vararia, Xylobolus
11. XPR-3s. Hymenophore poroid, stratified. Examples: some resupinate species of Phellinus
12. XPR-3ol. Hymenophore daedaleoid, lamellate, poroid, one-layered. Examples: resupinate species of Antrodia, resupinate forms of Daedalea, Lenzites

Lateral

13. XPL-1s. Hymenium smooth, stratified. Examples: Stereum s. lato
14. XPL-2s. Hymenophore hydnoid, stratified. Examples: Sclerodon
15. XPL-3s. Hymenophore poroid, stratified. Examples: Fomes, Fomitopsis
16. XPL-3ol. Hymenophore poroid, daedaleoid, lamellate, one-layered. Examples: Trametes, Daedalea, Lenzites

Stipitate

17. XPS-3s. Hymenophore poroid, stratified. Examples: some species of Ganoderma

TABLE 4. Continued

AEROHUMUSOMYCETES

ANNUAL, WINTERING

Resupinate

18. HAR-1. Hymenium smooth. Examples: *Scytinostroma galactinum*, *Ypsilonidium sterigmaticum*

Stipitate

19. HAS-1. Hymenium smooth, wrinkled. Examples: Ramaria, Clavariadelphus, Craterellus
20. HAS-2. Hymenophore hydnoid. Examples: Phellodon, Hydnellum, Hydnum.
21. HAS-3. Hymenophore poroid, lamellate. Examples: Boletopsis, Scutiger, *Cantharellus cibarius*

TABLE 5. Life-forms of Aphyllophoraceous Fungi of Leningrad Region

Life-form	% (absolute number)	Sporophore form % (a.n.)
	ANNUAL XYLOPHILOUS	
XAR-1	31.2(92)	resupinate
XAR-2	3.8(11)	50.2%(148)
XAR-3	15.2(45)	
XAL-1	2.7(8)	lateral
XAL-2	1.7(5)	19.6%(58)
XAL-3	15.2(45)	
XAS-1	2.4(7)	stipitate
XAS-2	0.7(2)	6.5%(19)
XAS-3	3.4(10)	
	PERENNIAL XYLOPHILOUS	
XPR-1s	0.3(1)	resupinate
XPR-3s	3.8(11)	4.1%(12)
XPL-2s	0.3(1)	lateral
XPL-3s	5.8(17)	7.2%(21)
XPL-3ol	1.0(3)	
	ANNUAL HUMICOLOUS	
HAR-1	0.3(1)	resupinate 0.3%(1)
HAS-1	5.4(16)	stipitate
HAS-2	5.1(15)	12.2%(36)
HAS-3	1.7(5)	
Total	100(295)	

TABLE 6. Life-forms of Aphyllophoraceous Fungi in Main Groups of Forests of Leningrad Region % (absolute number)

Life-form	Group of Forests				
	Fir-types	Pine-types	Mixed forests	Flood lands	Cleared space
XAR-1	28.8(47)	24.2(15)	28.8(21)	10.0(3)	14.2(6)
XAR-2	3.7(6)	1.6(1)	6.9(5)	3.3(1)	2.4(1)
XAR-3	18.4(30)	19.4(12)	11.0(8)	23.4(7)	4.8(2)
XAL-1	3.1(5)	1.6(1)	6.9(5)	6.7(2)	4.8(2)
XAL-2	—	1.6(1)	—	—	2.4(1)
XAL-3	18.4(30)	17.7(11)	24.6(18)	36.7(11)	33.3(11)
XAS-1	1.8(3)	—	1.4(1)	—	2.4(1)
XAS-2	0.6(1)	—	—	—	—
XAS-3	3.1(5)	—	2.7(2)	3.3(1)	4.8(2)
XPR-3s	4.3(7)	1.6(1)	4.1(3)	3.3(1)	2.4(1)
XPR-3ol	—	1.6(1)	—	—	—
XPL-2s	0.6(1)	—	—	—	—
XPL-3s	4.9(8)	—	9.5(7)	13.3(4)	11.9(5)
XPL-3ol	2.4(4)	3.2(2)	2.7(2)	—	4.8(2)
HAS-1	6.2(10)	9.7(6)	1.4(1)	—	9.5(4)
HAS-2	3.1(5)	12.9(8)	—	—	—
HAS-3	0.6(1)	4.9(3)	—	—	2.3(1)
Total	100(163)	100(62)	100(73)	100(30)	100(42)

TABLE 7. Distribution of Sporophore Forms in Different Groups of Forest-types in Leningrad Region (%)

Sporophore type	Groups of Forest				
	Fir-types	Pine-types	Mixed forests	Flood lands	Cleared space
Resupinate annual	50.9	45.2	46.7	36.7	21.4
Lateral annual	21.5	20.9	31.5	43.4	40.5
Stipitate xylophilous	5.5	—	4.1	3.3	7.2
Resupinate perennial	4.3	3.2	4.1	3.3	2.4
Lateral perennial	7.9	3.2	12.2	13.3	16.7
Stipitate humicolous	9.9	27.5	1.4	—	11.8
Total	100	100	100	100	100

TABLE 8. Morphological Features of Polypores from Different Territories of the Former USSR

Feature	Species % (absolute number)						
	Uzbekistan (92)	Kazakhstan (111)	Turkmenistan (45)	Leningrad region (127)	Estonia (122)	Bielaruss (155)	Ural (Russia) (161)
SPOROPHORE							
resupinate	19.6(18)	19.8(22)	28.9(13)	35.4(45)	39.3(48)	35.5(55)	33.5(54)
lateral	69.5(64)	73.0(81)	55.6(25)	55.2(70)	51.7(63)	56.8(88)	59.0(95)
stipitate	10.9(10)	7.2(8)	15.5(7)	9.4(12)	9.0(11)		
CONTEXT							
fragile	26.1(24)	31.5(35)	31.2(14)	40.2(51)	39.4(48)	50.3(78)	41.6(67)
coriaceous suberous	52.2(48)	54.1(60)	51.0(23)	52.7(67)	54.1(66)	44.8(71)	51.6(83)
woody	21.7(20)	14.4(16)	17.8(8)	7.1(9)	6.5(8)	3.9(6)	6.8(11)
HYPHAL SYSTEM							
monomitic	26.1(24)	27.9(31)	26.7(12)	40.2(51)	41.0(50)	47.1(73)	38.5(62)
dimitic	41.3(38)	46.0(51)	44.4(20)	39.3(50)	39.3(48)	34.8(54)	41.6(67)
trimitic	32.6(30)	26.1(29)	28.9(13)	20.5(26)	19.7(24)	18.1(28)	19.9(32)
ROT							
white	72.8(67)	81.1(90)	75.6(34)	76.4(97)	77.1(94)	73.6(114)	70.8(114)
brown	26.1(24)	18.9(21)	24.4(11)	23.6(30)	22.9(28)	26.4(41)	29.2(47)
unknown	1.1(1)	—	—	—	—	—	

Note: Data from Bondarceva et al., 1992.

TABLE 9. Life-forms of Polypores in Different Geographical Regions
% (absolute number)

Life-form	Territory			
	N.Americas	E.Africa	CUBA	Leningrad Region (RUSSIA)
XAR	29.2(109)	17.3(58)	12.2(19)	34.1(47)
XAL	35.4(132)	29.2(98)	34.8(54)	32.6(45)
XAS	9.7(36)	18.2(61)	11.6(18)	7.2(10)
XPR	11.2(42)	6.3(21)	7.7(12)	8.0(11)
XPL	9.4(35)	26.6(89)	30.4(47)	14.5(20)
HAS	5.1(19)	2.4(8)	1.9(3)	3.6(5)
Total	100(373)	100(335)	100(155)	100(138)

Note: Date on polypores of N. America from Gilbertson & Ryvarden, 1986, 1987; E. Africa — Ryvarden & Johansen, 1980; Cuba & Leningrad Region — Bondarceva, unpublished data.

Stenlid, J. (1993). Community and Population Biology of Boreal Decay Fungi. In D.N. Pegler, L. Boddy, B. Ing & P.M. Kirk (Editors). Fungi of Europe: Investigation, Recording and Conservation, pp. 171–180. Royal Botanic Gardens, Kew.

14. COMMUNITY AND POPULATION BIOLOGY OF BOREAL DECAY FUNGI

J. STENLID

Department of Forest Mycology & Pathology, Swedish University of Agricultural Sciences, Box 7026, S-750 07 Uppsala, Sweden

Abstract

External and intrinsic factors that influence the development of communities and populations of decay fungi are reviewed. Boreal forests have developed under the influence of ice ages, forest fires and most recently, forestry. Different colonization processes of various decay fungi, and the interactions among them and with the abiotic environment result in successional sequences that also strongly influence the composition of the decay flora in the boreal forests. Among the basidiomycetes, certain species, or intersterile groups within species complexes, are specialized for growth on particular tree species. Regional colonization patterns following the last ice age can be inferred from data on genetic variation in the fungi, and compared with those of their particular host species obtained from pollen records. Data on genetic variation can also be used to study questions regarding gene flow among local populations today and how gene flow is restricted by topographic barriers. Examples of approaches to studying these matters in boreal fungi are presented.

Introduction

This paper focuses on decay fungi in boreal forests. Such forests, which are found in northern Scandinavia, northern Russia and Siberia and in the northern parts of North America, cover over 14.7 million km², or 11% of the earth's terrestrial surface (Bonan & Shugart, 1989). Conifer forests are of the taiga type and typically include one or several species of *Picea*, *Pinus*, *Larix* and *Abies*. In Scandinavia, Scots pine (*Pinus sylvestris*) and Norway spruce (*Picea abies*) are the two native conifers, but deciduous broadleaf trees are also present, e.g. species of *Betula*, *Alnus*, *Populus*, *Sorbus* and *Salix*. Within these systems decay fungi play an important role as decomposers of woody substrata as well as needle and leaf litter. Frankland (1982) estimated that basidiomycete mycelia in stumps and dead roots constitute about 80% of the fungal biomass in temperate forests, and the proportion is probably similar in boreal systems. Wood-decay fungi may also act as pathogens of living trees (Rayner & Boddy, 1988), thereby gaining access to the large volume of biomass unavailable to non-pathogens. Some decay fungi are economically important since they substantially decrease the quality and quantity of commercial timber.

A comprehensive list of the fungi responsible for decay in boreal forests will not be given here. For such information see Käärik & Rennerfelt (1957); Strid (1975); Ryvarden (1976, 1978); Jahn (1979); Gilbertson (1980); Niemälä & Kotiranta (1982); Ryman & Holmåsen (1984); Breitenbach & Kränzlin (1986); Koski-Kotiranta & Niemälä (1987); Hjortstam, Larsson & Ryvarden (1987, 1988); Renvall, Renvall & Niemälä (1991).

Community Biology

A number of factors influence the composition of the decay community in boreal forests. They can be divided into external factors of non-fungal origin, either biotic or abiotic, and factors associated with the fungi, i.e. their ecological abilities.

External Factors

Of the external processes, the succession leading to changes in stand species composition is one of the most important. The boreal regions have been subjected to a number of ice ages which have influenced the development of their forests (Martin & Wright, 1976). Since decay fungi are heterotrophs, and some are restricted to certain tree species, the development of decay communities is influenced by forest development. The latter depends, amongst other factors, on recolonization following the last ice age; the geographical distributions of fungi are thus dependent on the migration history of their host tree species. On a shorter time scale, adaptation to seral or climax host tree species is of importance. Catastrophic disturbance caused by fire has historically played the most important part in restarting this successional cycle (Zackrisson, 1977; Bradshaw & Zackrisson, 1990; Bergeron, 1991). Therefore knowledge of how decay fungi survive fire and the ways in which they have recolonized and exploited conditions on the burnt sites is necessary.

Uncontrolled forest fires are uncommon today. Instead, human disturbance associated with forestry is the most significant external factor (Hansson, 1992). Many boreal decay-fungi may have adapted to an old forest landscape that has largely been replaced with forests designed for sustained timber and pulp production. There are four main differences between fire- and forestry-driven changes in boreal forests that affect decay fungi: 1) forestry is more apt to reduce the mean age of the stand, 2) forestry leaves less debris including dead trees, logs and brush, 3) clear-cut areas are typically smaller than areas razed by large forest fires; and 4) seral broadleaved trees (pines on poorer sites) are overlooked in favour of climax species in forestry (Hansson, 1992). As a result of the forestry-induced changes in stand structure, many decay-fungi are considered endangered species, e.g. *Amylocorticium laceratum* (Litsch) Hjortst. & Ryv., *Antrodia crassa* (Karst.) Ryv., *A. crustulina* (Bres.) Ryv., *Atheloderma mirabile* Parm., *Chaetoporellus curvisporus* (Eriks. & Hjortst.) Eriks. & Hjortst., *Perenniporia subacida* (Peck) Donk and *Radulodon erikssonii* Ryv., in Sweden alone (Anonymous, 1991).

Some decay fungi are clearly influenced by seral stages of plant communities, and some fungi actually influence plant succession. For example, the root rots of conifers along the Pacific coast of North America influences succession of forest trees. Differences between the tree species in their susceptibilities to *Phellinus weirii* (Murr.) Gilberts., *Heterobasidion annosum* Bref., and *Armillaria ostoyae* Romagn. may shift the balance towards pioneer or climax species (Goheen & Hansen, 1993). How this balance changes depends on the species composition of the stand and the relative susceptibility of component trees.

Intrinsic Factors

Intrinsic factors include spreading strategies, competition and differential substrate requirements etc. One important feature is the relative extent to which spores and mycelia effect fungal establishment. Spores can travel long distances, but have limited internal resources and sexually produced spores typically show genetic dissimilarities, hence different colonies are unable to act synergistically (Rayner & Boddy, 1988). On the other hand, mycelial spread is local, but established mycelia can import water and nutrients, and genetically identical aggregates can cooperate (Rayner & Boddy, 1988). Another important factor is the degree to which the community is still open for primary colonization, i.e. the proportion of suitable sites that have not already been colonized (Cooke & Rayner, 1984).

Succession Among Species

Succession occurs where species colonize substrata at different times or when one species replaces another species (whether arriving at the same or different times) or otherwise dominates the community during different stages of decomposition. It has been proposed that succession in the decomposition of standing trees starts with discoloration caused by non-decay organisms (Shigo, 1967), but this is frequently not the case (Boddy & Rayner, 1983).

An important conceptual division when discussing succession is between primary and secondary resource capture (Rayner & Webber, 1984). Primary resource capture, i.e. colonizing an unoccupied site in an open community, is achieved by fungi which are adapted to arrive rapidly, e.g. as airborne spores or latent invaders. However, many decay fungi have to rely on secondary resource capture, i.e. they colonize substrata that have already been colonized by other organisms. This can be accomplished through non-species specific combative mechanisms or through selective, species-specific parasitism (Rayner & Boddy, 1988). Examples of the latter include *Bjerkandera adusta* (Fr.) Karst. and *Coriolus versicolor* (L.: Fr.) Quél. which are mycoparasitic on *Pseudotrametes gibbosa* (Pers.) Bond. & Singer and *Lenzites betulina* (Fr.) Fr. (Rayner et al., 1987). Other examples of secondary colonizers which tend to follow specific earlier colonizers are; *Junghuhnia collabens* (Fr.) Ryv. that fruits on trunks decayed by *Phellinus ferrugineofuscus* (Karst.) Bourdot, *Piloporia sajensis* (Parm.) Niem. follows *Trichaptum laricinum* (Karst.) Ryv., and *Antrodiella citrinella* and *A. americana* Ryv. & Gilbertson follow *Fomitopsis pinicola* and *Hymenochaete tabacina*, respectively (Renvall & Niemälä, poster presented at this symposium). In fact, the development of these fungi which develop late in succession probably only occurs in forests left relatively undisturbed for long periods. Consequently, such species can be used as indicators of undisturbed forest ecosystems. Because they have specific habitat requirements and in view of the fragmented distribution of undisturbed forests, such decay species are also likely to be rare or endangered.

Succession in aerial parts of trees differs from that occurring in the roots (Table 1). Aerial parts are typically reached by fungal spores while the soil may act as a supporting matrix for fungal vegetative organs such as rhizomorphs and cords, e.g. *Marasmius androsaceus* (L.) Fr., *Resinicium bicolor* (Alb. & Schwein. Fr.) Parm., *Hypholoma spp.*, *Peniophora spp.*, *Coniophora spp.*, and *Merulius spp.* (Rayner & Boddy, 1988; Kirby, Stenlid & Holdenrieder, 1990; Holmer & Stenlid, 1991; Kile et al., 1991). Mycelial spread by rhizomorphs through the ground is an important means of primary resource capture in some fungi, such as pathogenic *Armillaria* species. Since rhizomorphs and cords have a high infection potential (Redfern & Filip, 1991) and can facilitate water and nutrient transport (Brownlee & Jennings, 1982; Wells & Boddy, 1990) mycelial spread is also important in secondary resource capture.

Fungi can be roughly categorized according to the successional stages at which they are found. They include: actively pathogenic fungi, species that usually infect standing trees but seldom kill them; fungi that usually decay damaged trees; species that establish in the wood of trees that have recently died or fallen; and decay fungi that only invade old logs. Parasitic fungi must have ways of penetrating the living bark to gain access to internal tissues, and a range of mechanisms are evident. (1) Root-rot fungi may cause lesions on roots, often occurring at sites of contact with previously colonized wood, e.g. *H. annosum* (Hodges, 1969), *Armillaria ostoyae* and *A. borealis* Marm. & Korh. (Morrison et al., 1991), and *Coltricia tomentosa* (Fr.) Murr. (Lewis & Hansen, 1991). (2) Some enter through wounds, e.g. *Stereum sanguinolentum* (Alb. & Schwein.: Fr.) Fr. (Isomäki & Kallio, 1974), or through suppressed or dead twigs, e.g. *Phellinus pini* (Brot.: Fr.) Amer. (Haddow, 1938). (3) Others, e.g. *Inonotus obliquus* (Pers.: Fr.) Pilàt, cause cankers on stems or branches,

TABLE 1. Typical means of establishment of decay fungi in aerial parts of trees and roots

	Aerial parts	Roots
Living trees	Pathogens arriving typically by spores	Pathogens arriving typically by mycelium
Recently dead trees	Latent establishment	Cord-formers, primary invasion.
	Saprotrophs arriving by spores	Development from minor lesions.
Old logs	Secondary invasion spores, cords	Secondary invasion by cord-formers. Invasion of soil fungi, at late stages, following invertebrate invasion.

thereby also opening the wood for subsequent exploitation (Sinclair et al., 1987). (4) The fungi which cause latent infections of broadleaved trees must have means of accessing xylem, although the mechanism(s) is unknown (Chapela & Boddy, 1988). Fungal structures may, following establishment, lie dormant for many years until suitable environmental conditions activate them. For this type of development to take place the fungus has to be able to establish itself within the xylem, initially by penetrating/traversing the cambium. For example, *Phellinus pini* remains in twig stubs within pine sapwood until it has turned into heartwood, and the fungus is 'activated' by the right conditions (Haddow, 1938; Sinclar *et al.*, 1987). *Phellinus tremulae* (Bond.) Bond. & Borisov. seems to have a similar strategy for establishing in *Populus tremula* in Sweden (Holmer, Nitare & Stenlid, unpublished). *Hypoxylon mammatum* (Wahl.) Karst. which has been found on aspen in boreal regions, has also been shown to be a latent invader in American aspen (Chapela, 1989).

There are ways in which certain weakly pathogenic decay fungi can enhance their probability of being the first organism to exploit dead or dying wood. *Hymenochaete tabacina* (Sow.: Fr.) Lév. often connects individual branches of *Corylus avellana* and other tree species through mycelial pads (Stenlid & Holmer, 1991; cf. *H. corrugata* (Fr.) Lév. in temperate forests; Ainsworth & Rayner, 1990). Fruit-bodes developing on an infected, dead branch form connections between this branch and nearby healthy branches, and a canker slowly develops on the healthy branch at the point of contact. When the healthy branch eventually becomes suppressed by shading etc. the fungus is in a perfect position for colonization. Contact may also form between branches of different stools. This type of contact spread resembles the strategy used, below ground, by root infecting fungi such as *H. annosum*. Some wood-decay fungi may also be able to spread by contact between logs lying on the ground. Some fungi can spread between spatially separate logs through soil by the means of mycelial cords (Rayner & Boddy, 1988).

Fruiting bodies of decay fungi often appear on trees that have recently died. This has been attributed to an inability of these fungi to attack vigorous trees. However, the fact that extensive decay columns are often associated with the sporocarps suggests that the fungus was already present in the tree before it died. Wood of conifers that have recently died contains a characteristic flora of decay fungi, e.g. *Fomitopsis pinicola* (Sow.: Fr.) Karst., *Trichaptum abietinum* (Pers.: Fr.) Ryv. and *Gloeophyllum sepiarium* (Wulfen: Fr.) Karst. Later on, other species will usually take over and dominate the decay process. In Sweden, decay fungi with different substratum requirements ranging from recently dead trees to old logs e.g. *Phellinus nigrolimitatus* (Rom.) Bourdot & Galzin, *Fomitopsis rosea* (Fr.) Karst., *Amylocystis lapponicus* (Rom.) Singer, and *Skeletocutis tschulymica* (Pilàt) Keller, have been used to indicate sites with increasing elements of primary forests untouched by modern foresty ("Steget före", Karstöm, 1992).

The size of individual mycelia strongly influences the outcome of fungal interactions and is likely to be particularly important in non-selective replacement. In an experimental system with differently sized mycelia, Holmer & Stenlid (unpublished) demonstrated that *Resinicium bicolor* was able to replace *Heterobasidion annosum* when the former was larger or the two mycelia were equally sized, but not when the *H. annosum* mycelium was substantially larger. The same was true for the combinations of *R. bicolor* and *Phanerochaete sanguinea* (Fr.) Pouzar or *Coniophora sp.* This indicates that small mycelia that recently originated from spores may have difficulties in invading and replacing established mycelia in a decay community unless the former exhibit a selective secondary resource capture strategy, as mentioned above.

Population Biology

Although many records on the geographical distribution of decay fungi are available, much less is known about their population biology. Moreover, little is known about the extent to which sporocarps reflect the distribution of vegetative mycelia. Nevertheless, three intraspecific levels of variation can be used in characterizing decay communities: (A) the subspecies or intersterility group level, (B) local population structures and dynamics, and (C) the regional scale.

A. Intersterility Group Level

The host ranges of intersterility groups within a species are generally dissimilar. This might in fact be a prerequisite for their coexistence in the same geographic region (Brasier, 1987; Hallenberg, 1991). In some cases, the geographical distribution of a given intersterility group may be determined by that of its host species rather than by direct environmental effects on the fungus. For example, in *Peniophora nuda* (Fr.) Bres. one intersterility group is restricted to *Ulmus carpinifolia* on the islands Öland and Gotland in the Baltic Sea, while the other groups are found on various deciduous trees elsewhere (Hallenberg, 1991). A number of recent reviews cover topics related to intersterility groups in fungi (Burnett, 1983; Boidin, 1986; Ainsworth, 1987; Fischer, 1987; Brasier, 1987; Korhonen, 1987; Hallenberg, 1987, 1991; Chase & Ullrich, 1988).

B. Local Population Structure

Structure and dynamics of local populations are important aspects of fungal ecology, and impinge on areas such as S-C-R selection (see Cooke & Rayner, 1984), succession, pathogenicity, resource capture, etc. The individuals making up local populations are initiated through spore germination, and enlargement occurs through mycelial growth. Population structures have been characterized through studies of somatic incompatibility in a number of species of boreal decay-fungi (Barrett & Uscuplic, 1971; Korhonen, 1978;

Adams et al, 1981; Rayner & Turton, 1982; Stenlid, 1985; Kirby et al., 1990; Stenlid & Holmer, 1991; Lewis & Hansen, 1991). However, *Heterobasidion annosum* is the only species studied thoroughly enough to allow comparisons to be made of population structures in different forest types (Stenlid, 1985, 1987; Siepmann, 1989; Piri et al., 1990; Swedjemark & Stenlid, 1993). On average, older populations of *H. annosum* contain larger but fewer mycelia than populations in first-rotation plantations (Fig. 1). Before the first thinning operation, second rotation plantations contain fragmented individual mycelia that have survived the clear-cutting of the preceding forest generation (Stenlid, 1987). In *H. annosum* the importance of spore dispersal as a mode of spread is increased following disturbance. Decay species that are not able to exploit a disturbance with high spore production or other means of primary resource capture will tend to be excluded from such sites. In other words, forestry tends to favour selection in decay fungi. One interesting comparison in this respect can be made between *Coltricia tomentosa* and *H. annosum*. Both fungi cause death to groups of conifers in natural forests and spread via root contacts (Stenlid, 1985; Lewis & Hansen, 1991). In forest plantations, *H. annosum* because of its high spore production throughout most of the year, can effectively establish itself in stumps created by thinning operations. By contrast *C. tomentosa*, which does not have the capability of exploiting thinning stumps, does not prosper in managed plantations.

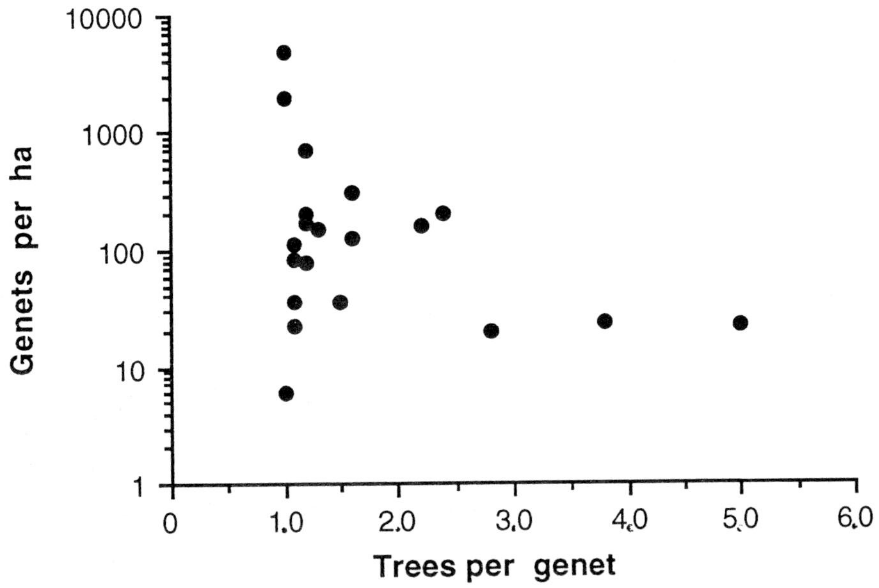

FIG. 1. Density and size of individual mycelia (genets) in populations of *Heterobasidion annosum*. Each point represents one population. Size is expressed as the average number of trees infected by a genet in the population. Populations with average genet sizes exceeding 2 trees were all from old forest sites. Average genet sizes less than 2 trees were found in first rotation stands. Density is expressed as the number of individual mycelia (genets) per hectare. Data from Stenlid (1985, 1987); Piri et al. (1990); Swedjemark & Stenlid (1993); and Stenlid (unpublished).

C. Regional Scale

Norway spruce was forced into refuges in at least two areas during the most recent ice age in northern Europe (Huntley & Birks, 1983). One area was in the Ural and Kaukasus mountain regions, and another was on the Balkan peninsula. Once the ice receded from northern Europe, Norway spruce colonized to the east and north of the two refuge areas, respectively. The two populations only merged about a thousand years ago (Silvertown, 1987). Interestingly, this intraspecific variation seems to be reflected in the regional population structure in *Heterobasidion annosum*. Strains in the S intersterility group, which in Europe is confined to spruce, differ in mating behaviour depending on from which side of the Alps they originate. Strains derived from Finland and Lithuania readily mate with single-spore strains of the F group which is found mostly on fir in Italy (Korhonen et al., 1992). By contrast, S group single-spore isolates from Italy are completely intersterile with the F group. The results suggest that the gene flow between the southern and northern S groups in Europe is very low. In accordance with this, large differences in DNA markers were found between Scandinavian and Italian S populations (Stenlid, Karlsson & Högberg, unpublished). Future work in population genetics should be able to determine whether any barriers exist to spore spread and gene flow in the boreal region. Does, for example, the Baltic sea serve as a barrier or are gene frequencies similar on both sides? Questions like these have just begun to be asked, some of the answers obtained could shed light on past climatic change, availability of host/substrata, and the movement of decay fungi over regions on a historical time-scale.

Acknowledgements

I am indebted to A. Dahlberg, L. Holmer, M. Johansson, and D. Tilles for comments on drafts of this manuscript. Financial support was received from the Swedish Council of Agricultural and Forestry Research.

References

Adams, T.J.H., Todd, N.K. and Rayner, A.D.M. (1981). Antagonism between dikaryons of *Piptoporus betulinus*. *Trans. Brit. Mycol. Soc.* **76**: 510–513.

Ainsworth, A.M. (1987). Occurrence and interactions of outcrossing and non-outcrossing populations in *Stereum*, *Phanerochaete* and *Coniophora*. In A.D.M. Rayner, C.M. Brasier and D. Moore (editors). Evolutionary Biology of the Fungi, pp. 285–299. Cambridge University Press, Cambridge.

Ainsworth, A.M. and Rayner, A.D.M. (1990). Aerial mycelial transfer by *Hymenochaete corrugata* between stems of hazel and other trees. *Mycol. Res.* **94**: 263–266.

Anon. (1991). Kommenterad lista över hotade svampar i Sverige. (An annotated red data list of fungi from Sweden.) *Windhalia* **19**: 87–130. (In Swedish with summary in English.)

Barret, D.K. and Uscuplic, M. (1971). The field distribution of interacting strains of *Polyporus schweinitzii* and their origin. *New Phytol.* **70**: 581–598.

Bergeron, Y. (1991). The influence of island and mainland lakeshore landscapes on boreal forest fire regimes. *Ecology* **72**: 1980–1992.

Boddy, L. and Rayner, A.D.M. (1983). Origins of decay in living deciduous trees: the role of moisture content and a re-appraisal of the expanded concept of tree decay. *New Phytol.* **94**: 623–641.

Boidin, J. (1986). Intercompatibility and the species concept in saprobic Basidiomycotina. *Mycotaxon* **26**: 319–336.

Bonan, G.B. and Shugart, H.H. (1989). Environmental factors and ecological processes in boreal forests. *Annual Rev. Ecol. Syst.* **20**: 1–28.

Bradshaw, R. and Zackrisson, O. (1990). A two thousand year history of a northern Swedish boreal forest stand. *J. Veg. Sci.* **1**: 519–528.

Brasier, C.M. (1987). The dynamics of fungal speciation. In A.D.M. Rayner, C.M. Brasier and D. Moore (editors). Evolutionary biology of the fungi, pp. 231–260. Cambridge University Press, Cambridge.
Breitenbach, J. and Kränzlin, F. (1986). Fungi of Switzerland 2. Non-gilled fungi. Verlag Mykologia, Lucerne.
Brownlee, C. and Jennings, D.H. (1982). The pathway of translocation in *Serpula lacrymans*. *Trans. Brit. Mycol. Soc.* **79**: 401–407.
Burnett, J.H. (1983). Speciation in fungi. *Trans. Brit. Mycol. Soc.* **81**: 1–14.
Chapela, I.H. (1989). Fungi in healthy stems and branches of American beech and aspen: a comparative study. *New Phytol.* **113**: 65–75.
Chapela, I.H. and Boddy, L. (1988). Fungal colonization of attached beech branches. II. Spatial and temporal organization of communities arising from latent invaders in bark and functional sapwood, under different moisture regimes. *New Phytol.* **110**: 47–57.
Chase, T.E. and Ullrich, R.C. (1988). *Heterobasidion annosum*, root- and butt-rot of trees. *Advances Pl. Pathol.* **6**: 501–510.
Cooke, R.C. and Rayner, A.D.M. (1984). Ecology of Saprotrophic Fungi. Longman, London.
Fischer, M. (1987). Biosystematische Untersuchungen an den Porlingsgattungen *Phellinus* und *Innonotus*. *Biblioth. Mycol.* **107**: 133 p.
Frankland, J.C. (1982). Biomass and nutrient cycling by decomposer basidiomycetes. In. J.C. Frankland, J.N. Hedger and M.J. Swift (editors). Basidiomycetes: Their Biology and Ecology, pp. 241–261, Cambridge University Press, Cambridge (British Mycological Society Symposium 4).
Gilbertson, R.L. (1980). Wood-rotting fungi of North America. *Mycologia* **72**: 1–49.
Goheen, D.J. and Hansen, E.M. (1993). Effects of rot disease fungi and associated bark beetles in conifer forests of western North America. In T.D. Schowalter and G.M. Filip (editors). Beetle-pathogen interactions in conifer forests. Academic Press.
Haddow, W.R. (1938). The disease caused by *Trametes pini* in White pine. *Trans. Roy. Canad. Inst.* **22**: 21–80.
Hallenberg, N. (1987). On speciation in Corticiaceae (Basidiomycetes). *Windahlia* **17**: 19–25.
Hallenberg, N. (1991). Speciation and distribution in Corticiaceae (Basidiomycetes). *Pl. Syst. Evol.* **177**: 93–110.
Hansson, L. (1992). Landscape ecology of boreal forests. *Trends Ecol. Evol.* **7**: 299–302.
Hjortstam, K., Larsson, K.-H. and Ryvarden, L. (1987). The Corticiaceae of North Europe, 1. Fungiflora, Oslo.
Hjortstam, K., Larsson, K.-H. and Ryvarden, L. (1988). The Corticiaceae of North Europe, 8. Fungiflora, Oslo.
Hodges, C.S. (1969). Modes of infection and spread of *Fomes annosus*. *Annual Rev. Phytopathol.* **7**: 247–266.
Holmer, L. and Stenlid, J. (1991). Population structure and mating system in *Marasmius androsaceus*. *New Phytol.* **119**: 307–314.
Huntley, B. and Birks, H.J.B. (1983). An atlas of past and present pollen maps of Europe, 0–13000 years ago, Cambridge University Press, Cambridge.
Isomäki, A. and Kallio, T. (1974). Consequences of injury caused by timber harvesting machines on the growth and decay of spruce (*Picea abies*). *Acta Forest Fenn.* **136**: 1–25.
Jahn, H. (1979). Pilze die an Holz wachsen. Busse, Herford.
Käärik, A.A. and Rennerfelt, E. (1957). Investigations on the fungal flora of spruce and pine stumps. *Meddeland. Statens Skogsforskningsinst.* **47**: 1–88.
Karström, M. (1992) Stegel före – en presentation. *Svensk Bot. Tidskr.* **86**: 103–114. (In Swedish with English summary.).
Kile, G.A., McDonald, G.I. and Byler, J.W. (1991). Ecology and disease in natural Forests. In C.G. Shaw and G.A. Kile (editors). *Armillaria* root disease, pp. 102–121, USDA Forest service, Washington.
Kirby, J.J.H., Stenlid, J. and Holdenrieder, O. (1990). Population structure and responses to disturbance of the basidiomycete *Resinicium bicolor*. *Oecologia* **85**: 178–185.
Korhonen, K. (1978). Interfertility and clonal size in the *Armillaria mellea* complex. *Karstenia* **18**: 31–42.

Korhonen, K. (1987). Breeding units in the forest pathogens *Armillaria* and *Heterobasidion*. In A.D.M. Rayner, C.M. Brasier and D. Moore (editors). Evolutionary biology of the fungi, pp. 301–310. Cambridge University Press, Cambridge.
Korhonen, K., Bobko, I., Hanso, S, Piri, T. and Vasiliauskas, A. (1992). Intersterility groups of *Heterobasidion annosum* in some spruce and pine stands in Byelorussia, Lithuania and Estonia. *Eur. J. Forest Pathol.* **22**: 384–391.
Koski-Kotiranta, S. and Niemelä, T. (1987). Hydnaceous fungi of the Hericiaceae, Auriscalpiaceae and Climacodontaceae in northwestern Europe. *Karstenia* **27**: 43–70.
Lewis, K. and Hansen, E.M. (1991). Vegetative compatibility groups and protein electrophoresis indicate a role for basidiospores in spread of *Innonotus tomentosus* in spruce forests of British Columbia. *Canad. J. Bot.* **69**: 1756–1763.
Martin, P.S. and Wright, H.E. (1976). Pleistocene extinctions. London.
Morrison, D.J., Williams, R.E. and Whitney, R.D. (1991). Infection, disease development, diagnosis, and detection. In C.G. Shaw and G.A. Kile (editors), *Armillaria* root disease, pp. 62–75, USDA Forest service, Washington.
Mukhin, V.A. (1987). Active main body of subarctic flora of xylotrophic basidiomycetes in west Siberia. *Mykol. Fitopatol.* **21**: 195–199.
Niemelä, T. and Kotiranta, H. (1982). Polypore survey of Finland 2. The genus *Phellinus*. *Karstenia* **22**: 27–42.
Piri, T., Korhonen, K. and Sairanen, A. (1990). Occurrence of *Heterobasidion annosum* in pure and mixed spruce stands in southern Finland. *Scand. J. Forest Res.* **5**: 113–125.
Rayner, A.D.M. and Boddy, L. (1988). Fungal Decomposition of Wood. John Wiley, Chichester.
Rayner, A.D.M. and Turton, M.N. (1982). Mycelial interactions and population structure in the genus *Stereum*: *S. rugosum*, *S. sanguinolentum*, and *S. rameale*. *Trans. Brit. Mycol. Soc.* **78**: 483–493.
Rayner, A.D.M. and Webber, J.F. (1984). Interspecific mycelial interactions — an overview. In D.H. Jennings and A.D.M. Rayner (editors). The ecology and physiology of the fungal mycelium. Cambridge University Press, Cambridge.
Rayner, A.D.M., Boddy, L. and Dowson, C.G. (1987). Temporary parasitism of *Coriolus spp.* by *Lenzites betulina*: A strategy for domain capture in wood decay fungi. *F.E.M.S. Microbiol. Ecol.* **45**: 53–58.
Redfern, D.B. and Filip, G.M. (1991). Inoculum and infection. In C.G. Shaw and G.A. Kile (editors). *Armillaria* root disease, pp. 48–61. USDA Forest service, Washington.
Renvall, P., Renvall, T. and Niemelä, T. (1991). Basidiomycetes at the timberline in Lappland: 2. An annotated checklist of the polypores of northeastern Finland. *Karstenia* **31**: 13–28.
Ryman, S. and Holmåsen, I. (1984). Svampar. Interpublishing, Stockholm.
Ryvarden, L. (1976). The Polyporaceae of North Europe. 1. Fungiflora, Oslo.
Ryvarden, L. (1978). The Polyporaceae of North Europe. 2. Fungiflora, Oslo.
Shigo, A.L. (1967). Succession of organisms in discolouration and decay of wood. *Int. Rev. Forest Res.* **2**: 237–299.
Siepmann, R. (1989). Intersterilitätsgruppen und Klone von *Heterobasidion annosum* in einem 31 jährigen Fichtenbestand. *Eur. J. Forest Pathol.* **19**: 251–253.
Silvertown, J. (1987). Introduction to plant population ecology. Longman, Harlow.
Sinclair, W.A., Lyon, H.H. and Johnson, W.T. (1987). Diseases of trees and shrubs. Cornell University Press, Ithaca & London.
Stenlid, J. (1985). Population structure in *Heterobasidion annosum* as determined by somatic incompatibility, sexual compatibility, and isoenzyme patterns. *Canad. J. Bot.* **63**: 2268–2273.
Stenlid, J. (1987). Controlling and predicting the spread of *Heterobasidion annosum* from infected stumps and trees of *Picea abies*. *Scand. J. Forest Res.* **2**: 187–198.
Stenlid, J. and Holmer, L. (1991). Infection strategy of *Hymenochaete tabacina*. *Eur. J. Forest Pathol.* **21**: 313–318.
Strid, Å. (1975). Wood-inhabiting fungi of alder forests in North-Central Scandinavia: 1. Aphyllophorales. *Wahlenbergia* **1**: 1–237.
Swedjemark, G. and Stenlid, J. (1993). Population dynamics of the root rot fungus *Heterobasidion annosum* following thinning of *Picea abies*. *Oikos* **66** 242–254.

Wells, J.M. and Boddy, L. (1990). Wood decay, and phosphorous and fungal biomass allocation, in mycelial cord systems. *New Phytol.* **116**: 285–295.
Zackrisson, O. (1977). Influence of forest fires on the north Swedish boreal forest. *Oikos* **29**: 22–32.

Svrcek, M. (1993). Discomycetes of Czechoslovakia: Their Occurrence & Distribution. In D.N. Pegler, L. Boddy, B. Ing & P.M. Kirk (Editors). Fungi of Europe: Investigation, Recording and Conservation, pp. 181–187. Royal Botanic Gardens, Kew.

15. DISCOMYCETES OF CZECHOSLOVAKIA: THEIR OCCURRENCE & DISTRIBUTION

M. SVRCEK

Národní Muzeum v Praze, Václavské nám. 68, 115 79 Praha 2, Czechoslovakia

Abstract

The occurrence and distribution of discomycetes in former Czechoslovakia (CSFR), based on the author's own observations during the last 50 years is discussed. These fungi were followed in various plant communities in many areas of Bohemia, Moravia, Silesia and Slovakia, particularly nature reserves, but also including natural and secondary woodlands from lowland to mountains. The results demonstrate a considerable qualitative and quantitative decline in these fungi, correlated with drastic environmental changes caused by sulphur dioxide pollution and acid rain over the past 30 years. This is analogous to the decline in lichens, though discomycetes are mostly less sensitive. In addition, during the past 10 years, fruiting has been affected, apparently by reduced rainfall, loss of swamps, marshes and small streams, and extinction of host plants. A decline of 50–80% in discomycetes during the past 40–50 years has occurred. However, mountain areas with a rich flora of vascular plants, mixed deciduous forests on calcareous soil, and alder woods are still relatively well-preserved habitats for discomycetes in CSFR. The decline is most conspicuous in Pezizales, originally representing about 30–35% of all discomycetes in CSFR. Many terrestrial operculates, especially those of acid soil, are now almost extinct, though lignicolous species are more frequent. In Helotiales, there is a large decrease in Sclerotiniaceae.

Geography and Vegetation

Czechoslovakia has a rich vascular plant flora. It occupies a biogeographically key position due to glaciation when a narrow passage was formed between the North European and Alpine areas of glaciation. It is positioned on the dividing line of the old Hercynian Bohemian Massif and the young Alpin-Carpathian mountain system, which have very different geological structure and relief. The Bohemian Massif is formed by Proterozoic and Paleozoic rocks, and the Carpathians formed by Mesozoic and Tertiary sediments. While the Carpathians consist of a system of distinct mountain ridges and deep river valleys, the Bohemian Massif is a peneplain diversified by sharp valley incisions.

The floristic composition of the various regions is determined primarily by the geological substratum, but also by relief and local climate. The climate is temperate, with a regular weather pattern and no extremes during winter or summer: 3–9°C mean annual temperature: 400–1200 mm mean annual average of precipitation. The vegetation is influenced by both oceanic and continental climate. The primeval plant cover consisted predominantly of deciduous forests and stands of mountain pine in the highest part of the mountains. Natural and semi-natural plant communities are now represented mainly in the mountains, and other inaccessible habitats, and are rare at lower altitudes. The primeval forests were largely replaced by spruce and pine monocultures. Mans activities are the main source of recent changes in vegetation.

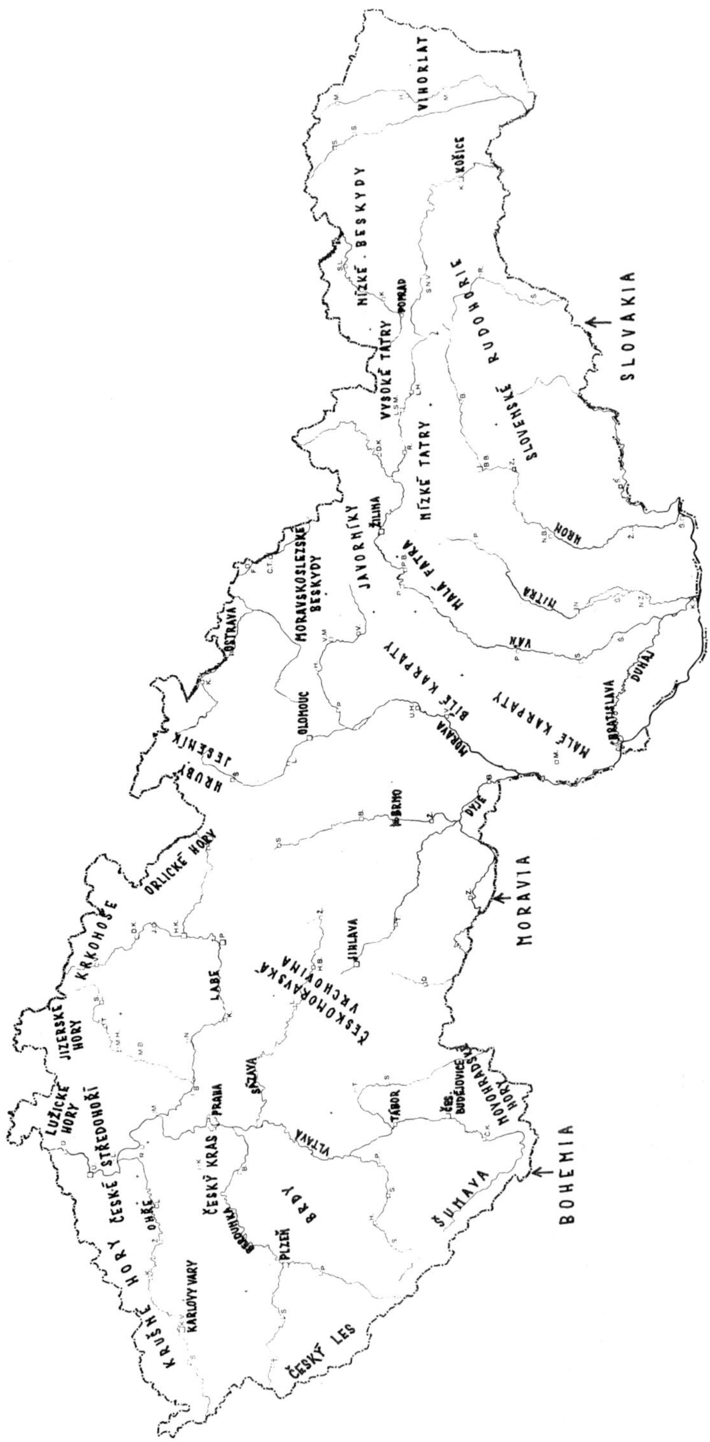

FIG. 1. Collecting localities in former Czechoslovakia.

Discomycetes most characteristic of major plant communities will be considered.

Communities on Wet Soils

Alnetea glutinosae are formed on boggy soil of alder (*Alnus*), poplar (*Populus*), willow (*Salix*), ash (*Fraxinus*) and elm (*Ulmus*). The richest associations for discomycetes are the smaller stands of *Alnus glutinosa* and *A. incana*, the latter mainly in the submontane and mountain belt. Undergrowth varies, e.g. *Carex brizoides, Lysimachia vulgaris, Urtica dioica* and *Rubus fruticosus* agg., and the diversity of hosts and substrata, and constant soil and air humidity are favourable for fungi. Discomycetes and other groups are less common than formerly as undergrowth is considerably reduced and altered in many places, due to mans activities, including extensive drainage, during the last 20 years. The mycoflora of the alder woods on mountain slopes, e.g. in the Giant Mts. (Krkonoše, Bohemia), and in Slovak Carpathians, is relatively well conserved.

Characteristic discomycetes are found on different components of alder litter: *Encoelia furfuracea* (Roth) Karst. on dead branches still attached to the tree; numerous species, mainly *Mollisia benesuada* (Tul.) Phill., *Orbilia luterorubella* (Nyl.) Karst., *Hymenoscyphus vernus* (Boud.) Dennis, *H. imberbis* (Bull.) Dennis on rotten branches lying on ground; *Mollisia uda* (Pers.) Gill., *Scutellinia spp.* on fallen trunks and rotten stumps; in the past, *Ciboria amentacea* (Balb.) Fuckel, *C. viridifusca* (Fuckel) Hoehn. on fruits, seeds, catkins; *Ciboria conformata* (Karst.) Svr.; on fallen leaves; *Disciotis venosa* (Pers.) Boud. on wet ground.

Similar habitats are found in stands of willows (*Salix aurita, S. cinerea*) forming order *Salicetalia auritae* and alliance *Salicion cinereae* on boggy ground, in Southern and West Bohemia, and the Šumava Mts. The most characteristic discomycetes are *Pezizella amenti* (Batsch) Dennis, *Hymenoscyphus salicellus* (Fr.) Dennis, *Ocellaria ocellata* (Pers.) Schroet. The foliicolous parasitic *Rhytisma salicinum* (Pers.) Fr. is now rare, but still found in the Krkonoše Mountains; its decline, as with discomycetes, is probably due to sulphur dioxide pollution. Several new species have been found on willows including: the foliicolous *Psilocistella cejpii* Svr. on *Salix cinerea*; *Parorbiliopsis salicis* Svr. on dead leaves of *Salix aurita*; *Amicodisca brdensis* (Vel.) Svr. on rotten trunks of *S. aurita*, partly submerged in water in swamps in Southern Bohemia.

Communities of the alliance *Alnion* occur in river source areas, in swamps and peat bogs. Plant communities of the class *Phragmiti-Magnocaricetea* are found, with the dominant reed (*Phragmites australis*) and high sedges (*Carex*). Altogether, 30–40% of all discomycetes in Czechoslovakia are found in associations of hygrophilous plants. Some host specific fungal species are found on living *Sphagnum. Conchatium airae* (Vel.) Svr. seems to be specific to dead leaves of *Deschampsia caespitosa. Tapesia kneiffii* (Wallr.) Kze., more than 30 other species of discomycetes (e.g. very rare *Dennisiodiscus cerberi* (Vel.) Svr., *Discocistella rudolphi* (Vel.) Svr. and *Lachnum winteri* (Cke.) Rehm.)] are found on *Phragmites. Cladium mariscus* (saw-sedge), a very rare plant in Bohemia, surviving in two small localities in nature reserves, is host of *Hyaloscypha cladii-marisci* Svr., a species close to *H. cladii* Nag Raj et Kendrick. It is found in Britain, where *Cladium* is more frequent. Specialized discomycetes, e.g. *Hymenoscyphus salmanovicensis* Svr. and *Myriosclerotinia dennisii* (Svr.) Schwegler occur on *Eriophorum vaginatum* (cotton-grass) in South Bohemia. *Schoenoplectus lacustris* (bulrushes), which often form large stands in ponds, has some characteristic species, two of which were described from South Bohemia asnew: *Dasyscyphus* (*Lachnum*) *pulchricolor* Svr. and *Discocistella calyptrata* Svr. *Discocistella albidolutea* (Feltg.) Svr. is very common on dead basal leaves of *Scirpus sylvaticus*, while the conspicuously blue-coloured *Cejpia amoena* (Boud.) Svr. is rare. Few discomycetes, e.g. *Unguicularia iridis* Svr., have been collected on *Iris pseudacorus*. On both species of Reedmace (*Typha angustifolia* and *T. latifolia*) three species, originally described from Bohemia, are found on dead stems and leaves: *Psilocistella fonticola* Svr., *Discocistella typhae* Svr. and *Orbilia arundinacea* Vel.

The peat-moss (*Sphagnum*) is widespread in wet areas, the most extensive ones being situated in Southern Bohemia and in mountainous regions. There are several rare host specific discomycetes on *Sphagnum*: *Sarcoleotia turficola* (Boud.) Dennis, known from three localities in Bohemia and Slovakia, *Hymenoscyphus sphagnisedus* (Vel.) Svr. (in Bohemia), and *Kubickia tatrensis* Svr. occurring at one small lake in Vysoké Tatry Mountains. *Equisetum limosum* is the host of *Gorgoniceps boltonii* (Phill.) Dennis and *Sclerotinia veselyi* Pil. & Svr. Rushes (*Juncus*), have many host specific discomycetes, e.g. *Olla costata* (Boud.) Svr., *Discocistella fugiens* (Phill. ex Bucknall) Svr. and *Myriosclerotinia curreyana* (Berk. ex Curr.) Buchw., the latter being almost extinct.

There is a group of aquatic discomycetes which grow on fallen twigs, branches and leaves lying submerged in streams. These fungi are not associated with particular plant communities but are affected by water conditions, communities being considerably altered where there is permanent eutrophication or pollution. Twelve species are found in CSFR, particuarly in running water: *Vibrissea truncorum* (Alb. & Schw.) Fr. and *Tapesia rivularis* Svr. only in mountains; *Apostemidium fiscellum* (Karst.) Karst., *A. guernisaci* (Crouan) Boud., *Mitrula paludosa* Fr., *Miladina lecithina* (Cooke) Svr., *Thecotheus rivicola* (Vacek) Kimbr. & Pfister, *Psilopezia nummularia* Berk., *Pachyella babingtonii* (Berk.) Boud. at lower altitudes; *Hymenoscyphus vernus* (Boud.) Dennis, *H. peruni* (Vel.) Svr. and *Hymenoscyphus imberbis* (Bull.) Dennis, are more tolerant of water quality.

Grassland and Herb Communities

The class *Molinieto-Arrhenatheretea* includes communities of meadows and grasslands on moist soil. The alliances *Calthion* and *Filipendulion* cover moist habitats where herbs and grasses are not cut regularly. Conditions are excellent for fruiting of micromycetes and discomycetes. In dense stands of *Filipendula ulmariae* there are many species on rotting, dead stems and leaves, one of the rarest being *Chlorosplenium aeruginellum* (Karst.) Karst. A new foliicolous species, *Aivenia tantula* Svr., was described on *Comarum palustre*, and the caulicolous *Mollisia potentillae-erectae* Svr. is found on *Potentilla erecta*. *Verpatinia calthicola* Whetzel, occurring on dead petioles of *Caltha palustris*, is known only from one locality, and *Hymenoscyphus callorioides* Rehm. occurs in the mountains. *Belonium hystrix* (de Not.) Hoehn. is specific to the grass *Molinia caerulea*.

Communities of acidophilous, short-grass meadows and grasslands, heath and peat bogs are grouped in the class *Nardo-Callunetea*. The extensive stands of *Nardus stricta*, present in mountains, are characterised by *Actinoscypha punctum* (Rehm) E. Müll. The large stands of *Calluna vulgaris* (heather), whose extent has been considerably reduced in the last 20 years, have several specific discomycetes, e.g. *Godronia callunigera* (Karst.) Karst. and *Trichobelonium obscurum* (Rehm) Rehm.

The class *Festuco-Brometea* consists of xerothermous grassland communities rich in herbs, both on acid and alkaline soil. In Central Bohemia, these communities are best preserved in the limestone region known as the Bohemian Karst, with xerothermous "steppe" grass associations (*Festucion vallesiaceae*) on calcareous rocky slopes. A similar steppe has developed on hills in the North Bohemia, the Southern Moravia and the Slovak Carpathians, where sunny slopes are occupied with associations of plants of Pontic and Panonian floristic elements. Discomycetes are scarce in these sunny, dry habitats, nevertheless some are characteristic of and restricted to these habitats, fruiting occurring after rain. For example *Belonidium procopii* Svr. occurs on dead stems of *Coronilla varia*, *Centaurea scabiosa*; *B. badium* (Rehm) Svr. on *Salvia verticillata*, *S. pratensis* and *Stachys recta*; *Rutstroemia iridis-aphyllae* Svr. on *Iris aphylla* subsp. *bohemica*. More discomycetes are found on xerothermous grasses (*Festuca sulcata*, *F. vallesiaca*, *Stipa ioannis*, *S. capillata*, *Bothriochloa ischaemum*, *Sesleria calcaria*, *Brachypodium pinnatum* etc.).

The class *Mulgedio-Aconitetea* embraces tall herb plant communities of the subalpine belt. I have studied the discomycetes of this group in the Slovak Carpathians, namely in the Belanské and the Nízke Tatry Mts., and also in Krkonose (Giant Mountains) in Bohemia. The rich flora has many host limited discomycetes. With regard to herbs (e.g. *Aconitum, Delphinium, Cicerbita (Mulgedium) alpina, Rumex alpinus, Senecio subalpinus, Veratrum lobelianum*), there are many conspicuous discomycetes, amongst which are: *Niveostoma leucostoma* (Rehm) Svr., *Lachnum (Trichopezizella) relicina* (Fr.) Karst., *Mollisia adenostylidis* Rehm, *Pezizellaster ochraceus* (Vel.) Svr., *Urceolella crispula* (Karst.) Boud., *Unguicularia aconiti* Svr. In the highest belt of mountains the class *Juncetea trifidi* often develops; *Lachnum calycioides* Rehm is specific for the rush *Juncus trifidus*. Another class, *Salicetea herbaceae*, formed of dwarf willows, has also some characteristic fungi, for example, a new species *Microscypha monticola* Svr. on fallen leaves of *Salix reticulata* was described from here.

The alliance *Epilobium angustifolii* (one of the order Atropetalia) consists of herbs of woodland clearings. Discomycetes are here present only when stands are dense, tall, and soil relatively moist. Most species are found on dead stems of *Chamaenerion (Epilobium) angustifolium, Rubus idaeus, R. fruticosus* agg., as well as on the ground. The community has been called "woodland-ecoton". These communities are classified as orders *Prunetalia* and *Trifolio-Geranietea*, with the alliance *Sambuco-Salicion*. Many discomycetes are found here not only on shrubs (*Crataegus, Rosa, Prunus spinosa, Cornus, Swida, Ligustrum, Berberis*), but also on various herbs and grasses (personal observation). The most interesting includes: *Zoellneria rosarum* Vel., *Velutarina rufo-olivacea* (Alb. & Schw.) Korf, *Haglundia sarmentosa* Svr., *Hymenoscyphus furinus* (Vel.) Svr., *Rutstroemia rubi* Vel., *Plectania melastoma* (Sow.) Fuckel.

Angiosperm Woodlands

There are several classes of predominantly angiosperm woodland including: *Alnetea glutinosae*, mentioned above; *Querco-Fagetea* representing mixed broad-leaved, xerothermous as well as hygrophilous communities, including the orders *Fagetalia silvaticae* and *Quercetea pubescentis*; acidophilous oak woods (*Quercetea roboris-petraeae*), often with *Betula* and *Pinus silvestris*. The natural structure of the vast majority of these woods was destroyed or alterend by mans activities. Despite this many natural woods are preserved in Bohemia, Moravia and Slovakia, and mostly protected as nature reserves. The highest frequency of discomycetes occurs on decaying stumps, fallen branches and leaves of oak (*Quercus*), hornbeam (*Carpinus betulus*), beech (*Fagus sylvatica*). Some species are host limited, e.g. of about 100 species known in CSFR on oak 20–30% are exclusive to it. *Quercetalium pubescentis* is a xerothermous community with several interesting discomycetes: *Lachnum claviculatum* Vel. and *L. radotinense* Vel., both on leaves of *Quercus pubescens*.

Associations of *Tilio-Acerion* and *Fraxino-Carpinion* are more widespread, covering western and northern slopes in hilly country and deep valleys of streams, typically developing on scree. *Acer pseudoplatanus, A. platanoides, Fraxinus excelsior, Ulmus glabra, Carpinus betulus, Tilia cordata* and *Fagus silvatica* are represented. This community is rich in lignicolous fungi, with discomycetes including: *Orbilia* spp., *Hyalinia crystallina* (Quél.) Boud., *Lanzia luteovirescens* (Rob. ex Desm.) Dumont & Korf, *Sarcoscypha coccinea* (Scop.) Lambotte, *Urnula craterium* (Schw.) Fr. Several discomycetes are specific on oak (*Quercus petraea* and *Q. robur*): *Hyaloscypha cupularum* Svr. on cupules; *Haglundia perelegans* Nannf., *Psilocistella quercina* (Vel.) Svr., *Hyaloscypha priapi* (Vel.) Svr., *H. luteopallida* Svr., *H. trapezispora* Svr., *Protounguicularia vandae* (Vel.) Svr., *Mollisia olivaceo-cinerea* Svr., *M. sericeomarginata* Svr. on wood, *Arachnopeziza aurelia* (Pers.) Fuckel was frequent in Central and Southern Bohemia, but is now almost extinct, its decline being attributed to high air pollution and acid precipitation. *Pseudombrophila fellneri* Svr. is ecologically interesting being found in three localities in Bohemia, in experimental mixed oak woods, on ground to which urea had been added to stimulate fungal fruiting. The single find of *Cenangiopsis quercicola* (Romell) Rehm is from Slovakia.

Carpinus betulus is often found in *Querco-Carpinetum* community. Formerly *Rutstroemia bolaris* (Batsch) Rehm was common on fallen twigs, but has become almost extinct in the last ten years. Discomycetes on *C. betulus* include *Ciboriopsis gemmigera* Svr. (on buds), *Encoelia glaberrima* (Rehm) Kirschst., *Hyaloscypha amyloideopilosa* Svr., *H. intacta* Svr. (described from Bohemia, but subsequently collected on other hosts in Europe and Canada). The semihypogeous discomyste, *Calyptellopsis reticulata* (Vacek) Svr., occurrences on decaying fine roots of trees, hidden under a high layer of fallen leaves or in small cavities in the ground.

Extensive beech forests (*Fagetalia sylvaticae*) are preserved in the Slovak Carpathians, several of them protected as virgin forests. Most of the discomycetes are found on fallen trunks, branches, stumps, dead leaves and fallen cupules. The most characteristic fagicolous species here are: *Bisporella citrina* (Batsch) Korf & Carp., *B. pallescens* (Pers.) Carp. & Korf, *Neobulgaria pura* (Fr.) Petrak, *Hymenoscyphus fagineus* (Pers.) Dennis, *H. rokebyensis* (Svr.) Matheis, *H. serotinus* (Pers.) Phill., various *Orbilia* spp. etc. One of the rarest is *Ionomidotis irregularis* (Schw.) Durand, found at the eastern boundary of Slovakia in a Carpathian virgin forest. *Chlorosplenium aeruginosum* (Pers.) auct. and *Ch. aeruginascens* (Nyl.) Karst. are also common on beech wood.

Coniferous Woods

Coniferous woods form about 60–70% of the CSFR woodland area. Many are planted spruce or pine monocultures, which have replaced deciduous or mixed forests. Spruce (*Picea abies*) is mainly natural in the mountains where it forms *Piceion excelsae*, *Athyrio-Picetalia* (rich in ferns) and *Sphagno-Piceeta* on moist or boggy ground. Discomycetes are less frequent in these spruce stands than in other coniferous woods, and the majority of species are limited to specific parts of the spruce. Surprisingly, in the planted spruce woods at lower altitudes discomycetes are often more frequent than in the primeval forests. Nevertheless, the decrease during the last 20 years is marked. Some of the commonly occurring species in the past are now rare or absent, e.g. *Spathularia flava* Pers., *Podophacidium xanthomelum* (Pers.) Kavina, *Pseudoplectania nigrella* (Pers.) Fuckel, *Sarcosoma globosum* (Schmidt.: Fr.) Rehm. Also *Tryblidiopsis picea* Vel. and *Piceomphale bulgarioides* (Rabenh.) Svr. have recently become uncommon. On fallen cones on moist soil, *Ombrophila janthina* (Karst.) Rehm, and *Hyaloscypha sulphureo-pilosa* Svr. occur; on wet wood, *Mollisia psilopezioides* Svr. occurs. The new species *Gorgoniceps hypothallosa* Svr. and *Ciliolarina corcontica* Svr. were described from natural spruce forests in mountains; *Orbilia cryptocarpa* and *O. roseofuscella* Svr. are found in lowland. *Hymenoscyphus ravus* Svr. was recently described on dead bud scales from Bohemia and found subsequently in Germany. *Ciboria subvillosula* (Rehm) Svr. is ecologically interesting, occurring in winter and early spring on fallen needles, and often fruiting on frozen ground. Some discomycetes are specific on resinous exudations on trunks: *Biatorella resinae* (Fr.) Th. Fr. [= *Sarea resinae* (Fr.) Kuntze] was common 20 years ago but is now almost extinct; *Hymenoscyphus resinae-piceae* Svr., *Cystopezizella venceslai* (Vel.) Svr., *Ciliolarina ligniseda* (Vel.) Svr.; *Pezizella nigrostipitata* Svr. on roots and stumps.

Pine (*Pinus silvestris*) is found as planted monocultures over much of CSFR on poor substrata at lower altitudes, but it is also often mixed with other trees, mainly oak, hornbeam and spruce. Relict pine woods are preserved in some submontane regions. In South Bohemia, *Pinus rotundata* (*P. uncinata*) forms a characteristic association on the peat bogs with *Ledum palustre*, *Oxycoccus*, *Vaccinium uliginosum*, *Andromeda polifolia*, *Eriophorum*, *Trichophorum* and *Sphagnum* in the undergrowth. In the high mountains, *Pinus mugho* forms uniform, large, dense growths at the border of forest. All these pine communities are rich in discomycetes, including *Desmazierella acicola* Lib., *Weinmannioscyphus messerschmidii* (Weinm.) Svr., *Hyalopeziza trichodea* (Phill. & Plowr.) Raitv., *Lachnum pulverulentum* (Lib.) Karst., *L. mughonicolum* (Svr.), *Antinoa* spp.,

Psilocistella lignatilis (Vel.) Svr., *Fuscoscypha acicularum* (Vel.) Svr. on needles; *Hymenoscyphus sazavae* (Vel.) Svr. and *H. coeruleo-annulatus* Svr. on cones. Some species, common in the past, have become extinct or almost so in the last 20 years, e.g. *Lophodermium pinastri* (Schrad.) Chev. and *Naemacyclus niveus* (Pers.) Sacc., and *Monilinia baccarum* (Schroet.) Whetzel on *Vaccinium myrtillus* (which formed extensive undergrowths in pine woods). *Pinus nigra* is not indigenous in CSFR, but is frequently planted predominantly on sunny, rocky, calcareous slopes and hills. It has a similar discomycete flora to *Pinus silvestris*, nevertheless two new discomycetes hitherto known only from Bohemia were described from needles — *Pseudohelotium vernale* (Vel.) Svr. and *Septopezizella oreadum* (Vel.) Svr.

On *Abies alba*, a conifer almost extinct in many regions but forming mixed forest with beech in the Carpathians, some specific, now rare discomycetes are found, e.g. *Ciboria rufofusca* (Weberb.) Sacc., *Rutstroemia elatina* (Alb. & Schw.) Rehm, *Pithya vulgaris* Fuckel. *Larix decidua*, forming natural stands in the Slovak Carpathians, is often planted elsewhere, and has several interesting species, e.g. *Ciliolarina laricina* (Raitv.) Svr. growing mostly on black resinous exudations and on short-shoots, and *Sarcotrochila alpina* (Fuckel) Hoehn. on fallen needles. The decline of the common *Lachnellula willkommii* (Hartig) Dennis and *L. hahniana* (Seaver) Dennis is evident. *Juniperus communis* fairly common in the past, is now missing in Bohemia; *Chloroscypha sabinae* (Fuckel) Dennis on juniper needles is now one of the rarest discomycetes in the country.

Habitats Strongly Influenced by Man

Discomycetes occur in fields, dumps, gardens, parks, plant communities associated with trampled down soils, burnt ground, bonfire sites, and dung. Operculate discomycetes (Pezizales), occur on ground or amongst moss, but their decline began 20–25 years ago as did species from all fungal groups in this country.

References

Dostal, J. (1989). Evětena ČSSR. 1.–2. Praha.

Svrček, M. (1974–1993). New or less known Discomycetes: I.–XXII. *Česká. Mykol.* **28**: 129–137 (1974); **29**: 129–134 (1975); **30**: 7–16 (1976); **31**: 8–14 (1977); **31**: 132–138 (1977); **31**: 193–200 (1977); **32**: 11–18 (1978); **32**: 157–160 (1978); **32**: 202–204 (1978); **33**: 193–206 (1979); **36**: 146–153 (1982); **37**: 65–71 (1983); **38**: 197–202 (1984); **40**: 203–217 (1986); **41**16–25 (1987); **41**; 88–96 (1987); **42**: 76–80 (1988); **42**: 137–148 (1988); **43**: 65–76 (1989); **43**: 215–226 (1989); **45**: 134–143 (1991); **46**: 27–33 (1993).

Svrček, M. (1979). A taxonomic revision of Velenovsky's types of Operculate Discomycetes (Pezizales) preserved in National Museum, Prague. *Sbor. Nár. Muz. v Praze, Rada B* **32**(2–4): 115–194, tab. 1–8 (1976 publ. 1979).

Svrček, M. (1981). Katalog operkulátních d iskomycetů (Pezizales) Československa: I.–III. (List of Operculate Discomycetes (Pezizales) recorded from Czechoslovakia: I.–III.) *Česká Mykol.*, **35**: 1–24, 64–89, 134–151

Svrček, M. (1985). A taxonomic revision of Inoperculate Discomycetes described by J. Velenovský in the genus Helotium, preserved in National Museum, Prague. *Sbor. Nár. Muz., v Praze Rada B* **40**: 129–215, tab. 1–20 (1984 publ. 1985).

Velenovský, (1934). Monographia Discomycetum Bohemiae: 1.–2. Pragae.

Wasser, S.P. & Grodzinskaya, A.A. (1993). Content of Radionuclides in Macromycetes of the Ukraine in 1990–1991. In D.N. Pegler, L. Boddy, B. Ing & P.M. Kirk (Editors). Fungi of Europe: Investigation, Recording and Conservation, pp. 189–210. Royal Botanic Gardens, Kew.

16. CONTENT OF RADIONUCLIDES IN MACROMYCETES OF THE UKRAINE IN 1990–1991

S.P. WASSER & A.A. GRODZINSKAYA

N.G. Kholodny Institute of Botany Ukrainian Academy of Sciences, Kiev, GSP, Repin str. 2, Ukraine.

Abstract

Radionuclide accumulation in the carpophores of macromycetes during the vegetation season of 1990–1991 in 44 locations of the Ukraine were studied. A total of 63 fungal species, soil and other substrates were gathered. Content of ^{137}Cs in carpophores was 1–2 orders of magnitude higher than in the substrata on which they were growing. Activity of ^{137}Cs in carpophores of *Amanita solitaria* (Bull.: Fr.) Mérat, *Suillus luteus* (L.: Fr.) S.F. Gray, and *Boletus edulis* Bull.: Fr. was 526, 300 and 55 times respectively greater than that of soil. Though the soils are contaminated by a set of radioactive isotypes (^{226}Ra, ^{228}Ra, ^{144}Ce, ^{125}Sb, ^{106}Ru, ^{154}Eu) most samples of macromycete carpophores mainly contained ^{134}Cs, ^{137}Cs and ^{40}K. The species belonging to the families Amanitaceae, Boletaceae and Russulaceae were mainly characterized by high contents of radiocesium, and may be considered as bioindicators of the radioactive contamination of the area. The tendencies of the increased radiocaesium accumulation in the series lignotrophs > saprotrophs > mycosymbiotrophs were observed.

Introduction

The scale of the ecological catastrophe in the Ukraine, resulting from the Chernobyl Atomic Power Station accident in April 1986, necessitated the constant monitoring of the uptake and distribution of radionuclides in all the links of natural ecosystems, including fungi. Radionuclide content in the carpophores of fungi has increased since the accident (Elstner *et al.*, 1987; Haselwandter *et al.*, 1988; Federov *et al.*, 1989; Haselwandter & Berreck, 1989a, b). Prior to this radiocaesium accumulation was monitored, in connection with the ground tests of nuclear weapons. Accumulation of radionuclides were first noted in lichens and mosses in 1961 and in fungus fruitbodies in 1963 (Oolbekkink & Kuyper, 1989). This temporal difference may be due to accumulation of atmospheric deposits in lichenes and mosses, but accumulation from the growth substrata, by fungi.

Accumulation of ^{134}Cs and ^{137}Cs depended on the level of radioactive contamination of territories following the Chernobyl disaster (Gans, 1986, 1987; Meijer *et al.*, 1988; Haselwandter *et al.*, 1988; Haselwandter & Berreck, 1989 a, b; Ritter, 1991). Bakken *et al.* (1990) found a positive correlation between accumulation in carpophores with the concentration of $^{134+137}$Cs in soil, but no correlation with pH. The concentration of stable Cs in the fungal fruitbodies positively correlated with the content of Cs radionuclides. Haselwandter *et al.* (1988) found a 3–4.8 fold increase of ^{137}Cs content in the fruitbodies of fungi in Austria post 1986.

Fruitbodies of some fungal species often have higher levels of radioactive contamination than plants and animals in the same exosystem (Fraiture *et al.*, 1990). Some fungal species are Cs accumulators, particularly in the *Cortinariaceae*, *Clavariaceae*,

Entolomataceae and *Strophariaceae* (Seeger & Schweinshaut cited in Oolbekkink & Kuyper, 1989). In contrast, low Cs contents are found in species of *Helvellaceae* and *Lycoperdaceae*.

In one of the first publications on radioactivity in fungi resulting from Chernobyl, Elstner *et al.* (1987) found high contents of ^{40}K in fungi, and these were always higher than in humus and soil. They also noticed considerable variability in the ^{137}Cs: ^{40}K ratio.

It should be noted that short-lived isotopes (e.g. ^{131}I) were major contributors of radioactivity of the fallen mixture immediately after the disaster; subsequently longer-lived isotopes, e.g. ^{90}Sr, ^{137}Cs and other products of uranium fission, predominated. Cs isotopes, when included into biological circulation, freely migrate along various biological food chains, and this determines their intake by humans and animals. As indicated by Zhuravlev (1990), the resorption coefficients of ^{137}Cs and ^{40}K in human lungs and the gastroenteric tract are 75–100%.

Despite the considerable variability in the content of caesium isotopes even within one species, it is evident that accumulation of radionuclides by fungi is species specific. Shcheglov *et al.* (1989) found 100-fold differences in ^{137}Cs content between fungal species and species specificity has been shown by others (Roemmelt *et al.*, 1988; Dietl, 1989; Elstner *et al.*, 1989; Nifontova & Kulikov, 1989; Reisinger, 1990; Tikhomirov *et al.*, 1990). *Paxillus involutus* (Batsch: Fr) Fr. and *Agaricus silvaticus* Schaeffer, had the highest levels of contamination; relatively low concentrations were found in tubular species, *Cantharellus cibarius* Fr., and species of *Lactarius* DC. (Schehglov *et al.*, 1989).

Tikhomirov *et al.*, (1990), when estimating the accumulating capacity of 30 species of fungi collected in a 30 km zone of the Chernobyl Atomic Power Station, found that *Clitocybe* (Fr.) Kummer species and *P. involutus* have the highest capacity for accumulation, and *Amanita phalloides* (Vaill.: Fr.) Link and *A. muscaria* (Linn.: Fr.) Hooker the lowest. Haselwandter & Berreck (1989 b) found that accumulation of ^{137}Cs by the fruitbodies of edible species was greater in wild *Cantharellus cibarius* than in cultivated *Agaricus bisporus* (J. Lange) Imbach, *Pleurotus ostreatus* (Jacq.: Fr) Kummer. *C. cibarius*, *Amanita rubescens* ([Pers.] Fr.) S.F. Gray and *Boletus edulis* accumulated amounts below the maximum permissible norms for European Economic Community (6Bq ^{137}Cs g^{-1} of dry weight).

Federov *et al.* (1989) determined the coefficients of radionuclide accumulation (C_a) in the fruitbodies of higher edible fungi in the territory of Belarus. The accumulation coefficient varied considerably from 0.06 to 56.5. Among widely gathered species of fungi, comparatively high C_a of radiocaesium were found for *Gyromitra esculenta* (Pers.) Fr., *Armillaria mellea* (Vahl: Fr.) Kummer — (0.6–1.0), and average levels for *Boletus edulis*, *Leccinum aurantiacum* [Bull.] S.F. Gray, and *L. scabrum* (Bull.: Fr.) S.F. Gray. *Agaricus campestris* L.: Fr. had 1.3–3.4, while *Lactarius deliciosus* (L.: Fr.) S.F. Gray, *L. torminosus* (Schaeff.: Fr.) S.F. Gray, *L. turpis* (Weinm.) Fr. were high — 5.8–13.8. Very high levels were noted for *Rozites caperatus* (Pers.: Fr.) Karsten, *Suillus luteus*, *Xerocomus badius* (Fr.) Gilb., *Lactarius rufus* (Scop.: Fr) Fr. between 31.5–56.5). The last group of fungi could be used as bio-indicators for environment contamination by radionuclides.

Ecological confinement apparently affects the degree of radionuclide accumulation, thus accumulation by parasites < saprotrophs < mycosymbiotrophs (Roemmelt *et al.*, 1988; Fedorov, *et al.*, 1989; Lambinon, 1989; Oolbekkink & Kuyper, 1989; Reisinger, 1990). Caesium accumulation also depends on time and place of gathering (Meijer *et al.*, 1988). Further, there is heterogeneity of radionuclide distribution in the fungal fruit body (Ernst & Van Rooij, 1987; Haselwandter *et al.*, 1988, Dietl, 1989).

The existence of various chemical forms of radionuclides, superposition of radioactive sediments at the same points, heterogeneity of their distribution in the ecosystems, migration, weather/climatic conditions, mosaic pattern of natural landscapes of the Ukraine, and a number of the other factors complicate the prediction of isotope redistribution and accumulation in macromycetes.

The following work aimed to study radionuclide accumulation by macromycetes of various regions of the Ukraine, distinguishing the indicator species and estimating the possibility of using wild edible fungi as food.

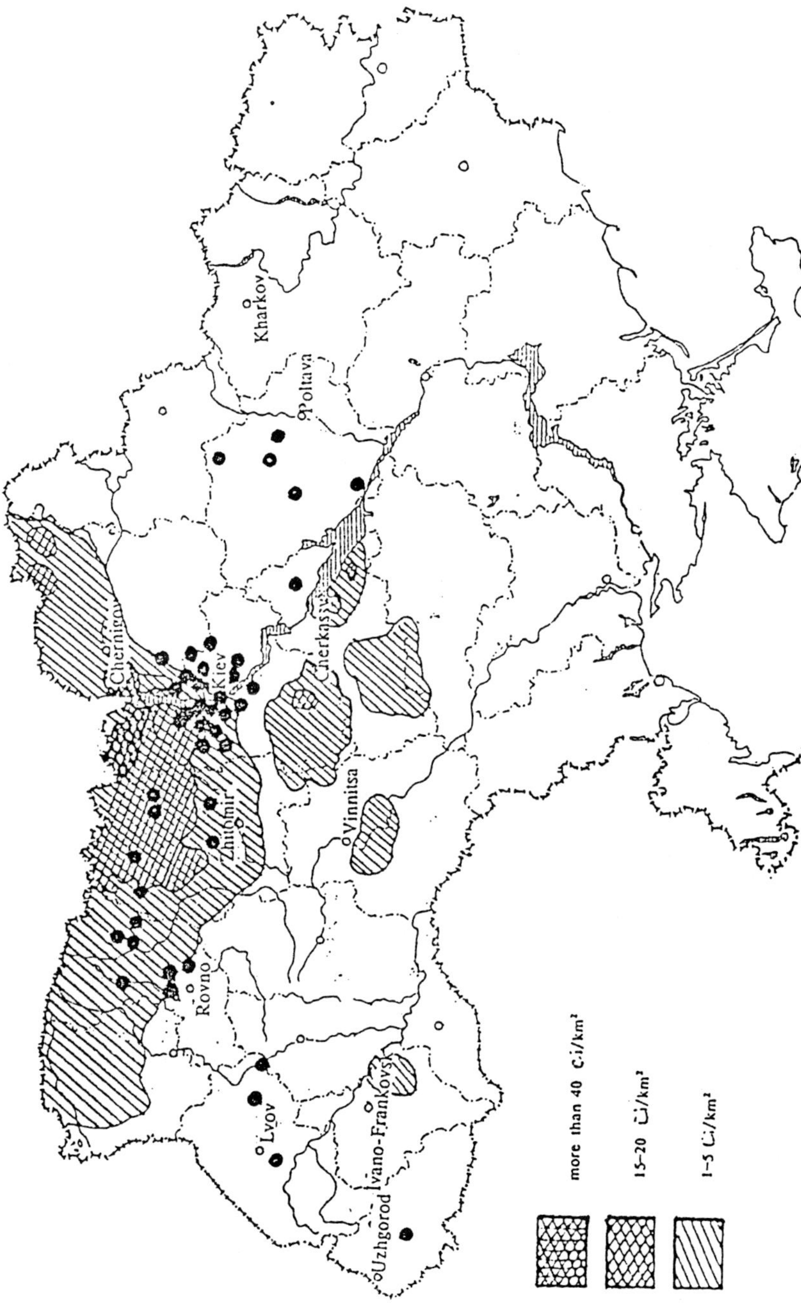

FIG. 1. Map of radioactive pollution of the Ukraine Territory and mushroom collection sites.

TABLE 1. Fungi species collected for radiological analysis in 1990–1991

A s c o m y c e t e s	
Rhizina inflata (Schaeff.) Karst.	litter saprotroph
B a s i d i o m y c e t e s	
Gasteromycetes	
Scleroderma aurantium Pers.	mycosymbiotroph
Scleroderma verrucosum Pers.	-"-
Lycoperdon perlatum Pers.	humus saprotroph
Aphyllophorales	
Laetiporus sulfureus (Bull.: Fr.) Murr.	lignotroph
Coltricia perennis (L.: Fr.) Murr.	litter saprotroph
Cantharellus cibarius Fr.	mycosymbiotroph
Piptoporus betulinus (Bull.: Fr) Karst.	lignotroph
Ramaria aurea (Fr.) Quel.	-"-
Agaricales s.l.	
Strobilomycetaceae	
Porphyrellus porphyrosporus (Fr.) Gill.	mycosymbiotroph
Boletaceae	
Boletus edulis Bull.: Fr	-"-
Boletus luridus Schaeff.: Fr	-"-
Leccinum aurantiacum (Bull.: St. Am.) S.F. Gray	-"-
Leccinum scabrum (Bull.: Fr) S.F. Gray	-"-
Suillus luteus (L.:Fr) S.F. Gray	-"-
Xerocomus badius (Fr.) Kuhn. ex Gilb.	-"-
Xerocomus chrysenteron (Bull.: St. Am.) Quel.	-"-
Paxillaceae	
Paxillus atrotomentosus (Batsch.: Fr	lignotroph
Paxillus involutus (Batsch.: Fr.) Fr.	mycosymbiotroph
Tricholomataceae	
Armillaria mellea (Vahl. *Fr.*) Kummer	lignotroph
Clitocybe gibba (Pers.: Fr.) Kumm.	litter saprotroph
Collybia butyraceae (Bull.: Fr.) Quel.	humus saprotroph
Collybia dryophila (Bull.: Fr.) Kumm.	litter saprotroph
Collybia maculata (Alb. et Schw.: Fr.) Quel.	humus saprotroph
Collybia peronata (Bolt. *Fr.*) Sing.	-"-
Laccaria laccata (Scot.: Fr.) Cke	-"-
Lyophyllum connatum (Schum.:Fr.) Sing.	-"-
Marasmius oreades (Bolt.: Fr) Fr.	-"-
Tricholoma saponaceum (Fr.) Kumm.	mycosymbiotroph
Tricholomopsis rutilans (Schaeff.: Fr.) Sing.	lignotroph

Materials and Methods

Macromycetes and soil samples (Fig. 1, Table 1) were gathered from 44 sites in summer and autumn of 1990–1991. One species, *Stropharia rugosoannulata* Farlow, was artificially grown on wheat straw substrate (taken from the Gadyach district of the Poltava which region was relatively uncontaminated). Fungi and soil samples were air dried (maximum 50 g).

Fungal samples were ground to a fine-dispersed state; all samples were analyzed by the standard technique of gamma-spectrometry used at the Institute of Nuclear Studies of the Academy of Sciences of the Ukraine. The method consists of detector fixation of γ-particles of the samples. Fungi and soil were analyzed using a multichannel gamma-analyzer AFORA LP 4900 B. The equipment resolution was 3 KeV.

Results and Discussion

Accumulated ^{137}Cs activity in 1990 was in the order: *Paxillus atrotomentosus* > *Amanita solitaria* > *Pholiota adiposa* > *Russula obscura* > *Boletus edulis* > *Suillus luteus* > *Laccaria laccata* > *Clitocybe gibba* > *Leccinus scabrum* > *Russula flava* > *Amanitopsis fulva* > *Cantharellus cibarius* > *Russula virescens* > *Coltricia perennis* > *Hypholoma sublateritium* > *Collybia dryophila* > *Ramaria aurea* > *Hypholoma fasciculare* > *Lycoperdon perlatum* > *Leccinum aurantiacum* > *Collybia peronata* > *Piptoporus betulinus* > *Russula cyanoxantha* (Table 2). The order for ^{134}Cs activity was slightly different. Radiocaesium activity did not exceed the background values in the species gathered in relatively uncontaminated regions: *Macrolepiota procera, Armillaria mellea, Xerocomus badius, Lycoperdon perlatum, Lactarius deliciosus, Tricholoma saponeum* (Poltava region, Gadyach district, vil. of Krasnaya Luka), *Lactarius pubescens* (Transcarpatian region, Mezhigorsky district, vil. of Kolochava, biological base of Uzhgorod University), *Armillaria mellea, Xerocomus badius, Porphyrellus porphyrosporus* (Lvov region, Skolevsky district, vil. of Smozhe), *Russula roseipes, R. aeruginea* (Lvov region, Stryi district, Lubenets Forestry). Caesium isotopes (^{134}Cs/604.7 KeV/ — 1384, ^{134}Cs/795.8KeV/ — 1519 and ^{137}Cs — 10000Bq/kg of dry weight) were found only in *Psathyrella candolleana* (Fr.) Maire (Lvov region, settl. of Kozevo).

Average content of ^{137}Cs was one or two orders of magnitude higher in the species which accumulate caesium than in the soil from the sampling site, and was variable within species. Thus, ^{137}Cs activity in the carpophores of *Amanita solitaria* was 526 times greater than soil, 243 times for *Paxillus atrotomentosus*, 131–300 times for *Suillus luteus*, and 3.7–55.7 times for *Boletus edulis.*

Though soils of the studied regions were contaminated, in 1991 by a complex of isotopes, which included ^{226,228}Ra, ^{144}Ce, ^{125}Sb, ^{106}Ru, ^{154}Eu (Table 3), only ^{134}Cs, ^{137}Cs and ^{40}K (1460.8KeV) were found in most fungal species analyzed. Relatively low levels of radium isotopes were fixed only in four specimens: *Piptoporus betulinus* (^{228}Ra, 583.1KeV, 31Bq kg^{-1} dry weight), *Pholiota destruens* (^{226}Ra, 351.9KeV, 22Bq kg^{-1}), *Marasmius oreades* (^{226}Ra, 609.3KeV, 38Bq kg^{-1}), and *Agaricus vaporarius* (^{144}Ce, 133.5KeV, 38Bq kg^{-1} dry weight). Reliability of these data is, however, doubtful because there are measurement errors and average values are close to the background level. Overall, concentrations of ^{134}Cs were similar in 1990 and 1991 samples (604.7 and 795.8KeV respectively) but ^{137}Cs was one order of magnitude higher. The activity of ^{134}Cs decreased in 1991 relative to ^{134}Cs due apparently to radioactive decay. Activity of ^{137}Cs in the soil samples ranged between 90Bq kg^{-1} (Feofania, Kiev) and 7870Bq kg^{-1} (near the settlement of Klavdievo-Tarasovo). The activity of ^{40}K in soil samples in 1991 ranged between 109Bq kg^{-1} (Borodyanka region, Khutor Poroskotel) and 692 Bq kg^{-1} (Feofania, Kiev). Minimum activity of ^{134}Cs (604.7 and 795.8KeV) was in wood inhabiting fungi (*Ph. squarrosa* – 13 and 9Bq kg^{-1} dry weight, *Hypholoma sublateritium* – 75 and 47 Bq kg^{-1}, no detectable activity in *Stropharia rugosoannulata, Coltricia perennis, Hypholoma fasciculare, Macrolepiota rhacodes* and

TABLE 2. Radioactivity of fungi and soils (1990)

Settlement	Fungus soil	^{134}Cs (E=604.7KeV) Bq kg^{-1}d.w.	C_a	^{134}Cs (E=795.8KeV) Bq kg^{-1}d.w.	C_a	^{137}Cs E=661.6KeV) Bq kg^{-1}d.w.	C_a	^{134}Cs:^{137}Cs
Kiev (Koncha Zaspa)	*Clitocybe gibba*	769	1.52	694	2.82	6167	1.52	0.237
	Laccaria laccata	968	1.91	1121	4.66	8156	2.01	0.266
	soil	507		246		4062		0.185
Kiev region Vasilkov dist. vil. of Malyutyanka	*Suillus luteus*	323		358		2743		0.248
	soil	BDL		BDL		BDL		
Makarov dist. vil. of Nokolaevka	*S. luteus*	1189		1225		8946	131.56	0.270
	soil	BDL		BDL		68		
Kiev-Svyatoshino region vil. of Ledevka	*Cantharellus cibarius*	549	1.97	561	1.25	4196	4.88	0.265
	C. cibarius	251	0.90	249	0.55	2078	2.42	0.241
	Piptoporus betulinus	BDL		BDL		152	0.18	BDL
	soil	279		450		859		0.849

TABLE 2 continued

Settlement	Fungus soil	^{134}Cs (E=604.7KeV) Bq kg^{-1}d.w.	C_a	^{134}Cs (E=795.8KeV) Bq kg^{-1}d.w.	C_a	^{137}Cs E=661.6KeV) Bq kg^{-1}d.w.	C_a	^{134}Cs:^{137}Cs
vil. of Pirnovo	*Leccinum scabrum*	2416		2689		18541		0.276
	Ramaria aurea	299		215		1887		0.272
Brovary dist. vil. of Kalita	*Russula cyanoxantha*	BDL		BDL		61		
Borodyanka dist. 5 km westwards town of Borodyanka	*Collybia dryophila*	120		85		704	4.07	0.291
	soil	BDL		BDL		173		
Near the settl. Klavdievo-Tarasovo	*C. dryophyla*	57	1.73	68	451	4.75	0.277	
	soil	33		BDL		95		0.347
Zhitomir region Radomyshl dist. vil. of Kocherov	*Hypholoma fasciculare*	99	7.62	126	14.0	967	12.89	0.233
	soil	13		9		75		0.293

TABLE 2 continued

Settlement	Fungus soil	^{134}Cs (E=604.7KeV) Bq kg^{-1}d.w.	C_a	^{134}Cs (E=795.8KeV) Bq kg^{-1}d.w.	C_a	^{137}Cs E=661.6KeV) Bq kg^{-1}d.w.	C_a	^{134}Cs:^{137}Cs
Korostyshev dist. 5 km eastwards town of Korostyshev	*Russula flava*	945	52.50	868	41.33	7143	39.46	0.254
	soil	18		21		181		0.215
	Collybia peronata	68		80		289	4.31	0.512
vicinity of t. of Krasnoarmeisk	*Amanita solitaria*	5036		6190		35263	526.31	0.318
	soil	BDL		BDL		67		
Korosten dist. 10 km eastwards town of Korosten near the route Korosten-Malin	*Collybia dryophila*	380	11.88	463	12.86	3019	9.80	0.279
	soil	32		36		308		0.221
Malin dist. vil. of Pirozhki	*Leccinum scabrum*	378	2.05	330	1.58	2394	1.85	0.296
	soil	184		209		1291		0.304

TABLE 2 continued

Settlement	Fungus soil	^{134}Cs (E=604.7KeV) Bq kg^{-1}d.w.	C_a	^{134}Cs (E=795.8KeV) Bq kg^{-1}d.w.	C_a	^{137}Cs E=661.6KeV) Bq kg^{-1}d.w.	C_a	^{134}Cs:^{137}Cs
near the Ubort river	*Boletus edulis*	2515		3537	73.69	21400	55.73	0.283
	soil	BDL		48		384		0.125
Luginsk dist. vil. of Letki	*Russula virescens*	361	1.25	490	1.54	3033	1.25	0.281
	R. flava	182	0.63	205	0.64	1707	0.70	0.226
	soil	288		319		2428		0.250
Olevsk dist. vicinity of vil. of Belokorovichi	*Suillus luteus*	1345		1155		10505	300.14	0.238
	Leccinum scabrum	209		350		1531	43.74	0.365
	L. aurantiacum	99		84		558	15.94	
	soil	BDL		BDL		35		0.328
Rovno region Kostopol dist. 15 km southwards t. of Kostopol	*Lycoperdon perlatum*	207	9.41	134	4.6	592	3.3	0.576

TABLE 2 continued

Settlement	Fungus soil	^{134}Cs (E=604.7KeV) Bq kg^{-1}d.w.	C_a	^{134}Cs (E=795.8KeV) Bq kg^{-1}d.w.	C_a	^{137}Cs E=661.6KeV) Bq kg^{-1}d.w.	C_a	^{134}Cs:^{137}Cs
	Ramaria aurea	BDL		BDL		99	0.9	
	Suillus luteus	3885	176.69	4325	149.1	27888	154.1	0.294
	Paxillus atrotomentosus	6592	299.64	6706	231.2	44012	243.2	0.302
	soil	22		29		181		0.282
5 km southwards of t. of Kostopol	*Amanitopsis fulva*	480	3.90	571	4.43	3537	3.47	0.297
	Boletus edulis	1607	13.07	1555	12.05	11153	10.96	0.284
	soil	123		129		1018		0.248
10 km northwards Kostopol	*Laccinum scabrum*	263	8.48	285	9.50	1866	8.80	0.294
	soil	31		30		212		0.288
Sarny dist. vicinity of Sarny	*Rusula obscura*	1864		2057		14110	131.87	0.278
	A. fulva	571		543	4176	39.03	0.267	
	soil	BDL		BDL		107		

TABLE 2 continued

Settlement	Fungus soil	^{134}Cs (E=604.7KeV) Bq kg^{-1}d.w.	C_a	^{134}Cs (E=795.8KeV) Bq kg^{-1}d.w.	C_a	^{137}Cs E=661.6KeV) Bq kg^{-1}d.w.	C_a	^{134}Cs:^{137}Cs
2 km eastwards of Sarny	*Boletus edulis*	662	2.97	878	3.70	6489	3.76	0.297
	Hypholoma fasciculare	310	1.39	206	0.87	1702	0.99	0.303
	soil	223		237		1725		
settl. of Klesov	*Coltricia perennis*	281	1.90	318	2.15	2249	2.14	0.266
	Pholiota adiposa	1971	13.32	2697	18.22	17360	16.50	0.269
	soil	148		148		1052		0.281

BDL — Below detection limit

Psathyrella candolleana). Activity of ^{137}Cs varied considerably between fungal samples, from a minimum in the cultivated species *S. rugosoannulata* (16Bq kg^{-1} oven dry weight) to maximum values in *Lactarius rufus* — 44904, *Amanita citrina* — 35080, *Xerocomus badius* —19574, *Scleroderma verrucosum* — 15190, *Suillus luteus* — 11900, *Xerocomus chrysenteron* —10749. It is of interest that this second group of species coincides with species-indicators noted by Fedorov *et al.* (1989).

Accumulated ^{137}Cs activity in 1991 was as follows: *Lactarius rufus* > *Amanita citrina* > *Xerocomus badius* > *Scleroderma verrucosum* > *Amanitopsis vaginata* > *Suillus luteus* > *Xerocomus chrysenteron* > *Lactarius helvus* > *Russula flava* > *Boletus edulis* > *Lactarius turpis* > *Amanita rubescens* > *Lyophyllum connatum* > *Russula foetens* > *Cantharellus cibarius* > *Paxillus atrotomentosus* > *Scleroderma aurantium* > *Tricholomopsis rutilans* > *Collybia maculata* > *Russula obscura* > *Amanita pantherina* > *Clitocybe gibba* > *Piptoporus betulinus* > *Ramaria stricta* > *Collybia butyracea* > *Rhizina inflata* > *Lycoperdon perlatum* > *Coltricia perennis* > *Agaricus bitorquis* > *Boletus luridus* > *Laetiporus sulfureus* > *Hypholoma sublateritium* > *Leccinum aurantiacum* > *Hypholoma fasciculare* > *Psathyrella candolleana* > *Marasmius oreades* > *Pholiota squarrosa* > *Agaricus vaporarius* > *Macrolepiota rhacodes* > *Stropharia rugosoannulata.*

Clearly, both in 1990 and 1991, mycosymbiotrophic species accumulated more of radiocaesium activity than saprotrophs and lignotrophs, confirming previous findings that among the other species of *Boletaceae*, a comparatively low activity of ^{137}Cs was recorded in *Leccinum aurantiacum* for these two years (1866 Bq/kg, 1990, Zhitomir region, Kostopol district, and 609 Bq/kg of dry mass 1991, Kiev region, Borodyanka district, Khutor Poroskoten).

Caesium is similar to potassium in chemical nature, and thus the basic element (potassium) can often be substituted by caesium. This cannot explain the whole phenomenon of caesium accumulation by fungal carpophores. Elstner *et al.* (1987) suggest that there probably exists a system of absorption (so-called "ionophore") which separates K cations and Cs cations and accelerates only the process of K absorption (including ^{40}K). It is known that ^{40}K is the most common radioactive element in nature. At the same time potassium is a macroelement in the mineral composition of fruitbodies of macromycetes, its absorption (including ^{40}K) from the substrate being, therefore, inevitable. High contents of ^{40}K, parallel with that of radiocaesium, in macromycetes of a number of the Ukraine regions, increases the dose load on the organisms and on people using fungi. It should be noted that in 1991 the following levels of ^{40}K were registered in soils of the studied regions: minimum — 109 Bq kg^{-1} (Borodyanka district, Khutor Poroskoten), maximum — 692 Bq kg^{-1} (Feofania, Kiev).

There were considerable differences in the ratio of $^{134+137}Cs$: ^{40}K — from 0.02 to 38.18. If the hypothesis of Elstner *et al.* 1987 on the prevalence of the process of ^{40}K absorption is accepted, such differences in $^{134+137}Cs$ and ^{40}K accumulation may apparently be connected with the specific radioecological situation obtained in the sampling sites.

A maximum of ^{40}K activity was recorded in *Xerocomus chrysenteron* — 7004 (from 4948 to 10842 Bq kg^{-1} d.w.), *Amanita citrina* — 6799, *Coltricia perennis* — 5180 Bq kg^{-1}, and a minimum in *H. capnoides* — 660 Bq kg^{-1}, *Scleroderma aurantium* —641, *Collybia maculata* —585 of dry weight (Table 3). The following lie between these extremes *Amanita pantherina* > *Hypholoma sublateritium* > *Amanitopsis vaginata* > *Boletus luridus* > *Scleroderma verrucosum* > *Agaricus bitorquis* > *Paxillus atrotomentosus* > *Psathyrella condolleana* > *Leccinum aurantiacum* > *Stropharia rugosoannulata* > *Russula foetens* > *Suillus luteus* > *Boletus edulis* > *Agaricus vaporarius* > *Tricholomopsis rutilans* > *Macrolepiota rhacodes* > *Pholiota destruens* > *Clitocybe gibba* > *Ramaria stricta* > *Lactarius turpis* > *Xerocomus badius* > *Russula obscura* > *Lactarius rufus* > *Lyophyllum connatum* > *Marasmius oreades* > *Cantharellus cibarius* > *Hypholoma fasciculare* > *Lycoperdon perlatum* > *Pholiota squarrosa* > *Rhizina inflata* > *Collybia butyracea* > *Russula flava* > *Laetiporus sulfureus* > *L. helvus.*

The highest coefficients of accumulation (C_a) of ^{137}Cs from soil to fruitbodies were in *Lactarius rufus* – 184, *Amanita citrina* – 143.7, *Xerocomus badius* – 80. C_a of ^{40}K was highest in

TABLE 3. Radioactivity of fungi and soils (1991) in "the zone of Chernobyl Atomic Power Station effect"

Settlement	Fungus soil	^{134}Cs	Ca	^{134}Cs	C_a	^{137}Cs	C_a	^{40}K	Ca	$^{134+137}Cs/^{40}K$	$^{134}Cs/^{137}$
Kiev (Feofania)	*Macrolepiota rhacodes*	BDL		BDL		13	0.14	1454	2.10	0.009	-
	M. rhacodes	BDL		BDL		23	0.26	1858	2.68	0.012	-
	soil	BDL		BDL		90		692		0.130	-
Kiev (Koncha Zaspa)	*Agaricus vaporarius*$_1$	11		22		66	0.94	1759	10.86	0.056	0.5
	Laetiporus sulfureus	7		25		461	6.58	680	4.2	0.725	0.07
	Hypholoma sublateritium	75		47		717	10.24	2951	18.22	0.284	0.170
	soil^2	BDL		BDL		70		162		0.432	
Kiev (Nauka Prospect)	*Boletus luridus*	68		96		932		2520		0.435	0.176
	Pholiota destruens$_3$	BDL		BDL		94		1648		0.057	-
Kiev (Chkalov St.)	*Agaricus bitorquis*	124		93		1029		2223		0.561	0.211
	Psathyrella candolleana	BDL		BDL		147		2041		0.072	
	Stropharia rugosoannulata$_4$	BDL		BDL		16		1972		0.008	
Kiev Region Obukhov dist. settl. of Kozin	*Marasmius oreades*$_5$	BDL		BDL		100	3.45	1303	3.43	0.077	

TABLE 3 continued

Settlement	Fungus soil	^{134}Cs	Ca	^{134}Cs	C_a	^{137}Cs	C_a	^{40}K	Ca	$^{134+137}Cs/^{40}K$	$^{134}Cs/^{137}$
	Macrolepiota rhacodes	BDL		BDL		259	8.93	2735		0.095	
	M. procera	BDL		BDL		199	6.86	2111		0.094	
	Xerocomus badius	110		167		1838	63.38	1489		1.420	0.15
	soil[6]	BDL		BDL		29		380		0.076	
Borispol	*Pholiota squarrosa*	13		9		78	1.5	1142	3.97	0.088	0.282
	Agaricus silvaticus	BDL		BLD		10	0.19	5846	20.3	0.002	
	Tricholomopsis rutilans	32		32		415	7.98	1465	5.09	0.33	0.154
	soil[7]	BDL		BDL		52		288			
Borodyanka dist.											
Khutor Poroskoten	*Lactarius turpis*	779	33.87	861	35.88	8739	35.82	1384	12.70	7.499	0.188
	L. helvus	831	36.13	997	41.54	10285	42.15	832	7.63	14.559	0.178
	L. rufus	3698	160.78	3623	150.96	44904	184.03	1368	12.55	38.176	0.163
	Scleroderma aurantium	431	18.739	510	21.25	4759	19.50	641	5.88	13.573	0.198
	Laccinum aurantiacum	135	5.870	68	2.83	609	2.50	2039	18.71	0.398	0.333
	Boletus edulis	617	26.826	1012	42.17	9123	37.39	1789	16.41	6.010	0.179
	Xerocomus chrysenteron	51	2.217	107	4.46	1059	4.34	4948	45.39	0.25	0.149
	X. badius	1755	76.30	1977	82.38	19574	80.22	1377	12.63	16.92	0.191

TABLE 3 continued

Settlement	Fungus soil	^{134}Cs	Ca	^{134}Cs	C_a	^{137}Cs	C_a	^{40}K	Ca	$^{134+137}Cs/^{40}K$	$^{134}Cs/^{137}$
	Amanitopsis vaginata	976	42.43	949	39.54	12820	52.54	2625	24.08	5.617	0.150
	Amanita citrina	3629	157.78	2983	124.29	35080	143.77	6799	63.38	6.132	0.188
	Lyophyllum connatum	472	20.52	545	22.71	5669	23.23	1324	12.15	5.150	0.179
	Russula flava	1066	46.35	1373	57.21	12887	52.82	846	7.76	18.12	0.189
	soil[8]	23		24		244		109		2.670	0.193
t. of Brovary	*Cantharellus cibarius*	967		1009		10205		1394		8.739	0.194
Brovary dist.											
vil. of Letki	*C. cibarius*	305		236		2643		905		3.528	0.204
vil. of Zazamie	*Paxillus atrotomentosus*	353	1.71	356	1.76	3868	1.50	1258		3.638	0.183
	soil	207		202		2582		400		7.478	0.158
vill of Bozhny	*Laetiporus sulfureus*	41		44		443		1085		0.487	0.19
t. of Vyshgorod	*Ramaria stricta*	135	7.11	213	9.68	1867	7.91	2267	28.34	0.977	0.186
	soil	19		22		236		80		0.463	0.174

TABLE 3 continued

Settlement	Fungus soil	^{134}Cs	Ca	^{134}Cs	C_a	^{137}Cs	C_a	^{40}K	Ca	$^{134+137}Cs/^{40}K$	$^{134}Cs/^{137}$
Vyshgorod dist.											
vil. of Lyutezh	*Amanita pantherina*	361	2.15	348	1.77	3162	1.69	3216	29.24	1.204	0.224
	A. rubescens	520	3.10	426	2.16	5677	3.03	5310	48.27	1.247	0.167
	Suillus luteus	902	5.37	1142	5.80	11900	6.35	1789	16.26	7.794	0.172
	Xerocomus chrysenteron	1465	8.72	2027	10.29	20054	10.71	10842	98.56	2.172	0.174
	Scleroderma verrucosum	860	5.12	1011	5.13	15190	8.11	2388	21.71	7.144	0.123
	Paxillus atrotomentosus	353	2.10	356	1.81	3868	2.07	1258	11.44	3.638	0.183
	Coltricia perennis	-	-	-	-	1099	0.59	5180	47.09	0.212	-
	Clitocybe gibba	284	1.69	312	1.58	3397	1.81	780	7.01	5.119	0.175
	Lycoperdon perlatum	87	0.52	141	0.72	1107	0.59	1161	10.55	1.150	0.206
	Russula foetens	420	2.50	497	2.52	5397	2.88	1833	16.66	3.445	0.170
	R. obscura	256	1.52	254	1.29	3199	1.71	1369	12.45	2.709	0.159
	Rhizina inflata	108	0.64	117	0.59	1276	0.68	1113	10.12	1.349	0.176
	Ramaria stricta	102	0.61	139	0.71	1368	0.73	576	5.24	2.793	0.176
	Tricholoma rutilans	978	5.82	1248	6.34	11598	6.19	1771	16.10	7.806	0.192
	soil[11]	168		197		1873		110		20.345	0.195
Kiev-Svyatoshino dist.											
vil. of Gurovshchina	*Hypholoma fasciculare*	BDL		BDL		167	2.78	1166	2.30	0.143	
	soil[12]	BDL		BDL		66		506	0.130		

TABLE 3 continued

Settlement	Fungus soil	^{134}Cs	Ca	^{134}Cs	C_a	^{137}Cs	C_a	^{40}K	Ca	$^{134+137}Cs/^{40}K$	$^{134}Cs/^{137}$
Chernigov region Bobrovitsy dist. vil. of Kobyzhcha											
	Agaricus campestris	BDL		BDL		7	0.1	2845	4.65	0.002	
	Collybia maculata	310		336		3306	47.29	585	0.96	6.756	0.195
	C. butyracea	151		86		1364	19.49	1082	1.78	1.480	0.174
	Hypholoma capnoides	149		85		1155	16.5	660	1.08	2.105	0.203
	Tricholomopsis rutilans	113		113		1154	16.49	1781	2.91	0.775	0.196
	soil	BDL		BDL		70			612		0.114
Rovno region Kostopos dist. vil. of Rakitnoe	*Cantharellus cibarius*	277		248		2926		1413		2.44	
Zhitomir region Luginsky dist. vil. of Letki	*Piptoporus betulinus*	192		185		1947					

TABLE 3 continued

BDL — Below detection limit

1 — the sample also includes ^{144}Ce (133.5 KeV) – 38 Bq/kg d.w.;
2 — the sample also includes ^{226}Ra (295.2 KeV) – 3 Bq/kg, ^{228}Ra (583.1 KeV) – 15 Bq/kg, ^{226}Ra (351.3 KeV) – 30 Bq/kg d.w.;
3 — the sample also includes ^{226}Ra (351.9 KeV) – 22 Bq/kg d.w.;
4 — cultivated species;
5 — the sample includes ^{226}Ra (609.3 KeV – 48 Bq/kg d.w.;
6 — the sample includes ^{226}Ra (351.9) – 21 Bq/kg, ^{228}Ra (583.1 KeV) – 21 Bq/kg ^{228}Ra (238.6 KeV – 5 Bq/kg, ^{226}Ra (609.3 KeV) – 6 Bq/kg d.w.;
7 — the sample includes ^{228}Ra (238.6 KeV) – 19 Bq/kg, ^{226}Ra (351.9 KeV) - 38 Bq/kg, ^{226}Ra (609.3 KeV) – 14 Bq/kg d.w.;
8 — the sample includes ^{144}Ce (133.5 KeV) – 20 Bq/kg, ^{226}Ra (351.9 KeV) – 3.3 Bq/kg d.w.;
9 — the sample includes ^{226}Ra (351.9 KeV) – 110 Bq/kg, ^{125}Sb (427.9 KeV) – 285 Bq/kg, ^{125}Sb (635.9 KeV) – 417 Bq/kg, ^{106}Ru (621.8 KeV) – 686 Bq/kg d.w.;
10 — the sample includes ^{144}Ce (133.5 KeV) – 366 Bq/kg, ^{106}Ru (621.8 KeV) – 554 Bq/kg, 154EU (127 KeV) – 70 Bq/kg, ^{154}Eu (1211 KeV) – 26 Bq/kg, ^{125}Sb (427.9 KeV) – 152 Bq/kg, ^{154}EU (722 KeV) – 95 Bq/kg d.w.;
11 — the sample includes ^{144}Ce (133.5 KeV) – 89 Bq/kg d.w.;
12 — the sample includes ^{226}Ra (295.2 KeV) – 30 Bq/kg, ^{226}Ra (351.9 KeV) – 8 Bq/kg, ^{228}Ra (238.6 KeV) – 9 Bq/kg d.w.;
13 — the sample includes ^{226}Ra (295.2 KeV) – 71 Bq/kg d.w.;
14 — the sample includes ^{228}Ra (238.6 KeV) – 13 Bq/kg, ^{226}Ra (351.9 KeV) – 11 Bq/kg, ^{228}Ra (583.1 KeV) – 46 Bq/kg, ^{226}Ra (609.3) – 12 Bq/kg d.w.;
15 — the sample includes ^{228}Ra (583.1 KeV) – 31 Bq/kg d.w.

TABLE 4. Radioactivity of fungi and soils outside the zone of Chernobyl Atomic Power Station effect

Settlement	Fungus soil	^{134}Cs Bq/kg d.w.	Ca	^{134}Cs Bq/kg d.w.	C_a	^{137}Cs Bq/kg d.w.	C_a	^{40}K bq/kg d.w.	Ca	$^{134+137}Cs/^{40}K$	$^{134}Cs/^{137}$
Poltava region Shishatsky dist. vil. of Yareski	*Suillus luteus*	62	7.75	60	7.50	881	8.99	1226	6.52	0.818	0.138
	Macrolepiota procera	BDL1		BDL		21	0.21	1580	8.40	0.013	
	Tricholomopsis rutilans	16	2.0	11	1.38	146	1.49	1093	5.81	0.158	0.185
	soil[2]	8		8		98		188		0.606	0.163
vil. of Kovankovka	*S. luteus*	25		36		764	9.32	1302	4.13	0.634	0.080
	Xerocomus badius	13		21		311	3.79	1448	4.60	0.238	0.109
	M. procera	BDL		BDL		16	0.20	1226	3.89	0.013	
	Marasmius oreades	BDL		BDL		12	0.15	2122	6.74	0.006	
	T. rutilans	8		19		203	2.48	1105	3.51	0.208	0.133
	soil[3]	BDL		BDL		82		315		0.260	
Kremenchug	*S. luteus*	128		127		1450	53.70	1167	2.98	1.460	0.176
	M. rhacodes	5		BDL		20	0.74	1262	3.28	0.020	0.25
	M. oreades	6		BDL		29	1.07	1416	3.62	0.025	0.207
	soil[4]	BDL		1		27		391		0.072	0.037

TABLE 4 continued

Settlement	Fungus soil	^{134}Cs Bq/kg d.w.	Ca	^{134}Cs Bq/kg d.w.	C_a	^{137}Cs Bq/kg d.w.	C_a	^{40}K bq/kg d.w.	Ca	$^{134+137}Cs/^{40}K$	$^{134}Cs/^{137}$
Cherkassy region Zolotonosha dist. vil. of Korobovka	*X. badius*	43		62		797	199.25	1393	4.10	0.648	0.132
	M. procera	12		13		61	15.25	692	2.04	0.124	0.410
	T. rutilans	BDL		BDL		122	30.50	1402	4.12	0.087	
	soil[5]	BDL		BDL		4		340		0.012	

1 — BDL — Below detection limit;
2 — soil sample also induced ^{226}Ra (351.9 KeV) – 13 Bq/kg, ^{226}Ra (295.2 KeV) – 20 Bq/kg, ^{226}Ra (609.3 KeV) – 17 Bq/kg, ^{228}Ra (238.6 KeV) – 7 Bq/kg, ^{228}Ra (583.1 KeV) – 14 Bq/kg of dry weight;
3 — soil sample included ^{226}Ra (295.2 KeV) – 29 Bq/kg, ^{228}Ra (238.6 KeV) – 9 Bq/kg of dry weight;
4 — soil sample included ^{226}Ra (361.6 KeV) – 7 Bq/kg, ^{226}Ra (609.3 KeV) – 8 Bq/kg of dry weight;
5 — soil sample included ^{226}Ra (609.3 KeV) – 7 Bq/kg, ^{228}Ra (238.6 KeV) – 11 Bq/kg of dry weight.

X. chrysenteron – 98.6, *A. citrina* – 62.4, and *A. rubescens* – 48.3. Considerable variations of C_a for the same species gathered from different habitats were observed. The highest C_a of ^{137}Cs and ^{40}K were on average found in samples gathered in the Borodyanka district of Kiev.

Much lower levels of radiocaesium accumulation were found in carpophores from relatively uncontaminated regions where, according to data of the Ministry of Forestry of the Ukraine, the activity of ^{137}Cs does not exceed 1 Ci km-2 and the use of forestry products, including fungi, is not limited (Table 4). C_a of ^{137}Cs in *X. badius* was within the limits of 3.79 and 199; in *Suillus luteus* – 8.99 to 53.7; in *Macrolepiota procera* – 0.20 to 15.25; *Tricholomopsis rutilans* – 1.49 to 30.50; *Marasmius oreades* – 0.15 to 1.07; *M. rhacodes* – 0.74. Accumulation coefficients of ^{40}K in these species were as follows: in *X. badius* – 4.1 to 4.6; *S. luteus* – 2.98 to 6.5; *M. procera* – 2.04 to 8.4; *M. rhacodes* – 3.2; *M. oreades* – 3.6 to 6.7; and *Tricholomopsis rutilans* – 3.5 to 5.8. *S. luteus* and *X. badius* (*Boletaceae*) accumulated, overall, greater amounts of radiocaesium, but activity did not exceed maximum permissible levels (6000 Bq kg^{-1} d.w.), which allows the use of these edible fungi. However, radiocaesium in samples from Kiev, Zhitomir and Rovno regions analyzed in 1990–1991 varied considerably and often exceeded maximum permissible levels.

It is worth noting that mycosymbiotrophic species, including many *Boletaceae*, *Amanitaceae*, *Russulaceae*, accumulate caesium isotopes. These species can be considered indicators of contamination of territories as the result of the accident at Chernobyl Atomic Power Station. With regard to toxic effect on humans, the joint effect of accumulated radioisotops and heavy metals must be considered. To be certain that food is not contaminated, the production of cultivated edible fungi on previously checked substrates is especially urgent for the Ukraine.

Summary

1. Radiological analysis of 63 species of macromycetes gathered in different regions of the Ukraine has shown that fungi are accumulators of caesium and potassium isotopes.

2. The degree of accumulation depends on several factors including the level and character of radioactive contamination of soils, ecological confinement and systematic position of the species.

3. Species-indicators of radioactive contamination of the territories are mycosymbiotrophs from families *Boletaceae*, *Amanitaceae* and *Russulaceae*.

4. Levels of radioactive contamination of wild edible fungi were close to or often exceeded permissible doses, and thus it is expedient to stop gathering fungi in a number of regions (Zhitomir, Kiev, and Rovno) which were contaminated by the accident at Chernobyl Atomic Power Station.

References

Bakken, L.R. and Olsen, R.A. (1990). Accumulation of radiocaesium in fungi. *Canad. J. Microbiol.* **36**: 74–710.

Dietl, G. (1989). Verteilung radioactiver Casiumnuklide im Pilzfruchtkor. *Z. Mycol.* **55**: 131–134.

Elstner, E.F., Fink, R., Holl, W., Lengfelder, E. and Ziegler, H. (1987). Natural and Chernobyl-caused radioactivity in mushrooms, mosses and soil-samples of defined biotops in SW Bavaria. *Oecologia* **73**: 553–558.

Elstner, E.F., Fink, R., Holl, W., Lengfelder, E. and Ziegler, M. (1989). Radioactivity in mushrooms, mosses and samples of defined biotops in SW Bavaria — two years after Chernobyl. *Oecologia* 80: 173–177.

Ernst, W.H.O. and Van Rooij, L.F. (1987). $^{134/137}Cs$ fall-out from Chernobyl in Dutch forests. In. S.E. Lindberg and T.C. Hutchinson, (editors). Heavy metals in the environment, pp. 284–286. CEP Consultants Ltd., Edinburgh.
Federov, V.N., Mironov, V.P. and Makarevich, V.I. (1989). Study of radionuclides accumulation in the fruitbodies of edible fungi in Belorussian SSR. In. First All-Union Radiobiol. Congr. Moscow, August 21–27, 1989. **2**, pp. 540–542.
Fraiture, A., Guillite, O. and Lambinon, J. (1990). Interest of fungi as bioindicators of the radiocontamination in forest ecosystems. In G. Desmet, P. Nassimbeni and M. Belli, M. (editors). Transfer of radionuclides in natural and semi-natural environments. pp. 477–484. Elsevier applied Science.
Gans, I. (1986). Radionuklidkonzentrationen in Berliner Pilzen. *Z. Mycol.* **52**: 446–454.
Gans, I. (1987). Radionuklidkonzentrationen in Berliner Pilzen. *Z. Mycol.* **53**: 151–154.
Haselwandter, K. and Berreck, M. (1989a). Radiocaesium accumulation in fruitbodies of Basidiomycetes collected in the province of Parma, Italy. In. Fungi atque loci Natura, Atti del IV Convegno Intern. di Micol. del 27–30 IX. 1987., Borgo Val di Taro, Italy, pp. 89–92.
Haselwandter, K. and Berreck, M. (1989b). Accumulation of 137Cs in fruitbodies of edible fungi — a comparison between wild and cultivated mushrooms. *Mushrooms Sci.* **12**(2) (Proc. of the Twelfth Intern. Congr. of the Science and Cultivation of Edible Fungi. Braunschweig, Germany): 587–592.
Haselwandter, K., Berreck, M. and Brunner, P. (1988). Fungi as bioindicators of radiocaesium contamination: pre- and post-Chernobyl activities. *Trans. Brit. Mycol. Soc.* **90**: 171–174.
Lambinon, J. (1989). Les champignons de la pollution atmospherique. *Sci. & Cult.* **34**: 12–16.
Meijer, R.J., Aldenkamp, F.J. and Jansen, A.E. (1988). Resorption of caesium radionuclides by various fungi. *Oecologia* **77**: 268–272.
Nifontova, M.G. and Kulikov N.V. (1989). Radioecology of spore plants. In. First All-Union Radiobiol. Congress. August, 21–27. 1989. Pushchino. **2**, pp. 487–488.
Oolbekkink, G.T. and Kuyper, T.W. (1989). Radioactive caesium from Chernobyl in fungi. *Mycologist* **3**: 3–6.
Reisinger, A. (1990). Correlation between radiocaesium activity of macrofungi and their systematic and ecological positions. In. Fourth Intern. Mycol. Congr., 28th August–3rd September, 1990. Regensburg. Germany. p. 339.
Ritter, G. (1991). Zur Radioaktivitat in Speisepilzen. *Micolog. Mitteilungsbl.* **34**: 24–29.
Roemmelt, R., Hiersche, L. and Wirth, E. (1988). Uptake of $^{134+137}Cs$ by higher fungi (Basidiomycetes) in terrestrial ecosystems. Biological aspects of the caesium transfer. 4-e Symp. Int. Radioecol. Cadarache "Impact accident orig. nucl. environ.", March, 14–18 1988, Cadarache. **1**, pp. D151–D161.
Shcheglov, A.I., Tikhomirov, F.A. Sidorov, V.P., Markin, M.V. Tchumak, V.K. and Ejdlin, V.Z. (1988). Radionuclides intake by forestry products under radioactive contamination. 4-e Symp. Int. Radioecol. Cadarache "Impact accident orig. nucl. environ.", March, 14–18 1988, Cadarache **2**, pp. 555-557.
Tikhomirov, F.A., Shcheglov, A.I. and Tsvetnova, O.B. (1990). Fungi as bioindicators of ^{137}Cs accessibility in the soils of the zone of radioactive contamination. Third All-Union Radiobiol. Conference on Agricult. Radiol., July, 2–7, 1990, Obninsk **1**, pp. 45–46.
Zhuravlev, V.F. (1990). Toxicology of radioactive substances. 336 pp. Energoatomizdat, Moscow.

E. Arnolds & B. de Vries (1993). Conservation of Fungi in Europe. In D.N. Pegler, L. Boddy, B. Ing & P.M. Kirk (Editors). Fungi of Europe: Investigation, Recording and Conservation, pp. 211–230. Royal Botanic Gardens, Kew.

17. CONSERVATION OF FUNGI IN EUROPE

E. ARNOLDS & B. de VRIES[1]

Biological Station, Centre for Soil Ecology, Kampsweg 27, 9418 PD Wijster, The Netherlands

Abstract

A concise survey is given of historical development of fungal conservation in Europe, including the role of the European Council for Conservation of Fungi. Main problems concerning fungal conservation are discussed at the hand of an analysis of Red Lists of threatened macrofungi from 11 European countries. Data from southern Europe are completely lacking. Considerable differences exist concerning numbers of species and proportions of various taxonomic groups, which are mainly caused by differences in criteria, taxonomic knowledge and consulted specialists. Groups with large, appealing sporocarps are overrepresented. The proportions of several niche-substrate groups of fungi in Red Lists of different countries are analyzed. The proportions of wood-inhabiting fungi, ectomycorrhizal species and grassland-inhabiting saprotrophs are strongly variable and related to the main problems concerning environmental conservation in these countries. It is striking that the numbers of basiphytic fungi are higher than those of acidophytic fungi in all countries. Almost half of the species is listed in only one of the 11 lists and only 13 species are included in 8 or 9 lists. Examples of species with high and low frequencies are presented. Their status as threatened species at a European level appears to be variable. The interpretation of many data is hampered by the lack of accurate data on distribution and ecology. It is suggested that far more research is needed on these and other topics before a reliable picture of threatened fungi in Europe can be given. It is concluded that all efforts in the scope of fungal conservation had some positive local effects, but that nevertheless the overall situation for the mycoflora is rapidly deteriorating in large parts of Europe.

Introduction

The topic "fungal conservation" has only recently attracted the attention of mycologists. In the 1970s hardly anyone worried about the fate of fungi, except perhaps about the effects of mass collecting on some popular edible mushrooms. In the 1980s reports started to appear on the prominent decrease of some groups of macrofungi in some countries, for instance a general reduction of species diversity in Saarland (Derbsch & Schmitt, 1987), the decline of many ectomycorrhizal fungi in the Netherlands (e.g. Arnolds, 1988) and parts of Germany (Derbsch & Schmitt, 1984, 1987) and Czechoslovakia (Fellner, 1989; this symposium) and the disappearance of grasslands with a varied mycoflora in Germany (Winterhoff, 1978), Sweden (Nitare, 1988), Denmark (e.g. Rald, 1985) and the Netherlands (Arnolds, 1980). The influence of mushroom picking appeared to be small in comparison with the effect of changes in the environment.

[1] Comm. no. 481 of the Biological Station, Wijster

Systematic mapping in some countries revealed essential information on the geographical distribution of macrofungi and showed the very limited occurrence of many species (e.g. Krieglsteiner, 1991; Nauta & Vellinga, 1992), hence the need to protect these sites. Ecological research in both the field and the laboratory stressed the essential functions of these organisms in ecosystems as decomposers of complex organic compounds (e.g. Frankland et al., 1982) and ectomycorrhizal symbionts of woody plants (e.g. Harley & Smith, 1983). Field experiments were carried-out to study the influence of various environmental factors on the mycoflora, such as acidification and nitrogen enrichment of forest soil due to acid precipitation (e.g. Høiland, 1986; Jansen & Dighton, 1991; Termorshuizen, 1990), changes caused by fertilizer application in forests (e.g. Kuyper, 1989) and grasslands (e.g. Lange, 1982; Arnolds, 1989a). As a consequence, research was initiated to study the impact of various management treatments, such as mowing, grazing and digging, on the mycoflora (see contributions by Nitare, Keizer and Baar). The foundation of the European Council for Conservation of Fungi in Oslo in 1985 has contributed considerably to exchange of information and cooperation on this topic between mycologists throughout Europe. The workshops in Lodz in 1988 and on the island of Vilm in 1991 were very successful in this respect. The proposal for a European Red List, to be discussed by Ing, has been the initiative of this group.

These efforts have led to the nowadays widespread notion that many fungi are threatened indeed and that efforts for their conservation are necessary (Arnolds, 1991). Information on threatened fungi has been summarized in a number of so-called Red Lists for different European countries. This paper is based on an analysis of these lists in order to present the actual state of the art of fungal conservation, to signalize gaps in our knowledge and to formulate suggestions for future activities.

A Comparison of Red Lists of Various European Countries

A. Methods

Our analysis is based on Red Lists from eleven European countries. A survey of these lists is presented in Table 1; full citations of authors and titles are given in the References. The Red List of Czechoslovakia exists only as unpublished manuscript. The former German Democratic Republic and German Federal Republic are in this case still treated as separate countries.

TABLE 1. Red lists of macrofungi in European countries in order of appearance, considered in the present paper (mnscr.: only as manuscript, not yet published). Full titles are given in the References

Year	Authors	Country
1982	Benkert	German Democratic Republic
1984	Winterhoff	German Federal Republic
1986	Krisai	Austria
1987	Rassi & Väisänen	Finland
1989	Arnolds	The Netherlands
1990	Vesterholt & Knudsen	Denmark
1991	Floravårdskommittén	Sweden
1992	Ing	Great Britain
1992	Wojewoda & Lawrynowicz	Poland (second edition)
1992	Bendiksen & Høiland	Norway
mnscr (1992)	Anon.	Czeckoslovakia

Quite recently a new Red List for entire Germany was published (Deutsche Gesellschaft für Mykologie, 1992). This list is not considered here. Regional lists, published for several states of Germany and regions in Czechoslovakia, were not taken into account. Red Lists have also been published in the former U.S.S.R. and various union states, now independent countries. However, these lists have a different character, comprising relatively low numbers of species, protected by law. We thought that it was not meaningful to use them for our calculations and conclusions.

It is striking that Red Lists were composed so far only in countries in northern, western and central Europe, but that information from southern Europe is completely lacking. When we use "Europe" in this text, we refer to the part of Europe covered by Red Lists, unless stated otherwise.

We have restricted our analysis to the macrofungi because only from few countries information was available on threatened microfungi. For instance, the list from Great Britain contains sizable numbers of Uredinomycetes, Ustilagomycetes and Myxomycotina (from a strict taxonomic point of view the latter group should rather be placed on a Red List of threatened animals!). Helfer will emphasize the problems in these groups in his contribution.

The categories of threatened species are in accordance to the IUCN classification (Table 2). In some lists a category of unclassified species was used, which therefore was added as a fifth category. In some countries class 2 (endangered species) was divided into groups of species threatened with extinction and of strongly threatened species. They are not distinguished in the present survey.

An alphabetical survey of species on all Red Lists was made, using the computer program SUPERCALC. The complete survey will be published separately together with the results of some calculations (Arnolds & De Vries, in prep.). One of the main difficulties was the tracing of synonyms, since nomenclatural and taxonomic concepts were often not identical. Author citations or references to taxonomic works were often lacking. Decisions on synonymy were made on the basis of "best professional judgement", but some incorrect conclusions may have been drawn. These decisions may result in small discrepancies between numbers of species reported in this publication and the original Lists.

B. Numbers of Species in Red Lists

The eleven Red Lists studied comprise a total of 2984 macrofungi: 2658 species of basidiomycetes and 326 ascomycetes. We have tried to compare these figures with the

TABLE 2. Categories of threatened macrofungi used in the present paper

1. Exinct species: Species which have not been found for several decades (in spite of intensive fieldwork).

2. Endangered species: Rare species which have strongly decreased in this century and/or occupying habitats which are endangered and declining.

3. Vulnerable species: Species which are likely to become endangered in the near future if the casual agents of their decline are not reduced or removed.

4. Rare (potentially threatened) species: Species with small populations or few sites which are not at present threatened, but may become at risk without adequate protection.

5. Indeterminate: Care demanding species which cannot be placed in one of the above categories due to lack of information.

TABLE 3. Numbers of species of different taxonomic groups of macrofungi mentioned in European Red Lists in comparison with the numbers described in standard flora

NF = number of species in flora. (For Agaricales Moser, 1983; for other basidomycetes Jülich, 1984). NRL = number of species mentioned in Red Lists of Europe. % = percentage of species number, included in Red Lists.

	NF	NRL	%
Basidiomycetes	**5106**	**2658**	**52**
Agaricales sensu lato	**3150**	**1747**	**55**
Agaricales sensu stricto	2790	1408	53
Russulales	249	173	69
Boletales	110	87	79
Gasteromycetes	**266**	**125**	**47**
epigeous	133	81	61
hypogeous	133	44	33
Aphyllophorales	**1455**	**674**	**46**
cantharelloid	11	12	100
clavarioid	137	96	70
corticioid	1000	298	30
hydnoid	48	43	90
polyporoid	259	198	76
Heterobasidiomycetes	**235**	**112**	**48**
Ascomycetes	**?**	**326**	**?**
Total	?	2984	?

numbers of species, described from northern and central Europe (Table 3). There is no complete, up to date survey of this area, but the figures for Agaricales in Moser (1983) and other higher basidiomycetes in Jülich (1984) were considered as representative (although too low) estimations of numbers of species for these groups. The European Red Lists contain no less than 52% of the number of basidiomycete species, mentioned in these books. A similar estimation for the ascomycetes was virtually impossible since some lists are confined to Pezizales and a few other groups whereas others include small Helotiales or Erisyphales. The number of species included in Red Lists (further indicated as "listed species") varies from 123 in Czechoslovakia to 1032 in West Germany. There is no correlation with either the geographical situation or the size of the countries (Fig. 1). The numbers are in our opinion only in part representative of the degree of threat to the mycoflora. For instance, the relative low numbers in Finland and Norway may reflect indeed the relatively undisturbed environments which are still widespread in these countries, whereas the high numbers in the Netherlands and West Germany may be indicative of the high pressure on (semi-) natural areas in these densely populated and highly industrialized countries. Differences in numbers of listed species are also caused by subjective factors, such as the criteria used for listing, the available knowledge on distribution and ecology and the access to information from specialists for various groups (see C).

The numbers and proportions of different categories of listed species are presented in Table 4. These data are influenced by differences in the applications of the distinguished

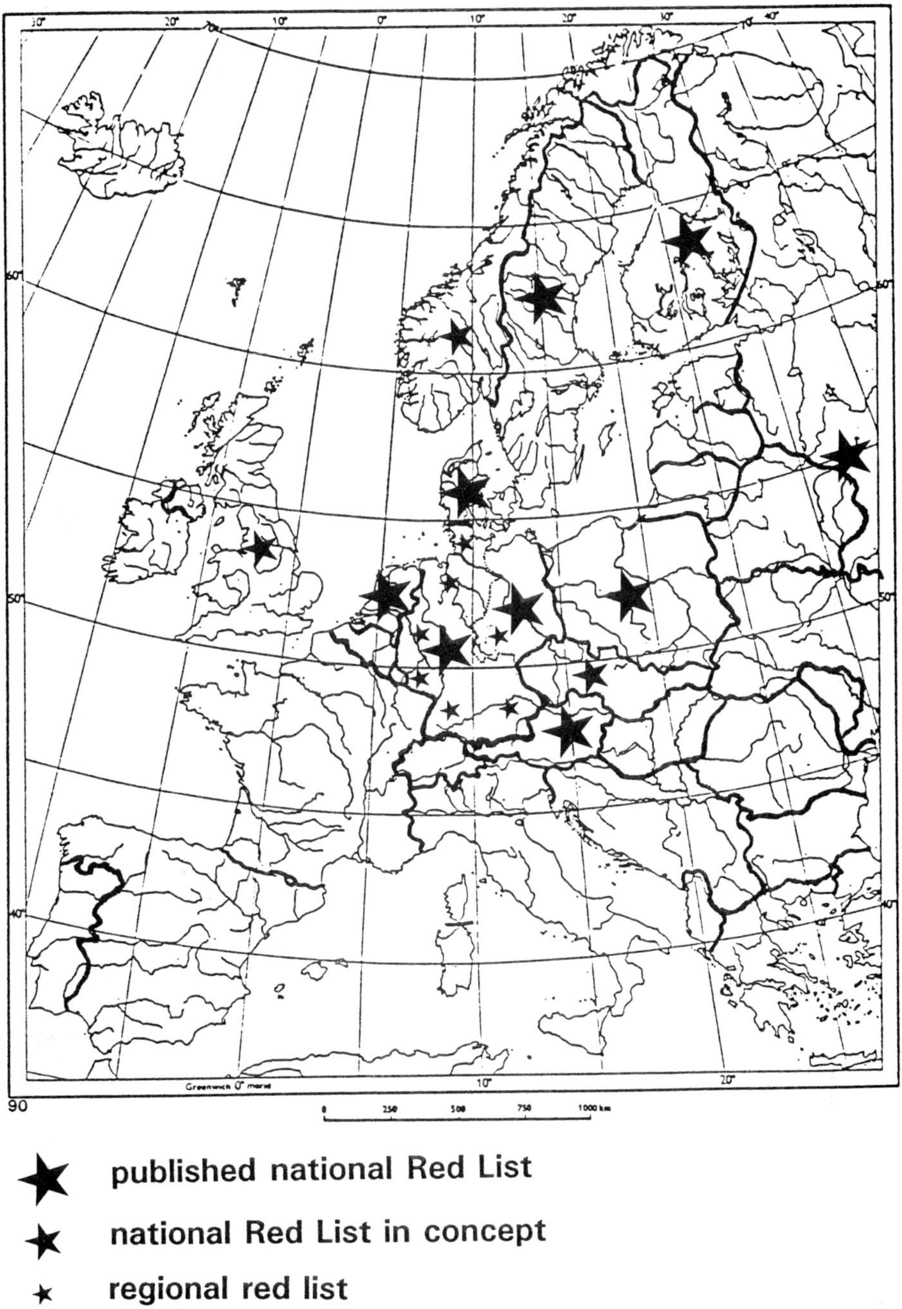

FIG. 1. Available Red Lists of threatened macrofungi in Europe.

categories, for instance the criteria used to determine whether a species has become extinct. In some countries part of the species are listed as "class indetermined". It is evident that the percentages of extinct and endangered species are highest in strongly industrialized and/or densely populated countries, such as the Netherlands, the two parts of Germany and Czechoslovakia. In the nordic countries and Great Britain a larger proportion is considered as vulnerable or rare and potentially threatened.

TABLE 4. Numbers of species in different categories of threat in Red Lists of macrofungi of 11 European countries

No = Norway　　Sw = Sweden　　Fi = Finland
DK = Denmark　　GB = Great Britain　　NL = Netherlands
GW = Germany (west)　　GE = Germany (east)　　Po = Poland
CS = Czechoslovakia　　Au = Austria

Category of threat:	Countries										
	No	Sw	Fi	DK	GB	NL	GW	GE	Po	CS	Au
Numbers of species:											
1. (Presumably) extinct	4	16	5	56	21	91	37	26	70	0	1
2. Endangered	45	106	29	143	56	355	406	178	169	81	41
3. Vulnerable	283	113	26	206	259	245	399	58	187	30	83
4. Rare (potentially threatened)	317	200	84	393	117	253	190	47	299	12	86
5. Indetermined	0	79	17	0	0	0	0	0	287	0	0
Total	649	514	161	898	453	944	1032	309	1032	123	211
Percentages:											
1. (Presumably) extinct	1	3	3	6	5	10	4	8	7	—	0
2. Endangered	7	21	18	16	12	37	39	58	17	66	19
3. Vulnerable	44	22	16	34	57	26	39	19	18	24	39
4. Rare (potentially threatened)	49	39	52	44	26	27	18	15	30	10	41
5. Indetermined	—	15	11	—	—	—	—	—	28	—	—

C. Taxonomic Groups in Red Lists

The proportions of various broad taxonomic groups in the European Red Lists, compared with the reference books used, varied less than we expected: from 46% of the Aphyllophorales to 55% of the Agaricales (Table 3). Within these groups more pronounced differences are observed. For instance, 61% of the epigeous Gasteromycetes are mentioned on at least one Red List and only 33% of the hypogeous species. Among the Aphyllophorales, cantharelloid and hydnoid species are present with 90–100% of their species, but corticioid fungi with only 30%. Such data raise a lot of questions, for instance: are these differences realistic reflections of the threatened position of various groups in Europe or perhaps mainly caused by differences in popularity among mycologists?

The proportions of various taxonomic groups in Red Lists of various countries show large variation (Table 5). The proportion of the agarics varies from 38% in Sweden to 78% in West Germany, of Gasteromycetes from 4% in Denmark to 13% in

Ceskoslovakia; of Aphyllophorales from 9% in West Germany to 39% in Sweden; of Heterobasidiomycetes from nil in Austria to 7% in Denmark; of Ascomycetes from 7% in West Germany to 17% in Finland. Within the Aphyllophorales, corticioid species score very high in Sweden, Norway and Denmark (Table 6) apparently due to the extensive knowledge on this group in the nordic countries (e.g. Eriksson & al., 1973–1988). The most striking difference is found in the Heterobasidiomycetes, a group with very few listed species (0–10) in most European countries, but in Denmark and Poland represented with 60 and 59 species, respectively. Apparently specialists on this group were involved in the compilation of these lists and not in others.

Epigeous and hypogeous gasteromycetes have approximately similar numbers of species in Europe (Table 3), but in all lists epigeous species are in the majority. Do these figures mean that epigeous fungi are more endangered than hypogeous species or that species of the former group are more widely distributed in Europe and therefore reported from more countries? Or are these figures merely caused by lack of knowledge on hypogeous fungi?

TABLE 5. Numbers of species belonging to different taxonomic groups in Red Lists of macrofungi of 11 European countries. (For abbreviations of the countries, see Table 4)

Taxonomic group	Countries											
	No	Sw	Fi	DK	GB	NL	GW	GE	Po	CS	Au	Eur
Number of species:												
Agaricales	320	197	78	447	290	707	799	162	499	76	137	1747
Gasteromycetes	36	39	15	31	25	45	64	27	60	16	20	125
Aphyllophorales	213	203	39	237	88	117	98	72	306	14	38	674
Heterobasidiomycetes	11	10	2	60	4	2	1	2	59	1	0	112
Ascomycetes	69	65	27	123	46	73	70	46	88	16	16	236
Total	649	514	161	898	453	944	1032	309	1012	123	211	2984
Percentages:												
Agaricales	49	38	48	50	64	75	78	53	49	62	65	59
Gasteromycetes	6	8	9	4	6	5	6	9	6	13	9	4
Aphyllophorales	32	39	24	26	19	12	9	23	30	11	18	23
Heterobasidiomycetes	2	2	1	7	1	0	0	1	6	1	—	4
Ascomycetes	11	13	17	14	10	8	7	15	9	13	8	11

The numbers of species of some important genera, mentioned on at least one European Red List, are given in Table 7. The 5 to 10 largest genera of each major taxonomic group are mentioned and in addition some smaller genera showing interesting features. Eight genera are presented with more than 80% of their European taxa: the agarics *Entoloma* (Fr.) Kummer, *Hygrocybe* (Fr.) Kumm. sensu lato, *Boletus* Fr. and *Lepiota* (Pers.) S.F. Gray, the gasteromycete genus *Geastrum* Pers. and the Aphyllophorales *Clavaria* Fr., *Hydnellum* Karsten and *Phlebia* Fr. *Antrodia* Karsten. The very high percentage of listed species in *Entoloma* is an overestimation, caused by the fact that the number of species described in Moser (1983) is not representative anymore since the revision of this genus by Noordeloos (1987), who distinguished 217 species in Europe. In *Clavaria* and *Boletus* the numbers of species in Red Lists are higher than the corresponding numbers in the reference books, indicating the incompleteness of the

TABLE 6. Numbers of species belonging to different groups of Aphyllophorales and Gasteromycetes in Red Lists of macromycetes of 11 European countries. (For abbreviations of countries, see Table 4)

Taxonomic group	Countries											
	No	Sw	Fi	DK	GB	NL	GW	GE	Po	CS	Au	Eur
Numbers of species:												
Clavarioid fungi	17	25	2	28	36	33	12	14	27	1	0	96
Cantharelloid fungi	5	2	0	3	5	5	2	1	3	0	2	12
Hydnoid fungi	17	14	1	24	22	25	18	11	19	1	3	43
Polyporoid fungi	68	64	33	50	18	30	54	32	136	9	28	198
Thelephora, Tomentella	0	1	0	22	0	4	1	0	2	0	0	26
Others (corticoid fungi)	106	97	3	110	7	20	11	14	118	3	5	298
Aphyllophorales total	213	203	39	237	88	117	98	72	305	14	38	674
Percentages:												
Clavarioid fungi	8	12	5	12	41	28	12	19	9	7	—	14
Cantharelloid fungi	2	1	—	1	6	4	2	1	1	—	5	2
Hydnoid fungi	8	7	3	10	25	21	18	15	6	7	8	6
Polyporoid fungi	32	32	85	21	20	26	55	44	45	64	74	29
Thelephora, Tomentella	—	0	—	9	—	4	1	—	—	4		
Others (corticioid fungi)	50	48	8	46	8	17	11	19	39	21	13	44
Numbers of species:												
Epigeous gasteromycetes	25	31	13	22	25	30	46	25	51	11	19	85
Hypogeous gasteromycetes	11	8	2	9	0	15	18	2	9	5	1	40
Gasteromycetes total	36	39	15	31	25	45	64	27	60	16	20	125
Percentages:												
Epigeous gasteromycetes	69	79	87	71	100	67	72	93	85	69	95	68
Hypogeous gasteromycetes	31	21	13	29	—	33	28	7	15	31	5	32

latter. These example reflect the relative value of such figures. Only 4 of the analyzed genera are represented with less than 20% of their species: *Hemimycena* (Singer) Singer, *Peniophora* Cooke, *Typhula* (Pers.) Fr. and *Exobasidium* Woronin.

There are considerable differences in the average frequencies of species of different genera in Red Lists (Table 7). All genera of heterobasidiomycetes have a very low score in this respect and it is striking that they are almost confined to lists of one country: *Dacrymyces* Nees to the list of Poland, *Exidiopsis* (Bref.) Möller and *Tulasnella* Schröter to the list of Denmark, *Tremella* Pers. to lists of these two countries. Also all ten *Tomentella* Pat. species are exclusively listed in Denmark. On the other hand, some genera have a score above 4, viz. *Geastrum* and *Tulostoma* Pers.: Pers., which implies that their Red List species are mentioned on the average in more than four European countries. Consequently, they are considered as threatened in large parts of Europe. Within the Agaricales, *Boletus* has a very high average (3.5), *Cortinarius* S.G. Gray emend. Fr. a low average (1.6). Does this mean that boletes are threatened in large parts of Europe and cortinarii only in some regions, or that boletes have larger distribution areas than *Cortinarius* species? Or is the attention of mycologists all over Europe focused on boletes and only in some countries on *Cortinarius*?

TABLE 7. Numbers of species of different genera of larger basidiomycetes in European Red Lists, in comparison with the numbers described in standard floras (for Agaricales after Moser, 1983; for other basidiomycetes after Jülich, 1984)

NF = number of species in flora.
NRL = number of species in European Red Lists.
% = percentage of species number, included in Red Lists
Av = average number of lists in which species of different genera are mentioned.

	NF	NRL	%	Av
Agaricales:				
Cortinarius (incl. Dermocybe)	502	264	53	1.6
Russula	159	111	70	1.9
Inocybe	158	76	48	2.0
Entoloma s. lato	146	142	97	2.2
Mycena	126	65	52	2.0
Psathyrella	100	42	42	1.6
Clitocybe	91	26	29	1.8
Coprinus	91	20	22	1.6
Lactarius	90	62	69	2.4
Hygrocybe (incl. Camarophyllus, Hygrotrama)	78	69	89	3.3
Tricholoma	66	49	74	2.4
Lepiota (incl. Cystolepiota)	58	55	95	2.5
Galerina	57	19	33	1.7
Pluteus	47	34	72	2.4
Melanoleuca	31	14	45	1.4
Boletus	27	29	>100	3.5
Hemimycena	18	3	17	1.0
Gasteromycetes:				
Hymenogaster	37	8	22	2.1
Geastrum	26	23	88	4.7
Lycoperdon	20	11	55	3.4
Tulostoma	17	7	41	4.4
Hysterangium	17	8	47	1.1
Aphyllophorales:				
Tomentella	49	10	20	1.0
Ramaria	41	29	70	1.7
Phlebia	33	27	82	1.6
Peniophora	32	5	16	1.6
Typhula	30	1	3	1.0
Phellinus	30	22	73	2.6
Clavaria	19	21	>100	1.9
Clavulinopsis	18	12	67	2.2
Hydnellum	17	14	82	3.8
Antrodia	15	13	80	1.8
Heterobasidiomycetes:				
Tulasnella	28	10	36	1.2
Exobasidium	26	3	12	1.0
Tremella	25	11	44	1.3
Exidiopsis	16	7	44	1.1
Dacrymyces	15	8	53	1.2

D. Ecological Groups in European Red Lists

The distribution of species among different broad niche-substrate groups in the analyzed lists is given in Table 8. A niche-substrate group is a group of species growing on a similar substrate in a similar (micro) habitat and exploiting the environment in a similar way. Ectomycorrhizal fungi contribute 31% to the species mentioned in Europe on one or more Red Lists (Table 7). The highest number of species (377) is listed in West Germany and the Netherlands, the highest proportion (47%) in Great Britain. Saprotrophic, soil-inhabiting fungi are the largest groups on combined European lists: 1034 species or 35%. The highest numbers of listed species are reported from the Netherlands (433) and western Germany (446); the highest proportions from the Netherlands (46%) and eastern Germany (48%). The numbers of dung-inhabiting species are very low in most European Red Lists, with noticeable exceptions of Poland (17 species), Denmark (14) and western Germany (11 species). The most striking differences are found in the proportions of wood-inhabiting fungi on various lists, ranging from 12% in the Netherlands (parasites and saprophytes combined) to 35% or more in Poland, Sweden and Finland.

TABLE 8. Numbers of species belonging to different niche-substrate groups in Red Lists of macrofungi of 11 European countries. (For abbreviations of the countries, see Table 4)

Niche-substrate group	Countries											
	No	Sw	Fi	DK	GB	NL	GW	GE	Po	CS	Au	Eur
Number of species:												
Ectomycorrhizal fungi	187	119	45	211	212	377	377	81	239	38	79	913
Saprotrophic fungi on soil	203	174	49	322	153	433	446	148	309	49	73	1034
Saprotrophic fungi on dung	7	1	1	14	1	3	11	4	17	0	1	31
Saprotrophic fungi on wood	210	188	50	284	62	99	164	55	375	26	39	823
Parasitic fungi on wood	16	20	9	19	9	18	20	11	37	4	16	52
Remaining/unknown	26	12	7	48	16	14	14	10	35	6	3	131
Percentages:												
Ectomycorrhizal fungi	29	23	28	23	47	40	37	26	24	31	37	31
Saprotrophic fungi on soil	31	34	30	36	34	46	43	48	31	40	35	35
Saprotrophic fungi on dung	1	0	1	2	0	1	1	1	1	—	0	1
Saprotrophic fungi on wood	32	37	31	32	14	10	16	18	37	21	18	28
Parasitic fungi on wood	2	4	6	2	2	2	2	4	4	3	8	2
Remaining/unknown	4	2	4	5	4	1	1	3	3	5	1	4

All species in Red Lists have been assigned to a certain habitat type, in which the species are supposed to occur optimally in Europe. The numbers of species in the distinguished habitat types are presented in Table 9. As to be expected the large majority of listed macrofungi (74%) are mainly found in forests. This proportion varies from 59% in eastern Germany to 81% in Norway. The numbers of listed species from frondose forests are in all countries higher than those of coniferous forests, although they are only slightly higher in eastern Germany and almost five times as high in Austria. There is no plausible explanation for these striking differences. The high percentage (26) of threatened fungi of coniferous forests in Great Britain may be linked to the small area of indigenous pine forests, restricted to Scotland.

TABLE 9. Numbers of species with optimum in certain habitat types in Red Lists of macrofungi of 11 European countries. (For abbreviations of the countries, see Table 4)

Habitat type	Countries											
	No	Sw	Fi	DK	GB	NL	GW	GE	Po	CS	Au	Eur
Numbers of species:												
Forests, not specified	96	60	19	152	79	93	107	28	136	12	18	442
Frondose forests on ± dry soil	240	211	74	347	149	375	399	69	382	41	94	1088
Boggy forests	17	10	6	35	17	52	35	17	27	6	16	101
Coniferous forests	171	118	30	140	111	145	183	68	239	27	23	611
Heathlands	3	1	2	9	3	22	10	4	10	0	4	34
Bogs, fens, wetlands	13	8	3	27	12	46	41	27	34	7	9	100
Alpine and arctic habitats	5	4	1	4	5	2	6	1	9	5	3	22
Grasslands	43	61	13	93	33	153	129	58	80	8	18	268
Sand dunes, steppe	14	17	4	14	15	25	30	18	23	10	8	61
Urban areas, parks, fields	19	13	4	26	11	26	58	12	48	2	13	101
Remaining/unknown	28	11	5	51	18	6	34	7	24	5	5	156
Percentages:												
Forests, not specified	15	12	12	17	17	10	10	9	13	10	9	15
Frondose forests on ± dry soil	37	41	46	39	33	40	39	22	38	33	45	36
Boggy forests	3	2	4	4	4	6	3	6	3	5	8	3
Coniferous forests	26	23	19	16	26	15	18	22	24	22	11	20
Heathlands	0	0	1	1	1	2	1	1	1	—	2	1
Bogs, fens, wetlands	2	1	2	3	3	5	4	9	3	6	4	3
Alpine and arctic habitats	1	1	1	0	1	0	1	0	1	4	1	1
Grasslands	7	12	8	10	7	16	12	19	8	6	9	9
Sand dunes, steppe	2	3	2	2	3	3	3	6	2	8	4	2
Urban areas, parcs, fields	3	3	2	3	2	3	6	4	5	2	6	3
Remaining/unknown	4	2	3	6	4	1	3	2	2	4	2	5

The numbers of threatened fungi characteristic of boggy forests (dominated by *Alnus*, *Salix*, sometimes *Populus*) are surprisingly low in most countries. The highest proportions (6%) are reported from the Netherlands and eastern Germany, where deep drainage of agricultural soils causes great damage to wet ecosystems, but also from Austria (8%). The percentage of fungi from bogs and other open wetlands are also high in eastern Germany (9%), the Netherlands (5%) and, in addition, in Czechoslovakia (6%). In the nordic countries, and Britain, still with vast areas of marshland, low numbers of this group are listed (1–3%). The largest difference between Red Lists is found in the grassland fungi, ranging from 6% in Czechoslovakia to 16% in the Netherlands and 19% in eastern Germany. Again, the high percentages in the latter countries are correlated with a more intensive agricultural practice.

Highest proportions of dune and steppe fungi occur in eastern Germany (6%) and Czechoslovakia (8%), where this habitat is the best developed and strongly threatened. The numbers of species from anthropogenic places are surprisingly high since these habitats are expanding everywhere and seem not to be threatened. However most species are classified among the potentially threatened or rare species. On the other hand the numbers of arctic and alpine habitats are lower than expected, taken into account the high number of endangered alpine plants and the damage done in central Europe by wintersport accommodations.

In this context it is tempting to wonder to which extent these figures really reflect the relative vulnerability of fungi in these habitats and to which extent they are influenced by the emphasis of mycological activities in forests.

We have also tried to make an estimation of the numbers of listed species, characteristic of forests and grasslands on basic or calcareous soils versus species from acidic soils, although it was in many cases difficult to find reliable data concerning this factor (Table 10). In all countries the number of threatened basiphytic species is considerably larger than that of acidiphytic species, in particular in Czechoslovakia (9 times larger) and Fennoscandia (4–7 times larger). The high proportions of threatened basiphytic species in Fennoscandia may be due to the rare occurrence of calcareous bedrocks and basic soils in these countries. Again, a number of questions arise: are basiphytic fungi more vulnerable than acidophytic fungi or is the number of basiphytic species higher? Or, perhaps, are the larger boletes, colourful *Phlegmacium* species, and attractive *Lepiota* species on basic soils more appealing than the less striking *Telamonia* species and *Galerinas* on acid soils?

TABLE 10. Numbers of species with optimum on basic (calcareous) or acidic soils in Red Lists of macrofungi of 11 European countries. (For abbreviations of the countries, see Table 4)

Habitat type	Countries											
	No	Sw	Fi	DK	GB	NL	GW	GE	Po	CS	Au	Eur
Forests on basic soil, not specified	8	4	3	6	9	6	8	2	3	1	2	20
Frondose forests on basic soil	65	76	25	88	50	137	125	20	69	20	23	253
Coniferous forests on basic soil	8	8	5	11	9	13	20	3	13	3	5	41
Grasslands on basic soil	14	26	4	32	8	41	32	13	16	3	6	52
Basiphytic species total	95	104	37	137	76	197	185	38	101	27	36	366
% of total listed species	15	22	23	15	17	21	18	12	10	22	17	12
Forests on acidic soil, not specified	2	1	0	3	2	10	7	0	8	0	1	16
Frondose forests on acidic soil	8	4	1	9	7	26	20	1	17	0	5	45
Coniferous forests on acidic soil	10	5	2	24	27	45	30	19	27	2	2	84
Grasslands on acidic soil	1	4	3	3	2	8	4	4	4	1	1	9
Acidophytic species total	21	14	6	39	38	89	61	24	56	3	9	154
% of total listed species	3	3	4	4	8	9	6	8	5	2	4	5
Ratio basiphytic/acidophytic species	4.5	7.4	6.2	3.5	2.0	2.2	3.0	1.6	1.8	9.0	4.0	2.4

E. Some Species in European Red Lists

The numbers of species, included in different numbers of Red Lists, are indicated in Fig. 2. It is striking that almost half of the species (49%) are listed in one country only. This might imply that there are many endemic species of fungi with small distribution areas in Europe. However, most species of fungi are known to have a rather wide geographical range. It might also mean that many species are threatened in only part of their distribution area, but this seems also unlikely in most cases since the factors affecting the

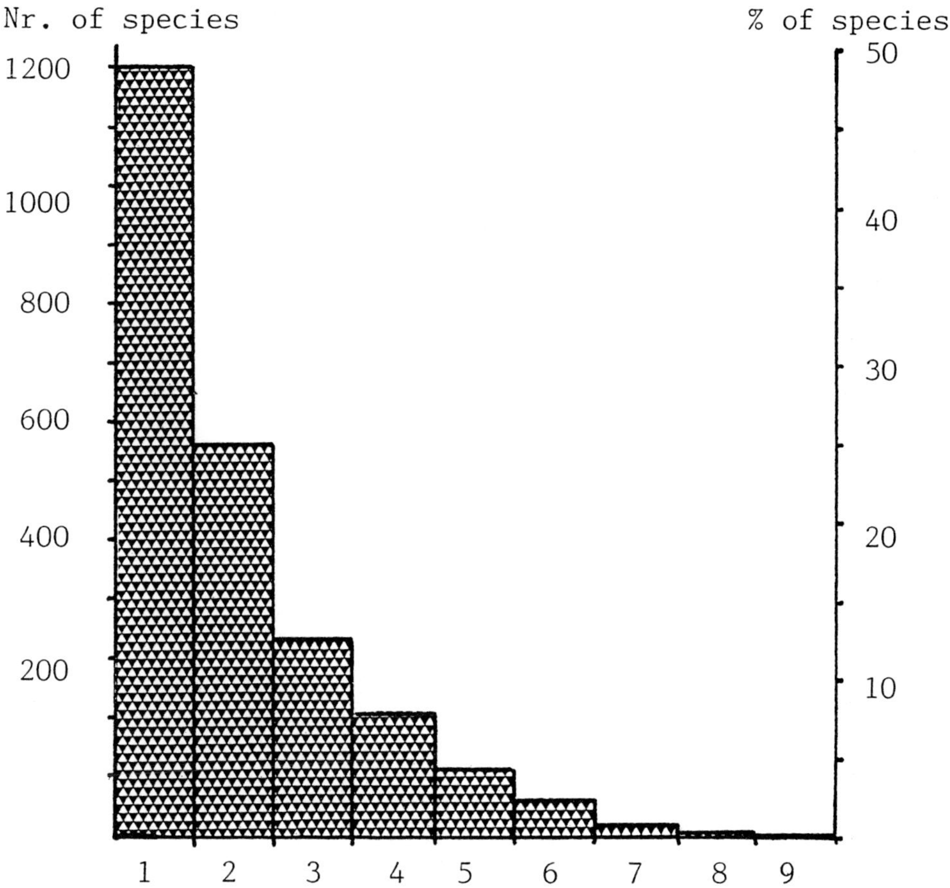

FIG. 2. Numbers of species of macrofungi listed in various numbers (1–9) of Red Lists in Europe.

mycoflora are rather similar in large parts of Europe. The high incidence of species on one list is in our opinion mainly caused by artefacts, such as the specialists working on a list (see previous example on heterobasidiomycetes) and available knowledge on certain groups. It is striking that the frequency of listing is highest in groups with striking sporocarps and well-known systematics.

The 23 species with the highest frequencies in Red Lists, mentioned 9 or 8 times, are presented in Table 11. These fungi are apparently widely distributed in Europe, but rare or decreasing in many places. In addition, most of them have striking and appealing sporocarps and belong to the most popular groups, such as earthstars and boletes. Do these species automatically belong to the most endangered European fungi? Certainly not. In fact, rare species with a local distribution are much more endangered. In addition, some species are mostly classified as rare or vulnerable, which is expressed in a relatively low average index of threat, e.g. *Tulostoma brumale* Pers. Some species, which are listed in many countries as threatened or extinct, indeed belong to the care demanding fungi, such as *Aurantioporus corceus* (Pers.: Fr.) Kotl. & Pouzar and *Poronia punctata* (L.: Fr.) Fr. The average index of threat (IT) is calculated by assigning 4 points to species class 1 (extinct), 3 to species in class 2 (endangered), 2 to species in class 3 (vulnerable) and 1 to species in classes 4 (rare) and 5 (indetermined). For each species these numbers are added and divided by the number of Red Lists in which that species is included. Consequently, high IT values indicate that the fungus is regarded as strongly endangered in many countries.

On the other hand, species which figure on a small number of Red Lists may have a variable status in Europe. Part of them are considered as threatened in countries at the border of their distribution areas (Table 12). For instance, *Fistulina hepatica* (Schaeffer) Fr., *Daedalea quercina* (L.) Fr., *Xylaria polymorpha* (Pers.) Grev. and *Collybia fusipes* (Bull.: Fr.) Quél. are regarded as threatened in Scandinavian countries. These species are exclusively or mainly found on the wood of *Quercus*, a tree genus with very limited distribution in southern Finland and Norway. They are widespread and common in large parts of central Europe. *Xeromphalina campanella* (Batsch: Fr.) Maire is a boreo-montane species, widespread in Fennoscandia and the central European mountains, but absent from the Netherlands and rare in Denmark and Great Britain, where it is considered as vulnerable. *Astraeus hygrometricus* (Pers.) Morgan is a southern continental species,

TABLE 11. Species, included in 9 or 8 (out of 11) Red Lists of macrofungi in Europe. (For abbreviations of countries, see Table 4)

1–5: categories of threat in Red Lists (see Table 5).

	Countries											
	No	Sw	Fi	DK	GB	NL	GW	GE	Po	CS	Au	Av
9 lists:												class
Melanophyllum eyrei	3	2	2	3	4	4		2	4	4		1.9
Geastrum badium	2	3	2	3	3	3	2	2			3	2.4
Aurantioporus croceus	3	3		1	2		2	2	1	2	3	2.9
Microglossum olivaceum	3	5	4	3	3	2	2	2	5			2.0
8 lists:												
Armillaria ectypa		3		2	3	2	2	2		2	2	2.8
Boletus fechtneri		3	5			2	2	4	2	2	3	2.3
Boletus impolitus	3	3		3	2	2	2		2		3	2.5
Boletus queletii	4	2	3	3		2	3	4	2			2.1
Boletus satanas		2	2	3	4	2	2		1		4	2.5
Entoloma euchroum	4	4	5	3		3	3		5		3	1.5
Russula pumila	4			3	4	3	3	2	5	4		1.6
Bovista paludosa	4	4	4		1		2	2	3	3		2.1
Geastrum berkeleyi		3		2	1		4	4	2	2	2	2.5
Geastrum corollinum		2		3	3	4	4	3	2		2	2.1
Geastrum fornicatum	2	2		2	3	4	1		2		3	2.6
Geastrum nanum	2	4	2			3	2	3	2		4	2.3
Geastrum pedicellatum	2	2		2	1	2	2	2	1			3.2
Tulostoma brumale	4	3		4		3	3		3		4	1.4
Gomphus clavatus	3	5	4	1	2		3	2	2			2.4
Polyporus umbellatus	3	3	3	3	4	4			3		3	1.7
Poronia punctata	2	2	2	1	2	2		2	5			2.9
Thuemenidium atropurpureum	4	5	4	3	4	2		2	4			1.6

TABLE 12. Examples of species of macrofungi, included in few Red Lists of European countries and widely distributed elsewhere. (For abbreviations of the countries, see Table 4)

—: absent in indicated country.
+: present in indicated country, not included in Red List.
1–5: categories of threat in Red Lists (see Table 4).

	No	Sw	Fi	DK	GB	NL	GW	GE	Po	CS	Au
Fistulina hepatica	3	+	4	+	+	+	+	+	3	+	+
Daedalea quercina	+	+	4	+	+	+	+	+	+	+	+
Xylaria polymorpha	3	+	2	+	+	+	+	+	+	+	+
Collybia fusipes	3	4	5	+	+	+	+	+	+	+	+
Xeromphalina campanella	+	+	+	4	3	—	+	+	+	+	+
Astraeus hygrometricus	—	—	—	—	3	2	3	+	2	+	+
Omphalotus olearius s. lato	—	—	—	—	3	4	4	1	—	+	+
Suillus bellini	—	—	—	—	3	—	—	—	—	—	—
Amanita caesarea	—	—	—	—	—	—	2	1	1	2	2
Auriscalpium vulgare	+	+	+	+	+	2	+	+	+	+	+
Suillus variegatus	+	+	+	+	+	2	+	+	+	+	+
Cantharellus tubaeformis	+	+	+	+	+	2	+	+	+	+	+
Boletus edulis	+	+	+	+	+	+	+	+	3	+	+
Tapinella panuoides	+	+	+	+	+	+	+	+	+	+	2
Rickenella setipes	+	+	+	+	+	+	+	+	+	+	3
Antrodiella semisupina	+	+	+	+	+	+	+	+	5	+	+
Bulgaria inquinans	+	+	+	+	+	+	+	+	5	+	+

lacking in northern Europe and regarded as rare and threatened in West European countries. *Omphalotus olearius* (DC: Fr.) Singer, *Suillus bellini* (Inz.) Watl. and *Amanita caesarea* (Scop.: Fr.) Grev. are examples of species with a mainly Mediterranean distribution, regarded as threatened in the northern parts of their range.

A second group of species are widespread in Europe, but a well-documented decrease has been observed in only part of the area. Examples are *Auriscalpium vulgare* S.F. Gray, *Suillus variegatus* (Swartz: Fr.) O. Kuntze and *Cantharellus tubaeformis* (Bull.) Fr. in the Netherlands (Arnolds, 1988). This decrease is mainly caused by air pollution and therefore it would be useful to pay special attention to their fate in other European countries in future. *Boletus edulis* is regarded as threatened in Poland only due to overcollecting for consumption.

A third group of species is widespread and often common in large parts of Europe, but they are for some unknown reason considered as threatened in one or two countries (Table 12), for instance *Tapinella* (= *Paxillus*) *panuoides* (Fr.: Fr.) Gilb. and *Rickenella setipes* (Fr.: Fr.) Raithelh. in Austria and *Antrodiella* (= *Antrodia*) *semisupina* (Berk. & Curtis) Ryv. & Johansen and *Bulgaria inquinans* (Pers.) Fr. in Poland. We can only guess whether such reports are based on an observed decrease, geographical factors or merely on an inaccurate estimation of its actual status by the authors of Red Lists.

The above groups of species cannot be regarded as threatened on a European level. However, some other species are included in only few Red Lists, which are known to have a restricted distribution at the same time. Some examples are given in Table 13. *Amylocystis lapponicus* (Rom.) Singer, *Haploporus suaveolens* (L.: Fr.) Donk, *Gloeoporus pannocinctus* (Rom.) J. Erikss. and *Hygrophorus inocybiformis* A.H. Smith have a restricted boreal or boreo-montane distribution. *Bondarzewia montana* (Quél.) Singer, *Battarraea stevenii* (Lib.) Fr. and *Suillus sibericus* Singer are confined to eastern Europe. It is not known whether these species are widely distributed outside the area covered by the analyzed Red Lists.

Some striking species are listed from and only known from one country. Six examples for different countries are given in Table 13. Do these species really have such limited distributions or are they more widespread than we know at present?

Many other examples could be given of species belonging to one of the groups discussed above. However, for the majority of species we can only guess what their actual status is at a European level because of lack of data on their distribution, frequency and possible changes in fruiting. For instance, are the hundreds of *Cortinarius* species, mentioned in one or two national Red Lists, really only threatened in these countries or are they common elsewhere or perhaps absent?

TABLE 13. Examples of species of macrofungi, included in few (<6) Red Lists of European countries, with restricted distribution. (For abbreviations of the countries, see Table 4)

— : absent in indicated country
+ : present in indicated country, not included in Red List
1–5 : categories of threat in Red Lists (see Table 4)

	No	Sw	Fi	DK	GB	NL	GW	GE	Po	CS	Au
Hygrophorus inocybiformis	+	4	4	—	—	—	—	—	—	—	—
Haploporus suaveolens	2	3	3	—	2	—	—	—	—	—	—
Amylocystics lapponicus	3	4	—	—	—	—	—	—	2	2	—
Gloeoporus pannocinctus	2	3	4	2	4	—	—	+	—	+	—
Bondarzewia montana	—	—	—	—	—	—	2	3	3	+	2
Battarraea stevenii	—	3	—	—	—	—	—	—	2	2	2
Piptoporus pseudobetulinus	—	—	2	—	—	—	—	—	—	—	—
Pleuroflammula ragazziana	—	—	—	—	3	—	—	—	—	—	—
Agaricus geesterani	—	—	—	—	—	4	—	—	—	—	—
Dentipranulum bialoviesense	—	—	—	—	—	—	—	—	2	—	—
Agrocybe stepposa	—	—	—	—	—	—	—	—	—	2	—
Amanita valens	—	—	—	—	—	—	—	—	—	—	2

Discussion

Our analysis of Red Lists enables us to draw the attention to some omissions and inconsistencies in our knowledge. It has become evident that fungi may be considered as threatened in a certain country for various reasons: they may be rare in their entire distribution area or only in some regions; they may have a very restricted distribution or occur at the borders of their range; they may show marked decline due to regional or more widespread factors; their habitat may be considered as threatened in one place but not in another; they may be merely listed, or more likely missed because of lack of knowledge. At present it is in most cases impossible to distinguish between these possibilities and consequently to evaluate the true status of a species in a European context. The following subjects may deserve attention:

Lack of information on microfungi

The gaps in our knowledge on macrofungi are even more pronounced in the microfungi except for some groups of economic importance. Conservational aspects of microfungi are more difficult to study due to unsolved methodological problems. Notable exceptions are Uredinomycetes and Ustilagomycetes.

Lack of information on parts of Europe

It is striking that Red Lists are entirely lacking in southern Europe. It is of primary importance to have information from these countries, also for the interpretation of data from central Europe. Many species from that area may in fact have their main distribution in more southern regions. For instance, after our own experience some species regarded as threatened in many countries appeared to be quite common in parts of Italy, e.g. *Cantharellus lutescens, Tricholoma atrosqumosum* (Chev.) Sacc. and *T. aurantium* (Schaeffer: Fr.) Ricken.

Which species are not regarded as threatened?

In fact each list contains a selection of indigenous species which are considered to be (potentially) threatened. It is not clear whether species which are not on these lists are common, lacking or just neglected because of lack of knowledge. For instance, large numbers of heterobasidiomycetes are only incorporated in lists of Denmark and Poland, but it seems unlikely that this phenomenon has a realistic basis. A good method to summarize this information is the production of annotated checklists, as were published in eastern Germany (Kreisel, 1987) and the Netherlands (Arnolds, 1984).

Distribution patterns and frequency in Europe

The lack of reliable information on distribution of fungi in Europe has repeatedly been indicated in the preceding paragraphs. Checklists, floras and revisions may provide valuable data but more accurate information is summarized in a distribution atlas, such as published from western Germany (Krieglsteiner, 1991). Other relevant subjects are the frequency of species in space (when present in a certain area, is it scarce or abundant?) and time (is it ephemeral or regular?), and changes in frequency and distribution during this century.

Information on ecology

The ecological data in floras and other taxonomic works on fungi are generally very superficial. There is a great need for accurate information on the ranges in habitat types, substrates and host specifity of species in different parts of Europe, by preference in a (semi)quantitative form. This information may be gathered by mycocoenological and autecological studies as well as during annotated floristical and geographical studies.

Special attention should be paid to experimental studies on the impact of various anthropogenic factors on the mycoflora, such as different kinds of forest and grassland management, various types of air pollution and future challenges, such as global warming. Monitoring in permanent plots is one of the most promising techniques to study natural and man-induced changes in the mycoflora.

Taxonomy

Floristical, geographical and ecological studies are impossible without thorough taxonomic and nomenclatural basis. On the other hand, geographical and ecological studies may provide valuable information for taxonomy. There is a continued need for both classical revisions of many critical groups of fungi and for advanced research on the genetical basis of our species concepts. Increasing discrepancies in the application of species concepts by various groups of workers in different parts of Europe are subject of concern.

Contacts with other ecological disciplines and nature conservationists

In order to be of any practical significance, the collected information on ecological functions of the mycoflora, threatened species, and the impact of human activities on fungi should spread from the mycological "incrowd" to ecologists, foresters, nature conservationists and environmentalists. It is necessary to improve contacts with other disciplines.

The gathering of the required information will demand for increasing cooperation between mycologists and mycological societies from different countries. Subjects such as mapping of fungi and the initiation of monitoring programs may demand a kind of central coordination. It is also important to improve contacts between professional and amateur mycologists. Successful mapping without participation of the latter category will be impossible.

Conclusions

In the past decade the interest for conservational aspects of macrofungi has increased enormously. Studies appeared on changes in the mycoflora and the influence of man-induced changes in the environment on fungi. Our collective knowledge has been summarized in a number of Red Lists and threatened mushrooms appeared in more general and popular journals, on television or even in departmental notes. Some local successes were obtained in the protection of areas with a valuable mycoflora and the realization of measures to benefit the mycoflora.

However, did all these efforts in fact lead to improving conditions for fungi in Europe? We are not so optimistic in this respect. Air pollution, threatening ectomycorrhizal fungi, is still a severe problem; EEC agricultural policy leads to a fast and irreversible loss of old, poorly managed pastures and their associated hygrocybes, entolomas and many other fungi; the few remains of old-growth forests with their characteristic wood-inhabiting fungi are still threatened by forestrial exploitation; peat bogs with their few, but strongly specialized fungi, are still drained and dug away; coastal sand dunes with their striking gasteromycetes are under strong pressure of recreation and expansion of towns, and so on. It is in our opinion essential that mycologists become increasingly aware of the precarious position of many fungi at present and of the man-induced environmental changes affecting their occurrence.

References

Anon. (1992). Fungi in Red Data Book of Czechoslovakia. 4 pp. Unpubl. mnscr.

Arnolds, E. (1980). De oecologie en sociologie van Wasplaten. *Natura* **77**: 17–44.

Arnolds, E. (1984). Standaardlijst van Nederlandse macrofungi. 364 pp. Nederlandse Mycologische Vereniging, Baarn. (*Coolia*, **26**, suppl.).

Arnolds, E. (1988). The changing macromycete flora in the Netherlands. *Trans. Brit. Mycol. Soc.* **90**: 391–406.

Arnolds, E. (1989a). The influence of increased fertilization on the macrofungi of a sheep meadow in Drenthe, the Netherlands. *Opera Bot.* **100**: 7–21.

Arnolds, E. (1989b). A preliminary Red Data List of macrofungi in the Netherlands. *Persoonia* **14**: 77–125.

Arnolds, E. (1991). Mycologists and nature conservation. In D.L. Hawksworth (editor), Frontiers in Mycology pp. 243–264. C.A.B. International, Wallingford, U.K.

Bendiksen, E. and Høiland, K. (1992). Red list of threatened macrofungi in Norway. In: Directorate for Nature Management, Report 1992–6: 31–42. Oslo.

Benkert, D. (1982). Vorläufige Liste der verschollenen und gefährdeten Grosspilzarten der DDR. *Boletus* **6**: 21–32.

Derbsch, H. and Schmitt, J.A. (1984). Atlas der Pilze des Saarlandes, Teil 1. Verbreitung und Gefährdung. 536 pp. Minister für Umwelt des Saarlandes, Saarbrücken. (Aus Natur und Landschaft im Saarland, 2).
Derbsch, H. and Schmitt, J.A. (1987). Atlas der Pilze des Saarlandes. Teil 2. Nachweise, Ökologie, Vorkommen und Beschreibungen. 816 pp. Minister für Umwelt des Saarlandes, Saarbrücken. (Aus Natur und Landschaft im Saarland, 3).
Deutsche Gesellschaft für Mykologie (1992). Rote Liste der gefährdeten Grosspilze in Deutschland. IHW-Verlag, Eching. 174 pp.
Eriksson, J., Hjortstam, K., Larsson, K.-H. and Ryvarden, L. (1973–1988). The Corticiaceae of North Europe. Vol. 1–8. 1631 pp. Fungiflora, Oslo.
Fellner, R. (1989). Mycorrhiza-forming fungi as bioindicators of air pollution. *Agric. Ecosyst. Environm.* **28**: 115–120.
Floravårdskommitén för svampar. (1991). Kommenterad lista över hotade svampar i Sverige. *Windahlia* **19**: 87–130.
Frankland, J.C., Hedger, J.N. and Swift, M.J. (editors) (1982). Decomposer basidiomycetes, their biology and ecology. 355 pp. Cambridge University Press, Cambridge. (British Mycological Society Symposium 4).
Harley, J.L. and Smith, S.E. (1983). Mycorrhizal symbiosis. 483 pp. Academic Press, London. New York.
Høiland, K. (1986). Rapport angående storsoppfloraens reaksjon overfor forsuring, med hensyn til mykorrhizasoppene. Report to the Norvegian Ministry of Environmental Affairs; Botanical Garden and Museum, Oslo. T-Serie, 62 pp.
Ing, B. (1992). A provisional Red Data List of British Fungi. *Mycologist* **6**: 124–128.
Jansen, A.E. and Dighton, J. (1990). Effects of air pollutants on ectomycorrhizas. 58 pp. EEC, Brussels. (Air Pollution Research Report, 30).
Jülich, W. (1984). Die Nichtblätterpilze, Gallertpilze und Bauchpilze. 626 pp. G. Fischer, Stuttgart, New York. (Kleine Kryptogamenflora II b/1).
Kreisel, H. (editor) (1987). Pilzflora der Deutschen Demokratischen Republik. 281 pp. G. Fischer, Jena.
Krieglsteiner, G.J. (1991). Verbreitungsatlas der Grosspilze Deutschlands (West). Band 1, 2. 1016 pp. Ulmer Verlag. Stuttgart.
Krisai, I. (1986). Rote Liste gefährdeten Grosspilze Österreichs. In H. Niklfield (editor), Rote Listen gefährdeten Pflanzen Österreichs, pp. 178–192. Bundesministeriums Gesundheit-Umweltschutz. Wien. (Grüne Reihe, 5)
Kuyper, Th.W. (1989). Auswirkungen der Walddüngung auf die Mykoflora. *Beitr. Kenntn. Pilze Mitteleuropas* **5**: 5–20.
Lange, M. (1982). Fleshy fungi in grass fields. Dependence on fertilization, grass species, and age of field. *Nordic J. Bot.* **2**: 131–143.
Moser, M. (1983). Die Röhrlinge und Blätterpilze. 532 pp. 5. Aufl. G. Fischer, Stuttgart, New York. (Kleine Kryptogamenflora II b/2. 5).
Nauta, M. and Vellinga, E.C. (1992). Towards a distribution atlas of macrofungi in the Netherlands. *Mycologist* **6**: 6–10.
Nitare, J. (1988). Jordtungor, en svampgrupp påtillbakegång i naturlige fodermarker. *Svensk Bot. Tidskr.* **82**: 341–368.
Noordeloos, M.E. (1987). *Entoloma* (Agaricales) in Europe. *Beih. Nova Hedwigia* **91**: 419 pp.
Rald, E. (1985). Vokshatte som indikatorarter for mykologisk vaerdifulde overdrevslokaliteter. *Svampe* **11**: 1–9.
Rassi, P. and Väisänen, R. (1987). Threatened animals and plants in Finland. 82 pp. Ministry of Environment, Helsinki.
Termorshuizen, A.J. (1990). Decline of carpophores of mycorrhizal fungi in stands of *Pinus sylvestris*. Ph.D. thesis Agricultural University Wageningen. 128 pp.
Vesterholt, J. and Knudsen, H. (1990). Truede storsvampe i Denmark- en rødliste. 64 pp. Foreningen til Svampekundskabens Fremme, København.
Winterhoff, W. (1978). Bemerkenswerte Pilze in Trockenrasen des nördlichen Oberrheingebietes. *Hess. Florist. Briefe* **27**(1): 2–8; **27**(3): 41–47.
Winterhoff, W. (1984). Vorläufige Rote Liste der Grosspilze (Makromyzeten). In J. Blab, E. Nowak, W. Trautmann, and H. Sukopp (editors). Rote Liste der gefährdeten Tiere und Pflanzen in der Bundesrepublik Deutschland. 4. Aufl., pp. 162–184. Kilda, Greven.

Wojewoda, W. and Lawrynowicz, M. (1986). Red list of threatened macrofungi in Poland. In K. Zarzyckiego, and W. Wojewoda (editors). List of threatened plants in Poland, pp. 47–82. Polsk Akademia Nauk, Warzawa.
Wojewoda, W. and Lawrynowicz, M. (1992). Red List of threatened macrofungi in Poland. In K. Zarzycki, W. Wojewoda & Z. Heinrich (editors). List of threatened plants in Poland, 2nd ed., pp. 27–56. Polish Academy of Sciences, Cracow.

Ing, B. (1993). Towards a Red List of Endngered European Microfungi. In D.N. Pegler, L. Boddy, B. Ing & P.M. Kirk (Editors). Fungi of Europe: Investigation, Recording and Conservation, pp. 231–237. Royal Botanic Gardens, Kew.

18. TOWARDS A RED LIST OF ENDANGERED EUROPEAN MACROFUNGI

B. ING

Department of Biology, Chester College of Higher Education, Cheyney Road, Chester CH1 4BJ, U.K.

Abstract

278 species of macrofungi are considered to be endangered throughout Europe, on the basis of published national and regional lists. They are arranged in four categories depending on the degree and geographical extent of endangerment. The lists are provisional as little data are available for southern Europe and the evidence of losses is biased towards central, western and northern Europe.

Introduction

Several countries have produced lists of endangered fungi: Austria (Krisai, 1986); the British Isles (Ing, 1992); Denmark (Vesterholt & Knudsen, 1990); Finland (Rassi & Väisänen, 1987); Germany (Benkert, 1982, Lettau, 1982, Runge, 1987, Schmitt, 1988, Winterhoff, 1984, Winterhoff & Krieglsteiner, 1984 and Wöldecke, 1987); the Netherlands (Arnolds, 1989); Norway (Bendiksen & Hiland, 1992); Poland (Wojewoda & Lawrinowicz, 1986) and Sweden (Ingelör, Thor & Hallingbäck, 1991). In addition information has been made available from Czechoslovakia, Lithuania, Russia, Switzerland and the Ukraine. So far no lists, published or otherwise, have been produced from Albania, Belgium, Bulgaria, Estonia, France, Greece, Hungary, Italy, Latvia, Portugal, Romania, Spain, Turkey or the former Yugoslav regions.

National lists are different in character, reflecting more local criteria based on the cultural importance of fungi and ecological priorities, and are necessarily parochial. A European list must try to reflect a continent-wide anxiety for species whose geographical ranges do not coincide with convenient political boundaries, and whose occurrence is as much dependent on suitable ecology as climate.

The absence of information from southern Europe and much of the east means that any conclusions drawn about decline or otherwise are at most tentative and could be completely misleading. Fungi at the edge of their natural range will always appear more vulnerable and this is especially true of the north-south boundary of many species regarded as endangered in the far north but "safe" in the Mediterranean, and vice-versa. Any list attempting to cover Europe as a whole must be a compromise and this is what has been attempted here. This is especially true during a period of rapid climatic change to which the highly mobile fungi are among the first organisms to respond. What follows is not claimed to be anything other than a first draft and will need to be frequently updated. It will probably be necessary to produce future editions as regional Red Lists associated with standard biogeographical zones but this is not yet possible. The current level of information may allow the Arctic, Boreal, Atlantic, Central European, Alpine and Coastal zones to be attempted, but no more.

What are Red Lists For?

There is little point in trying to conserve an individual species of fungus, either by legislation or by land management. What is needed is the safeguarding of important fungal habitats. This can be achieved by legislation but additionally involves the correct management procedures. The presence of one or more Red List species, whether at national or European level, in an area immediately raises the importance of that area for conservation. Sites may therefore be scored for national or European Red List species and ranked in importance. This may be a useful tool in selecting sites for reserve status and should also determine the appropriate management, particularly in relation to forestry practices and the disposal of fallen timber. Where a site has a high proportion of species on the European List it should carry more weight than those with only national Red List species. For example, during the compilation of this European list and the British list, it was noticed that the Caledonian pine forests of the Scottish Highlands have a fine representation of the most endangered European species and should therefore be graded in terms of European rather than national importance. Similarly the coniferous forests of northern Fennoscandia have a high proportion of endangered species and should be managed sympathetically, with the long-term survival of fungal habitats a major consideration. The Atlantic oak woodlands of the British Isles are just as important for fungi on a European scale as they are for bryophytes, lichenised fungi, angiosperms and beetles.

Red lists may be used to promote species-protection legislation, by making it illegal to collect them, but, in general, this is undesirable as it places an onus on mycologists to choose species which are easily identifiable by non-specialists, and also presupposes that small-scale collection is damaging, which is not proven. It may also conflict with the choice of species which are indicators of particular biotopes and which are therefore important in reserve selection. In the compilation of this list it is the ecological importance of the ecosystem which is used as a guide to species selection, as being both more practical and of more long-term conservation value.

Compilation of the European List

Species were chosen for this list as a result of two processes. The first took place at the conference of the European Council for the Conservation of Fungi at Vilm, Germany, in September 1991, and resulted in a list of some thirty species to be considered as special cases for the Bern Convention on endangered European organisms. All those present offered a selection of taxa for consideration. The second exercise has been a detailed analysis of those national and regional lists available. All species occurring in two mor more lists, including the Vilm list, were compiled into a database and the degree of endangerment given a simple numerical value. From the available data a 'Euro-index' was calculated for each species — ranging from 2 to 26 on the scale (the highest score was for *Poronia punctata*!). The species were grouped into the four categories, using the Euro-index, and based on the geographical spread of reports and the mean level of endangerment.

Group A: widespread losses, rapidly declining populations, many national extinctions, high-level concern expressed.

Group B: widespread losses, evidence of steady decline, some national extinctions, medium-level concern.

Group C: widespread, but scattered, populations, fewer extinctions, lower-level concern.

Group D: local losses, some extinctions but mainly at edge of geographical range.

When this had been done the nature of the fungi themselves was considered and a number of arbitrary, but practical, choices were made:

(a) no hypogeous species were included. It was felt that inadequate data, often confused taxonomy and irregularity of fruiting made it difficult to assess these species in conservation terms.

(b) no plant pathogens were included. They are too dependent on their host plants and their conservation depends on conserving the hosts and *their* habitats.

(c) no fungi in difficult or poorly resourced groups were included. There is little hope in using species which only a few specialists can distinguish and which are not included in readily available reference works.

(d) only fungi likely to be identified with some confidence by the non-specialist field mycologist to be included, thus omitting most small ascomycetes — indeed *macrofungi* to be the keyword.

(e) species should be conspicuous, distinctive and attractive. Red lists must be practical and be based on what non-specialists will find.

(f) no introduced species to be included. This includes mycorrhizal species introduced with exotic trees or litter species with hothouse plants. They have no place in the native vegetation and there is no justification for conserving them outside their natural range. This does not apply to species extending their range without human assistance as a result of climatic change.

(g) where possible species should be associated with ecosystems which are themselves endangered. This strengthens the argument that fungi are essential parts of ecosystems and should be conserved alongside their habitats. In general the majority are found in ancient woodland — both broad-leaved and coniferous, in unimproved grassland, in lowland bogs and sand dunes. Woodland mycorrhizal species are a high priority, including boletes, *Cortinarius*, *Hygrophorus*, *Tricholoma*, the stipitate hydnoids, etc. as well as *Geastrum*. From grassland comes a large assemblage of *Hygrocybe*, *Entoloma* and the Geoglossaceae. The gasteroid basidiomycetes are particularly well represented — their distribution is easier to study with long-lasting fruit-bodies and good literature.

After the weeding-out of non-eligible species in the above categories some 600 species remained. By adopting a more rigorous approach, namely allowing only species with three or more list entries to be retained, the present list of 278 species was achieved.

This is only a provisional list and must be improved as more data becomes available from countries not yet providing lists. Existing lists must be updated and the changes monitored. Lists of this kind must be ongoing — there is only one final list — when everything is extinct!

Acknowledgements

Thanks are due to Dr Ann-Elise Jansen, Dr Sigvard Svensson and Prof Solomon Wasser for providing copies of Red Lists and other relevant literature and to Dr Eef Arnolds and Prof Maria Lawrynowicz for valuable discussions. The final choice of species was mine and no-one else need be blamed!

References

Arnolds, E. (1989). A preliminary Red Data List of macrofungi in the Netherlands. *Persoonia*, **14**: 77–125.

Bendiksen, E. and Høiland, K. (1992). Red List of threatened macrofungi in Norway. *Report, Directorate for Nature Management*, **6**: 31–42.

Benkert, D. (1982). Vorläufige Liste der verschollenen und gefährdeten Grosspilzarten der DDR. *Boletus*, **6**: 21–32.

Ing, B. (1992). A provisional Red Data List of British fungi. *Mycologist* **6**; 124–128.

Ingelög, T., Thor, G. and Hallingbäck, T. (1991). Hotade växter i Sverige 1990. Databanken för hotade arter, Lund.

Krisai, I. (1986). Rote Liste gefährdeter Grosspilze Österreichs, In Rote Listen gefährdeter Pflanzen Österreichs. **5**, pp. 178–192. Grüne Reihe Bundesministerimus Gesundheit-Umweltschutz

Lettau, M. (1982). Vorläufige Liste verschollener und gefährdeter Grosspilze in Schleswig-Holstein. Schriftenreihe Landesamtes Naturschutz Landschaftplege Schleswig-Holstein. **5**: 58–71.

Rassi, P. and Väisänen, P. (1987). Threatened plants and animals in Finland. 82 pp. Ministry of Environment, Helsinki.

Runge, A. (1987). Vorläufige Rote Liste der gefährdeten Grosspilze (Makromyzeten) in Nordrhein-Westfalen. In Landensanstalt für Ökologie, Landschaftsentwicklung und Forstplanung NW, Rote Liste der in Nordrhein-Westfalen gefährdeten Pflanzen und Tiere, **2**, 18-03–18-15. Recklinghausen.

Schmitt, J.A. (1988). Die Pilze. In Rote Liste, Bedrohte Tier- und Pflanzenarten im Saarland pp. 77—106. Minister für Umwelt Saarbrucken.

Vesterholt, J. and Knudsen, H. (1990). Truede storsvampe i Danmark — en rødliste. Foreningen til Svampekundskabens Fremme, København.

Winterhoff, W. (1984). Vorläufige Rote Liste der Grosspilze (Makromyzeten), In J. Blab, E. Nowak, W. Trautman, and H. Sukopp, (editors) Rote Liste der gefährdeten Tiere und Pflanzen in der Bundesrepublik Deutschland, **4**, pp. 162–184. Greven.

Winterhoff, W. and Krieglsteiner, G.J. (1984). Gefährdete Pilze in Baden-Württemberg. *Beih. Verröff. Naturschutz Landschaftspflege Baden-Württemberg* **40**: 1–120.

Wojewoda, W. and Lawrinowicz, M. (1986). Red List of threatened macrofungi in Poland. In K. Zazyckiego and W. Wojewoda (editors) List of threatened plants in Poland pp. 47–82. Warsaw.

Wöldecke, K. (1987). Rote Liste der in Niedersachsen und Bremen gefährdeten Grosspilze, *Informationsdienst Naturschutz Niedersachsen* **7**: 1–28.

APPENDIX 1 A Provisional European Red List of Endangered Macrofungi

Group A: widespread losses, rapidly declining populations, many national extinctions, high-level concern.

Bankera fuligineoalba
Boletopsis leucomelaena
Boletus juquilleus
Boletus regius
Boletus rhodoxanthus
Boletus satanas
Cotylidia pannosa
Geastrum campestre
Geastrum fornicatum
Geoglossum peckianum (difforme)
Gomphus clavatus
Hapalopilus (Aurantioporus) croceus
Hygrocybe spadicea
Hydnellum ferrugineum
Phellodon confluens
Poronia punctata

Group B: widespread losses, evidence of steady decline, some national extinctions, medium-level concern.

Aleurodiscus disciformis
Armillaria ectypa
Boletus fechtneri
Boletus impolitus
Boletus queletii
Bovista limosa
Bovista paludosa
Bovistella radicata
Buglossoporus pulvinus
Catathelasma imperiale
Chamaemyces fracidus
Cortinarius bulliardii
Cortinarius violaceus
Disciceda bovista
Disciceda candida
Entoloma bloxamii (*madidum*)
Fomitopsis rosea
Geastrum berkeleyi
Geastrum corollinum (*recolligens*)
Geastrum elegans (*badium*)
Geastrum schmidelii (*nanum*)
Geoglossum simile
Hericium erinaceus
Hydnellum aurantiacum
Hydnellum caeruleum
Hydnellum peckii
Hydnellum scrobiculatum
Hygrocybe aurantiosplendens
Hygrocybe lacmus
Hygrocybe ovina
Hygrocybe perplexa (*sciophana*)
Hygrocybe radiata
Hygrophorus mesotephrus
Hygrophorus pudorinus
Hygrophorus russula
Lactarius mairei
Lactarius musteus
Leucopaxillus tricolor
Lopharia spadicea
Lycoperdon caudatum (*pedicellatum*)
Melanophyllum eyrei
Microglossum olivaceum
Myriostoma coliforme
Onnia romentosa
Phellodon niger
Phellodon tomentosus
Phylloporus rhodoxanthus
Polyporus umbellatus
Porpoloma spinulosum
Pseudorhizina sphaerospora
Pulveroboletus gentilis (*cramesinus*)
Pulveroboletus lignicolor
Ramariopsis crocea
Rhodotus palmatus
Russula pumila
Sarcodon fuligineoviolaceus
Sarcodon glaucopus
Sarcodon scabrosus
Sarcosoma globosum
Spathularia flavida
Suillus flavidus
Thuemenidium atropurpureum
Tricholoma acerbum
Tricholoma apium
Tricholoma aurantium
Tricholoma colossum

Group C: widespread, but scattered, populations, fewer extinctions, lower-level concern.

Agrocybe firma
Amanita franchetii (*aspera*)
Amanita friabilis
Astraeus hygrometricus
Boletinus cavipes
Boletus aereus
Boletus appendiculatus
Boletus fragrans
Boletus radicans (*albidus*)
Caloscypha fulgens
Camarophyllopsis schulzeri
Cantharellus cinereus
Camarops tubulina
Clavaria zollingeri
Climacodon septentrionalis
Clitocybe josserandii
Coltricia cinnamomea
Cordyceps capitatus
Coriolopsis gallica
Cortinarius anthracinus
Cortinarius canobarba
Cortinarius cinnabarinus
Cortinarius cyanites
Cortinarius orellanus
Cortinarius praestans
Cortinarius sebaceus
Cortinarius sodagnites
Cudonia circinans
Cyathus stercoreus
Dentipellis fragilis
Dichomitus campestris
Entoloma euchroum
Entoloma excentricum
Entoloma griseocyaneum
Entoloma prunuloides
Entoloma roseum
Funalia trogii
Galerina stagnina
Ganoderma pfeifferi
Ganoderma resinaceum
Geastrum coronatum
Geastrum floriforme
Geastrum melanocephalum
Geastrum minimum
Geastrum pectinatum
Geastrum quadrifidum
Geastrum rufescens (*vulgatum*)
Geastrum striatum
Geoglossum fallax
Geoglossum glutinosum
Geoglossum umbratile (*nigritum*)
Gomphidius maculatus
Grifola frondosa
Gyromitra gigas
Haploporus odorus
Hericium clathroides
Hydnellum compactum
Hydnellum concrescens
Hydnellum spongiosipes
Hygrocybe fornicata
Hygrocybe ingrata
Hygrocybe insipida
Hygrocybe intermedia
Hygrocybe lepida (*cantharellus*)
Hygrocybe obrussea (*quieta*)
Hygrocybe nitrata (*murinacea*)
Hygrocybe punicea
Hygrocybe subglobispora
Hygrocybe unguinosa
Hygrophorus arbustivus
Hygrophorus calophyllus
Hygrophorus camarophyllus
Hygrophorus gliocyclus (*ligatus*)
Hygrophorus hypothejus (*aureus*)
Hygrophorus nemoreus
Hygrophorus purpurascens
Hypocreopsis lichenoides
Inocybe godeyi
Inonotus hispidus
Ischnoderma resinosum
Lactarius acris
Lactarius aspideus
Lactarius controversus
Lactarius hysginus
Lactarius resimus
Lactarius violascens
Lariciformes officinalis
Lentinellus ursinus
Lentinus adhaerens
Lepiota cortinarius
Lepiota fuscovinacea
Lepiota grangei
Lepiota pseudofelina
Leucocortinarium bulbiger
Leucopaxillus gentianeus
Limacella glioderma
Limacella guttata
Lycoperdon mammaeforme
Lycoperdon marginatum
Lyophyllum transforme
Marasmius hudsonii
Microglossum viride

Group C continued

Mutinus caninus
Mycenastrum corium
Omphalina sphagnicola
Omphalotus olearius
Onygena equina
Otidea concinna
Oudemansiella (*Xerula*) pudens
Perenniporia fraxinea
Phellodon melaleucus
Phyllotus (*Pleurocybella*) porrigens
Pisolithus arhizus (*arenarius*)
Pluteus aurantiorugosus
Porphyrellus porphyrosporus
Porpoloma metapodium
Psathyrella caput-medusae
Pulcherricium (*Terrana*) caeruleum
Ramaria botrytis
Ramaria formosa
Ramariopsis pulchella
Russula minutula
Sarcodon imbricatus
Sarcodon leucopus
Scytinostroma portentosum
Sparassis laminosa *agg.*
Strobilomyces floccopus
Stropharia albocrenulata
Trametes suavolens
Trichoglossum hirsutum
Trichoglossum variabile
Tricholoma focale
Tricholoma sejunctum
Tricholoma squarrulosum
Tulostoma brumale
Tulostoma fimbriatum
Tulostoma melanocyclum
Tyromyces (*Aurantioporus*) fissilis
Urnula craterium
Urnula hiemalis
Verpa conica
Vibrissea truncorum
Volvariella bombycina
Volvariella caesiotincta
Xylobolus frustulatus

Group D: local losses, some extinctions but mainly at edge of geographical range.

Albatrellus ovinus
Albatrellus (*Scutiger*) pes-caprae
Amanita caesarea
Amylocystis lapponica
Antrodia crassa
Battarraea phalloides
Boletus torosus
Bovista graveolens
Clavariadelphus truncatus
Cortinarius cedretorum
Cortinarius scaurus
Cotylidia undulata
Dermoloma josserandii
Entoloma hirtum
Floccularia straminea
Geastrum hungaricum
Geastrum saccatum
Geastrum triplex
Geoglossum sphagnophilum (*glabrum*)
Gloeporus dichrous
Hohenbuehelia longipes
Holwaya mucida
Hygrophorus marzuolus
Hygrophorus poetarum
Inonotus dryophilus
Laurilia sulcata
Lentinellus vulpinus
Lepiota lignicola
Lindtneria trachyspora
Montagnea arenaria
Multiclavula mucida
Phellinus tremulae
Pholiota squarrosoides
Piptoporus pseudobetulinus
Pulveroboletus hemichrysus
Pycnoporellus alboluteus
Ramaria fagetorum
Russula aquosa
Russula borealis
Russula helodes
Sarcodon joeides
Sarcodon martioflavus
Sarcodontia crocea
Squamanita scheieri
Steccherinum robustius
Stereopsis vitellina
Thuemenidium (*Geoglossum*) arenarium
Torrendia pulchella
Tulostoma niveum
Tyromyces (*Loweomyces*) fractipes
Tyromyces placenta

R. Fellner (1993). Air pollution and mycorrhizal fungi in central Europe. In D.N. Pegler, L. Boddy, B. Ing & P.M. Kirk (Editors). Fungi of Europe: Investigation, Recording and Conservation, pp. 239–250. Royal Botanic Gardens, Kew.

19. AIR POLLUTION AND MYCORRHIZAL FUNGI IN CENTRAL EUROPE

R. FELLNER

Forestry and Game Management Research Institute, Strnady, 156 04 Praha 5 — Zbraslav, Czech Republic

Abstract

The decline of ectomycorrhizal fungi in Central Europe is described in terms of the disturbance of ectotrophic forest stability as a consequence of air pollution. This process includes latent, acute and lethal stages corresponding with both specific phases of the impoverishment of ectomycorrhizal mycocoenoses and the enrichment of lignicolous ones. Some results of mycological monitoring in former Czechoslovakia are given for exemplifying main trends.

Introduction

A rapid decrease of ectomycorrhizal fungi and many other macromycetes has become a well-known phenomenon in some parts of Europe in the last decade (Fellner 1983, 1985, 1988, 1989; Arnolds 1985, 1988, 1991; Derbsch & Schmitt 1987; Jakucs 1988). It has been argued that changes in populations of these fungi can be attributed to direct or indirect effects of air pollution (Dörfelt & Braun 1980; Arnolds, op. cit.; Gulden & Høiland 1985; Schlechte 1986, 1987, 1991; Fellner 1987; Termorshuizen & Schaffers 1987; Jansen & van Dobben 1988; Jakucs, op. cit.; Jansen 1991; Gulden & al. 1992). In view of the fact that in Central Europe these changes have preceded a visual damaging of forest or even its decline (Jakucs 1988; Fellner & Landa 1990; Fellner & Soukup in press), it seems that many ectomycorrhizal fungi are very suitable bioindicators of the disturbance of forest ectotrophic stability (Fellner 1986, 1989). It concerns not only particular species, but also whole assemblages (mycocoenoses) of ectomycorrhizal fungi (Fellner 1988).

Methods and Results

A mycocoenological approach in permanent plots is widely used in Central European studies of the decline of ectomycorrhizal fungi (Fellner 1985, 1987, 1988; Schlechte 1986, 1987, 1991; Fellner & Soukup in press). The size of permanent plots reaches mostly 2500 m^2 (Czechoslovakia) or even 3000 m^2 (Germany). It is supposed to be a sufficient and suitable size for a minimareal of macromycetes in Central European forest ecosystems (Fellner 1985; Schlechte 1991). In the case of smaller plots (e.g. 1000 m^2) it is recommended to reduplicate their number or even to cluster them in one cumulative "minimareal plot" in each locality (cf. Fellner 1992).

Investigations in Czechoslovakia described — in agreement with other European authors — a conspicuous decline of fungal species diversity, especially in ectomycorrhizal species in spruce mountain forests (Fellner 1985, 1991; Fellner & Soukup in press) (Fig. 1). As it was postulated by Fellner (1985) or Schlechte (1991) the number of ectomycorrhizal

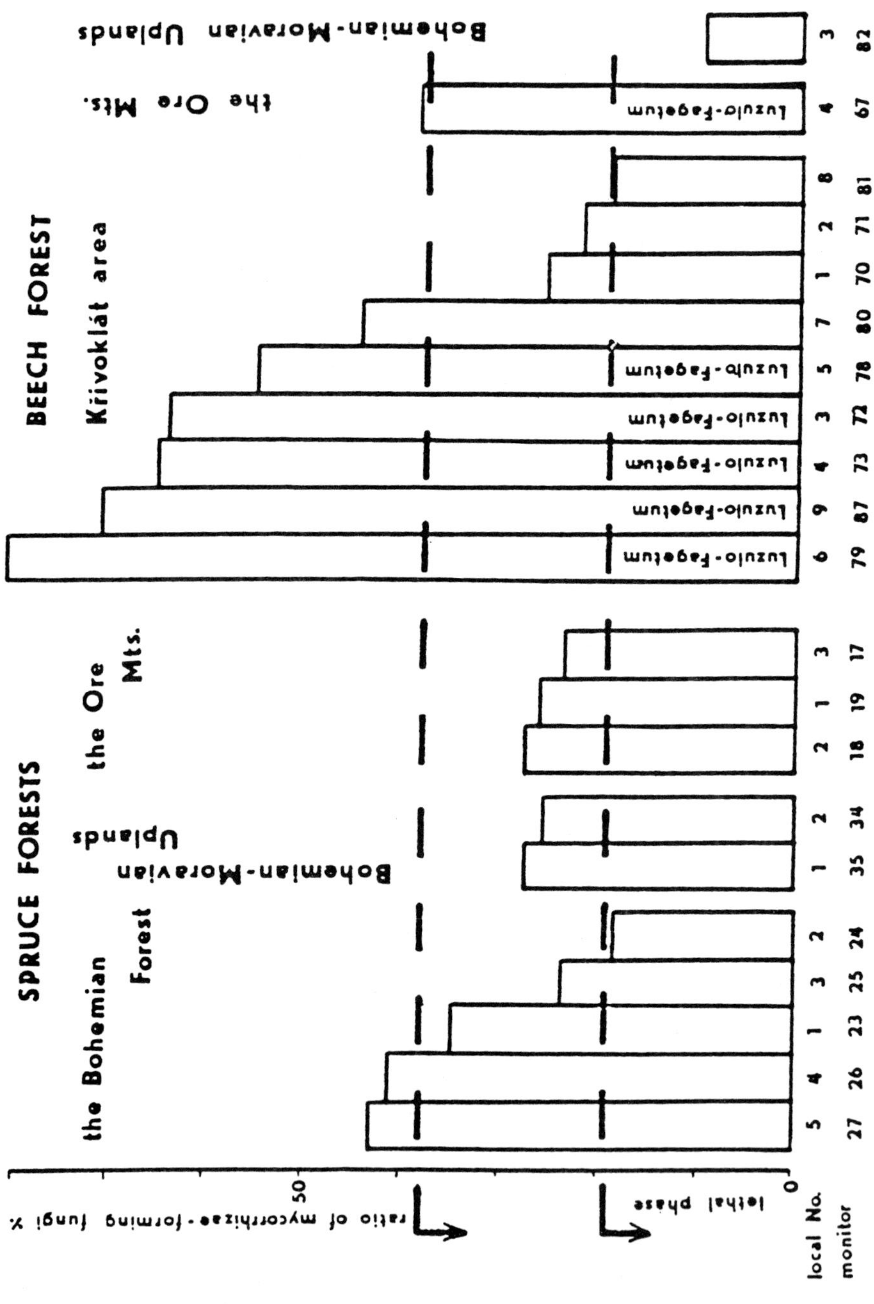

FIG. 1. The proportion of species of mycorrhizal fungi in permanent plots in spruce and beech forests in Bohemia (Czech Republic), 1986–1990.

species in permanent forest plots is ordinarily about 40–60% in proportion to all other macromycetes, at least in non-calcareous sites not damaged by air pollution. In air polluted areas, however, this ratio is much lower.

In contradiction to changes observed in ectomycorrhizal fungi, the number of lignicolous species is significantly increasing in air polluted areas (Fig. 2).

However, not only the number of species, but also the number of sporocarps is a factor significantly influenced by air pollution. While the total number of macromycete species is generally decreasing at a slow rate (at least in the beginning), the sporocarp production is, as a rule, influenced very quickly and considerably. This sporocarp production is strongly inhibited in most of the macromycetes with the exception of lignicolous fungi

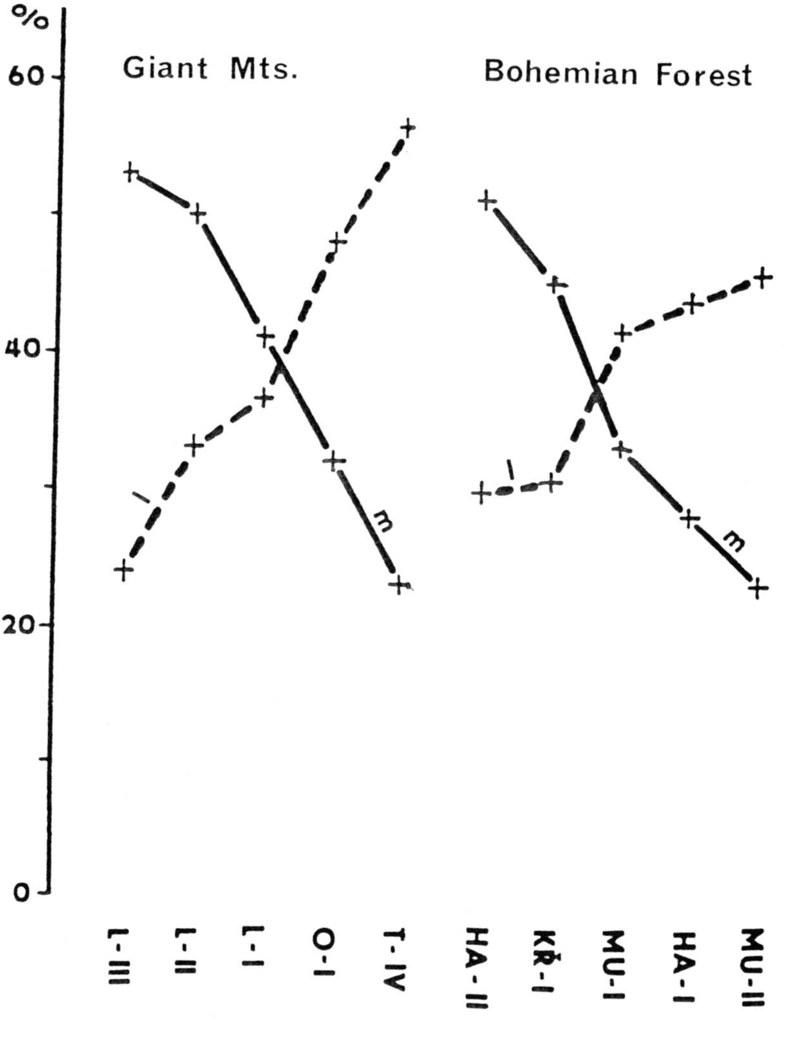

FIG. 2. The decreasing proportion of mycorrhizal species (m) and the increasing proportion of lignicolous species (1) in spruce permanent plots ecologically more protected or more exposed to air pollution (from the left to the right) in two mountain areas in Bohemia (1991).

whose production of sporocarps is at first slow, but then at a more rapid growth rate in air polluted sites (Arnolds 1985, 1991; Fellner & Soukup in press). The loss in the total sporocarp production is at first not able to be compensated by the enrichment in lignicolous fungal communities. In the last stage of forest deterioration, however, it is possible to observe a conspicuous expansion of lignicolous fungi (Fig. 3, 4).

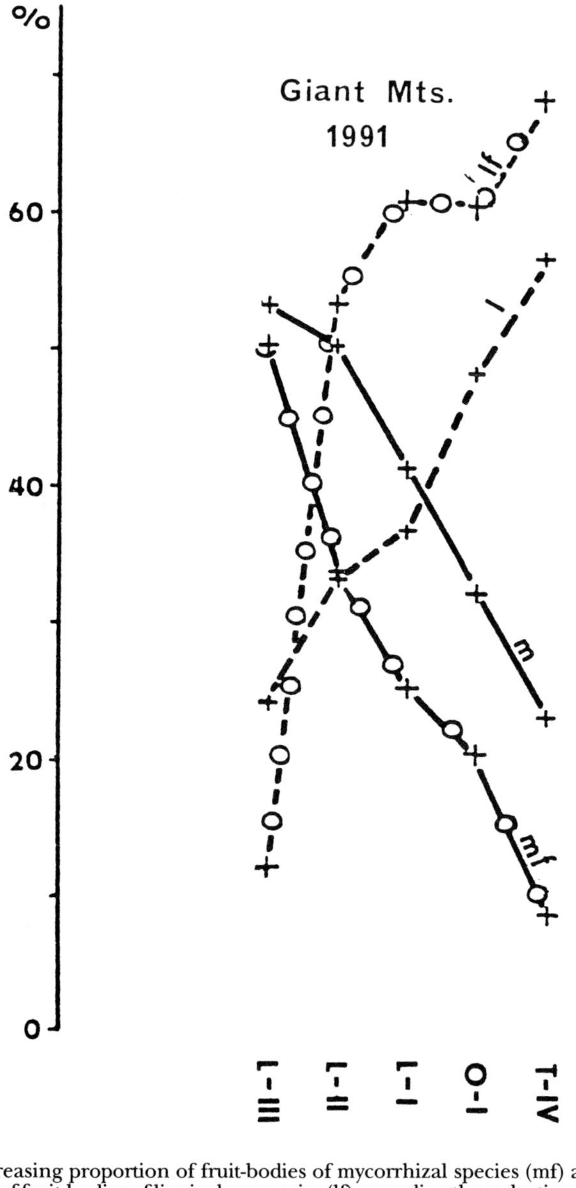

FIG. 3. The decreasing proportion of fruit-bodies of mycorrhizal species (mf) and the increasing proportion of fruit-bodies of lignicolous species (lf) preceding the reduction of species diversity of mycorrhizal fungi (m) and the enrichment of species diversity of lignicolous fungi (l) in spruce permanent plots ecologically more protected or more exposed to air pollution (from the left to the right) in the Giant Mts. (Krkonose), Northern Bohemia (1991).

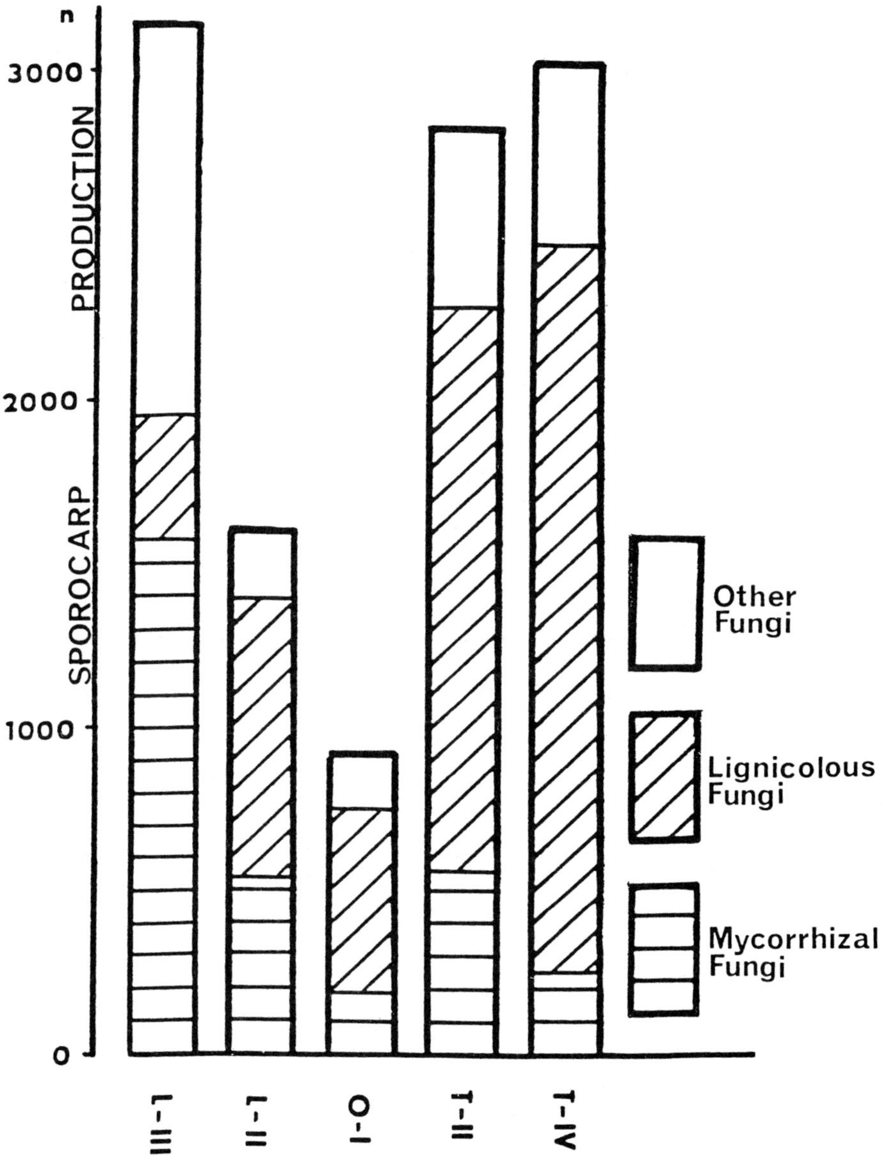

FIG. 4. The changing number of fruit-bodies of mycorrhizal, lignicolous, and other macromycete species in permanent plots exposed to lower or higher levels pollutants (from the left to the right) in spruce permanent plots in the Giant Mts. (1991).

The system of forest deterioration as a consequence of air pollution includes three stages of the disturbance of ectotrophic forest stability:

(a) latent disturbance, if the proportion of species of ectomycorrhizal fungi in the total count of macromycetes is decreasing to 40%, while the proportion of lignicolous species is tending to reach more than 30%;

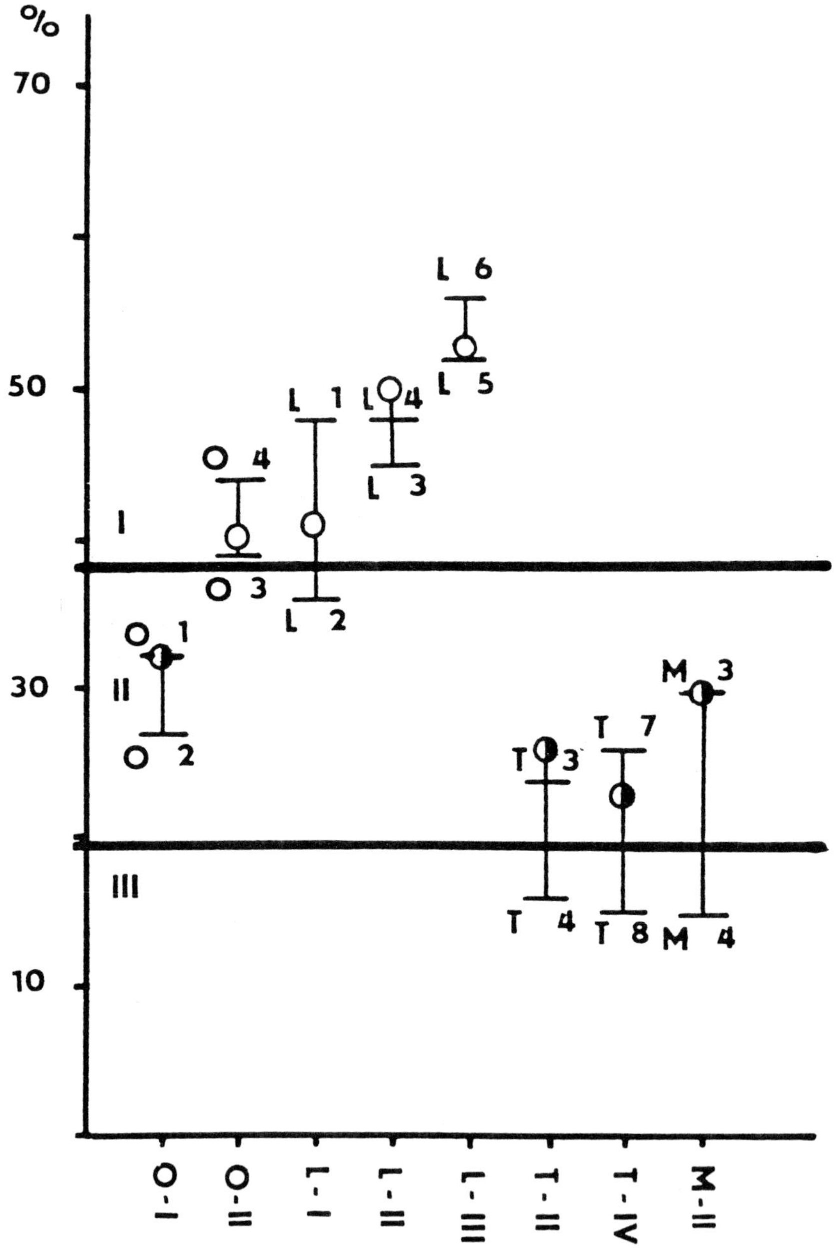

FIG. 5. The stages of the deterioration of ectotrophic forest stability in permanent spruce plots at the bottom of valleys (plots O–I, O–II, L–I, L–II, L–III), ecologically more protected, and on the top of hills (plots T–II, T–IV, M–II), ecologically more exposed to air pollution; I = latent, II = acute, III = lethal (Giant Mts., 1991).

(b) acute disturbance, if the ectomycorrhizal species constantly contribute less than 40% of the total number of macromycetes, while lignicolous species, as a rule, contribute more than 40%;

(c) lethal disturbance, if ectomycorrhizal species constantly contribute less than 20%, while lignicolous species, as a rule, contribute more than 55% of all macromycetes.

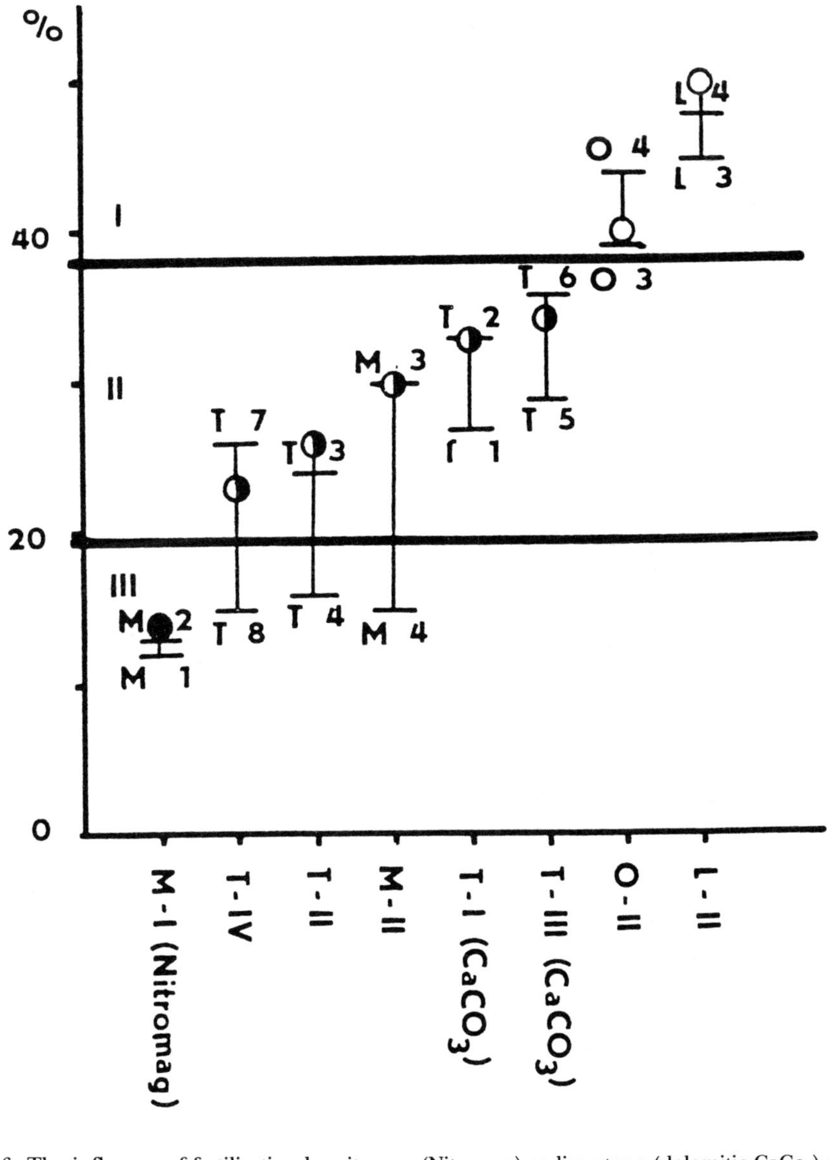

FIG. 6. The influence of fertilization by nitrogen (Nitromag) or limestone (dolomitic $CaCo_3$) on ectotrophic forest stability in experimental spruce plots (M–I, T–I, T–III, all on the top of hills) and in comparison with plots without any application (M–II, T–II, T–IV on the top of hills; L–II, O–II at the bottom of valleys); I = latent, II = acute, III = lethal (Giant Mts., 1991).

These three stages are linked directly to particular phases of impoverishment of ectomycorrhizal mycocoenoses and of enrichment of lignicolous mycocoenoses. The latent disturbance is, on the one hand, linked to the inhibition of sporocarp production of ectomycorrhizal fungi (accompanied by a decline of highly sensitive species), and on the other hand, to the stimulation of sporocarp production of wood destroying fungi. The acute disturbance is linked, on the one hand, to the reduction of species diversity of

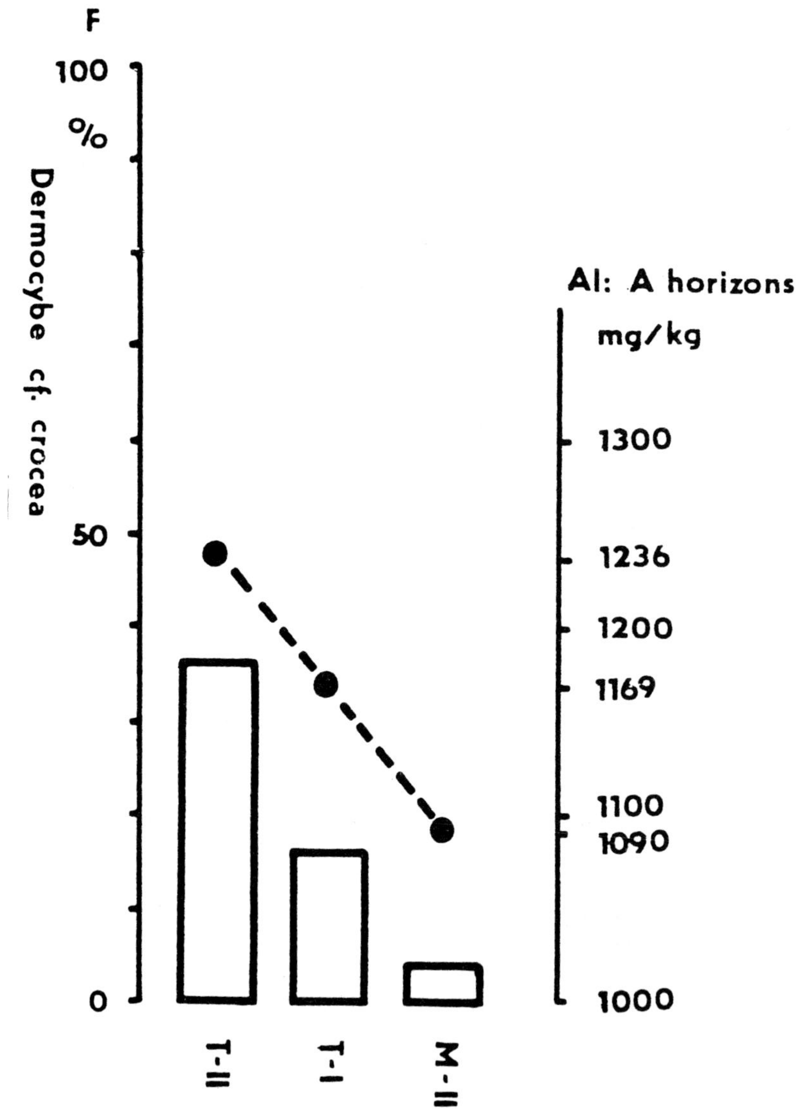

FIG. 7. The correlation between the frequency of occurrence of *Dermocybe* cf. *crocea* and the amount of aluminium in spruce permanent plots in the Giant Mts. (1991).

ectomycorrhizal fungi (with a continuous inhibition of sporocarp production), and on the other hand, to the increase of species diversity of wood destroying fungi (with a continuous stimulation of sporocarp production). The lethal disturbance is linked, on the one hand, to the partial to total destruction of mycorrhizal mycocoenoses, and on the other hand, to the expansion of lignicolous mycocoenoses — (cf. Fellner 1992).

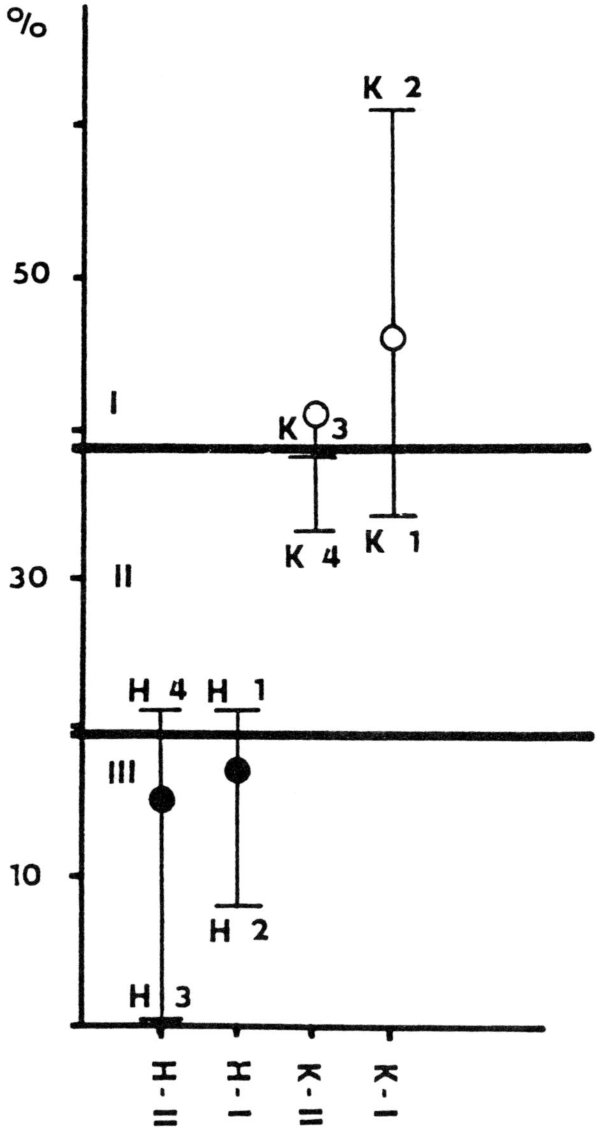

FIG. 8. The stages of deterioration of ectotrophic forest stability in permanent oak plots in the Slovakian Karst with higher (plots H-I, H-II) and lower (plots K-I, K-II) levels of SO_2 pollution, eastern Slovakia (1991); the mean annual concentration of SO_2 in H-I, H-II plots is more than 20 g.m^{-3}.

Evaluating annual results, it is possible to take into account separately data from partial segments and then from entire cumulative "minimareal plots" (e.g. the 1000 m^2 size and then the 2500 m^2 size, respectively). This allows inclusion of the first in the form of intervals as diagrammed in Fig. 5 describing mycorrhizal conditions in these minimareal units.

The same approach proved to be highly useful in the case of evaluating how the fertilization (by nitrogen or limestone) influences mycorrhizal mycocoenoses and the ectotrophic forest stability (Fig. 6). While the nitrogen application caused a strong loss of the species diversity of ectomycorrhizal fungi, corresponding to the lethal disturbance of ectotrophic forest stability, the limestone application surprisingly caused a slight increase in the percentage proportion of ectomycorrhizal species, indicating a trend to the reinforcement of ectotrophic stability. Consistent with some Schlechte's observations (Schlechte 1991), the limestone application (at least in the form of dolomite which less strongly influences soil pH) seems to have a stabilizing effect in acid spruce sites chronically affected by the air pollution (Fellner & al. 1991b), and more concretely in sites which already have been in the stage of the acute disturbance of ectotrophic forest stability. In such a case the dolomitic limestone application could be acceptable for a soil melioration but not before.

Høiland (1992) found that in acidified forests some species of *Cortinarius* subgenus *Dermocybe* are increasing. He suggested that these ectomycorrhizal fungi are able to detoxify aluminium ions which are released into soil-water due to the process of acidification. Indeed, we have found in permanent plots in the Giant Mountains (Northern Bohemia) that *Cortinarius* cf. *croceus* (Schaeff.: Fr.) Høiland occurs more frequently now. But its frequency was conspicuous only in plots containing at least 1000 mg kg-[1] of aluminium in A soil horizons (cf. Fellner & al. 1991b) (Fig. 7).

The proposed scheme of the disturbance of ectotrophic forest stability has been recognized and confirmed both in spruce and oak sites in Central Europe. For example in the Slovakian Karst Protected Landscape Area, the observed disturbance of ectotrophic forest stability in *Quercus virgiliana* localities was found in close correlation with known levels of SO_2 pollution (Fellner & al. 1991c) (Fig. 8).

Conclusions

A rapid decline of ectomycorrhizal fungi described from Czechoslovakia in terms of the disturbance of ectotrophic forest stability seems to be a European-wide problem. For the reason, a proposal has been made to monitor macromycetes in European spruce and oak forests at the Second Meeting of the European Council for the Conservation of Fungi held in Vilm, Germany, in 1991 (Fellner & Arnolds 1991). The main aim of the project would be to know the actual qualitative and quantitative composition of the mycoflora in spruce and oak forests, according to standard methods in various parts of Europe (enabling a further comparision) as well as to relate the composition of the mycoflora to some environmental factors (including air pollution levels). This project was proposed to be sponsored by an EEC programme stimulating cooperation between western and eastern Europe.

References

Arnolds, E. (editor) (1985). Veranderingen in de paddestoelenflora (mycoflora). *Wetensch. Meded. Kon. Ned. Natuurhist. Ver.* **167**: 1–101.

Arnolds, E. (1989). The changing macromycete flora in the Netherlands. *Trans. Brit. Mycol. Soc.* **90**: 391–406.

Arnolds, E. (1991). Decline of ectomycorrhizal fungi in Europe. *Agric. EcoSyst. Environm.* **35**: 209–244.

Derbsch, H. and Schmitt, J.A. (1987). Atlas der Pilze des Saarlandes, Teil 2. Nachweise, Ökologie, Vorkommen und Beschreibungen. 816 pp. Minister für Umwelt des Saarlandes, Saarbrücken. (Aus Natur und Landschaft im Saarland 3).

Dörfelt, H. and Braun, U. (1980). Untersuchungen zur Bioindikation durch Pilze in der Dübener Heide (DDR). In R. Schubert and J. Schuh (editors). Bioindikation 4, pp. 15–20. Martin Luther Universität, Halle (Saale).

Fellner, R. (1983). Mycorrhizae-forming fungi in climax forest communities at a timberline in Giant Mountains. *Ceská Mykol.* **37**: 109.

Fellner, R. (1985). Ektomykorrhizni houby klimaxovych lesnich spolecenstev pri horni hranici lesa v imisnich oblastech Krkonos (Ectomycorrhizae-forming fungi in climax forest communities at a timberline in air polluted areas of Giant Mountains). Ph.D. Thesis. Kostelec n C. 1. 300 pp. (unpubl.).

Fellner, R. (1986). The destruction of ectomycorrhizal conditions at a timber line in immission areas of Krkonose Mts. In 18th IUFRO World Congress, Lublan, Division 1, Vol. 2, p. 771.

Fellner, R. (1987). Monitorovani zmen v druhove diverzite mykorrhiznich hub na imisne ruzne exponovanych stanovistich (Monitoring of changes in species diversity of mycorrhizal fungi in sites exposed to different levels of air pollution). In R. Fellner (editor), Ekologie mykorrhiz a mykorrhiznich hub, pp. 93–103. DT CSVTS, Pardubice.

Fellner, R. (1988). Effects of acid depositions on the ectotrophic stability of mountain forest ecosystems in Central Europe (Czechoslovakia). In A.E. Jansen, J. Dighton and A.H.M. Bresser (editors). Ectomycorrhiza and Acid Rain. Proc. Workshop on Ectomycorrhiza, Berg en Dal 1987, The Netherlands, pp. 116–121. Commission Eur. Communities, Bilthoven.

Fellner, R. (1989). Mycorrhizae-forming fungi as bioindicators of air pollution. *Agric. EcoSyst. Environm.* **28**: 115–120.

Fellner, R. (1992). Monitoring of fungus in permanent plots. In K. Høiland and E. Ohenoja (editors). Methodological problems in the study of fungus flora, Lammi 1992, Finland (Proc. in prep.).

Fellner, R. and Arnolds, E. (1991). Proposal for monitoring of macromycetes in European spruce (*Picea*) and oak (*Quercus*) forests. In E. Arnolds (editor), Second meeting of the European Council for Conservation of Fungi, Vilm 1991, Germany, pp. 23–24. Biol. Station Wijster.

Fellner, R. and Landa, J. (1990). Poznatky z mykologickeho vyzkumu v CHKO Krivoklatsko (Notes to the mycological research in the Landscape Protected Area Krivoklatsko) In M. Rivola et al. (editors). Soucasny stav a cile botanickeho vyzkumu v CHKO Krivoklatsko, pp. 104–107. SSPPOP Stredoceskeho kraje, Praha.

Fellner, R. and Soukup, F. (in press). Mycological monitoring in air polluted areas of the Czech Republic. *Commun. Inst. Forest. Cech.* **17**.

Fellner, R., Bystrican, A., Tihounova, H., Soukup, F. and Javurek, M. (1991a). Monitorovani vlivu vapneni a kapalneho hnojeni na mykorrhizni pomery ve smrkovych porostech v Krkonosich — 1 (Monitoring the influence of liming and liquid fertilization on mycorrhizal conditions in spruce forests in Giant Mts. — 1.). Annual Res. Rep., Forestry and Game Management Research Institute Strnady, pp. 1–82 (unpubl.).

Fellner, R., Landa, J., Soukup, F. and Javurek, M. (1991b). Mykologicky monitoring ve smrkovych porostech na uzemi Krkonosskeho narodniho parku — 1 (Mycological monitoring in spruce forests in the Giant Mountains National Park — 1.) Res. Rep., Mykol. Monitoring Praha, pp. 1–50 (unpubl.).

Fellner, R., Landa, J. and Soukup, F. (1991c). Mykologicky monitoring v dubovych porostech na uzemi Chranene krajinne oblasti Slovensky kras — 1 (Mycological monitoring in oak forests in the Slovakian Karst Protected Landscape Area — 1). Res. Rep., Czech Union of Nature Conservation Local Organization, Strnady, pp. 1–36 (unpubl.).

Gulden, G. and Høiland, K. (1985). The role of ectomycorrhiza in a situation of air pollution and forest death. *Agarica* **6** (12): 341–357.

Gulden, G., Høiland, K., Bendiksen, K., Brandrud, T.E., Foss, B.S., Jenssen, H.B. and Laber, D. (1992). Fungi and air pollution. Mycocoenological studies in three oligotrophic spruce forests in Europe. *Biblioth. Mycol.* **144**: 1–81.

Høiland, K. (1992). Chemotaxonomy and its implication for ecology. In K. Høiland and E. Ohenoja (editors). Methodological problems in the study of fungus flora, Lammi 1992, Finland (Proc. in prep.).

Jakucs, P. (1988). Ecological approach to forest decay in Hungary. *Ambio* **17**: 267–274.

Jansen, A.E. (1991). The mycorrhizal status of Douglas fir in the Netherlands: its relation with stand age, regional factors, atmospheric pollutants and tree vitality. *Agric. EcoSyst. Environm.* **35**: 191–208.

Jansen, E.J. and Dobben, H.F. van (1987). Is decline of *Cantharellus cibarius* in the Netherlands due to air pollution? *Ambio* **16**: 211–213.

Schlechte, G. (1987). Ecological studies on mycorrhizae-forming fungi of forest stands exposed to different levels of air pollution. In R. Fellner (editor). Ekologie mykorrhiz a mykorrhiznich hub, pp. 82–92. DT CSVTS, Pardubice.

Schlechte, G. (1991). Zur Struktur der Basidiomyzeten-Flora von unterschiedlich immissionsbelasteten Waldstandorten in Südniedersachsen unter besonderer Berücksichtigung der Mykorrhizabildung, 224 pp. Jahn u. Ernst Verlag, Hamburg.

Termorshuizen, A.J. and Schaffers, A.P. (1987). Occurrence of carpophores of ectomycorrhizal fungi in selected stands of *Pinus sylvestris* in the Netherlands in relation to stand vitality and air pollution. *Pl. & Soil* **104**: 209–217.

Keizer, P.J. (1993). The influence of nature management on the macromycete flora. In D.N. Pegler, L. Boddy, B. Ing & P.M. Kirk (Editors). Fungi of Europe: Investigation, Recording and Conservation, pp. 251–269. Royal Botanic Gardens, Kew.

20. THE INFLUENCE OF NATURE MANAGEMENT ON THE MACROMYCETE FLORA

P.J. KEIZER[1]

Biological Station Wijster, Centre for Soil Ecology, Kampsweg 27, 9418PD Wijster, Netherlands.

Abstract

The paper aims to review the available information on nature management measures in relation to the mycoflora. Although much information is available for plants and animals, there is little quantitive knowledge available concerning the fungi in nature management methods. Consideration is given to maintenance management, restoration management and nature development with regard to forests, grasslands, tree-meadows and parkland.

Introduction

At the beginning of this century concern arose about the mass destruction of natural habitats. In that time nature conservation organisations were established whose primary aim was the acquisition of nature reserves. Later similar governmental organisations were set up, often as a part of a forestry department. The common practice was, roughly, fencing-in the nature reserves and banish trespassers.

In the Netherlands and the U.K. after World War II the observation was made that in some cases mere protection of reserves did not prevent a rapid decrease of diversity of species and communities by natural succession. Consequently, the insight was developed that for the protection of species and the maintenance of some communities active interference is necessary (Westhoff, 1945; Gabrielson, 1947; Westhoff & van Dijk, 1952). These observations were especially true for old, low productive but species-rich (green plants) grasslands which had an unchanged use for a long time. Van Leeuwen (1966) stated in his relation theory: "The best management practice is the continuation of the existing agricultural or other human practices".

The ideas described above on community (instead of species) protection and management measures were almost entirely based on the experiences among phytocoenologists in agricultural grassland ecosystems. Fungi were only rarely included as an argument for the protection of nature reserves (Arnolds, 1991a), because (1) no quantitative knowledge was available on the mycoflora in various ecosystems, nor on the proportion of rare or vulnerable species among them, (2) no knowledge was present on possible changes in numbers of macromycetes, i.e. the degree of threat among various species, (3) only few specialists were able to identify the large numbers of fungal species present in many habitats, and (4) many fungi occur in autumn, the period that many phytocoenologists start hibernating!

Early quantitative research on macromycetes was mainly focused on forest communities (e.g. Haas, 1932, Wilkins et al. 1937, 1938), and usually no information was

[1] Comm. no. 477 of the Biological Station Wijster.

supplied on the use or management of these forests. Only rarely the macromycetes in grasslands were studied (e.g. Neuhoff, 1949; Arnolds, 1981; Lange 1982).

Nowadays, among nature conservationists more attention is paid to other groups of organisms than traditionally to plants and large animals only (e.g. Siepel et al. 1987, Natuurbeschermingsraad, 1991), including fungi. For the latter, this was thanks to the concern of mycologists who had noticed a strong decline of many species of macromycetes and consequently have published regional and national "red data lists" of threatened fungi (e.g. Winterhoff (1984); Derbsch & Schmitt (1984); Hallingbäck (1988); Arnolds (1989a)). From these data, it appears that the proportion of the threatened species varies strongly between different ecological groups and habitats. In the Netherlands, for example, only 12% of the wood-inhabiting species are threatened, and most of them belong to classes of low degree of threat, whereas almost half (47%) of the ectomycorrhizal fungi are considered threatened (Arnolds, 1989a). On the other hand, in Sweden 39% of the threatened fungi are lignicolous species (Floravårdskommittén, 1991). In Table 1 the number of threatened species is presented for various habitats in the Netherlands.

TABLE 1. Survey of the number of fungal species and the proportion of threatened species in some habitats in the Netherlands.

N-sp = number of species, %TH = percentage threatened. MR, SA and LI indicate the ecological groups (in percentages) to which the threatened species belong. MR = ectomycorrhizal species, SA = terrestrial saprotrophic species, LI = lignicolous species (after Arnolds, 1989a).

Habitat	N-sp	%TH	MR	SA	LI
Dry grasslands	365	44	–	100	-
Grasslands (fertilized)	84	15	–	100	–
Grasslands (unfertilized)	158	78	–		
Heathlands	32	59	–	100	–
Deciduous forests	1565	33	54	30	16
Coniferous forests	439	30	80	9	11
Avenues/parks	126	78	91	6	3

In order to maximize the conservation efforts, protection of fungi and measures to maintain them should focus on the habitats where most threatened species occur. However, knowledge on effects of various nature management treatments on macromycetes is scanty and scattered in literature. The aim of this paper is to review some information on nature management measures and other treatments in various (semi)natural habitats in relation to the mycoflora and to present some own research on this subject.

It should be noted that large-scaled changes in the environment (e.g. air pollution, changed groundwater regimes and quality, eventual climatic changes) can greatly influence the environment in nature reserves as well. Many management measures are becoming insufficient to compensate for these (external) changes.

The survey presented below is ordered according to habitats, and not to management measures. On habitats that are not included here hardly any knowledge is available (e.g. fen communities) or the optimal management for the mycoflora (probably) agrees almost completely with that for most other organisms (e.g. peat bogs, drifting sand, etc.). Special attention is paid to the Dutch situation.

In western Europe, the vast majority of the ecosystems which are commonly denoted as "nature" are strongly influenced by human activity. Most forests are in use for timber production; grasslands usually serve for hay cutting or pasturage. *Seminatural* ecosystems are those systems where the arrangement of plants and species composition is spontaneous, but where the natural (phytosociological) formation is not reached as a consequence of human influence (Westhoff, 1949). This term is mostly used in relation to grasslands, but applies to several types of woodland (e.g. coppiced woodland) as well.

Nature management is (human) activity in (semi)natural habitats in order to create or maintain certain desired communities. If the nature management aims at maintaining a present community the term *maintenance management* can be used.

In the Netherlands, in some habitats managers try to return to a (former) situation, richer in species than the present state by active measures. For example, some grasslands, which have lost most of their diversity of green plant species by abundant use of fertilizers, are now no longer fertilized and regularly mown with removal of the hay in order to return to the original species-rich situation. This type of management is called *restoration management* (Bakker, 1987, 1989). *Nature development* is called the construction of a set of (abiotic) habitats meant to be colonized spontaneously by plants and animals.

Forests

Forests are the best known places where fungi are to be found. In western Europe, a strong decline has been observed among ectomycorrhiza-forming fungi (e.g. Arnolds, 1989a; Derbsch & Schmitt, 1984). This decline in the Netherlands is well-documented for the edible *Cantharellus cibarius* Fr. (Jansen & van Dobben, 1987) and it is likely that the decline of this species is representative for many more ectomycorrhizal species. Forests which are particularly rich in ectomycorrhizal fungi are often situated on hilly sites with a nutrient-poor soil with a thin or absent organic layer and a low coverage of vascular plants, e.g. the Dicrano-Quercetum (Barkman, 1974; Jansen, 1984), Carici-Fagetum (Jahn, 1986), the Cladonio-Pinetum (Lisiewska, 1983), the Luzulo-Fagetum (Darimont, 1973). These forest communities are threatened by eutrophication and/or acidification caused by air pollution. Management measures with the purpose to enhance these fungi should aim at such nutrient-poor circumstances and a thin to absent organic layer.

Another category of threatened fungi are the wood-inhabiting fungi which occur in natural, untouched forests. (Hallingbäck, 1988). These require large, old tree trunks and possibly also have special microclimatic requests that are only fulfilled in large, undisturbed areas. In Northern Scandinavia the last remains of natural forests including many characteristic fungi of this habitat are seriously endangered through exploitation by timber companies.

The following forest use or management practices are most commonly in use: (1) "traditional" use; (2) grazing; (3) fertilizing, including liming; (4) sods cutting and/or collecting of forest litter, and (5) no measures.

(1) **Traditional use**. Clear cutting of large areas and subsequent replanting or spontaneous regenerating from seed trees are most commonly practiced in forestry. Clear cutting causes drastic changes in the microclimate and the upper soil layer, even more so after soil scarification. A consequence is a strong increase in the decomposition of the organic layer. Clearly, many of the ectomycorrhizal fungi, dependent on living trees and other fungi characteristic of old, undisturbed soils will disappear. However, the inoculum potential of (some) ectomycorrhizal fungi remains present for several years in clearcuts (Dahlberg, 1991).

On the other hand, in young monoculture plantations of conifers a number of characteristic ectomycorrhizal species can be found, e.g. in the Netherlands *Suillus bovinus* (Linn.) O. Kuntze, *Inocybe lacera* (Fr.) Kummer, *I. brevispora* Huijsm., *Cortinarius fusisporus* Kühn., *Dermocybe croceocona* (Fr.) Maser, *D. semisanguinea* (Fr.) Moser, which decrease in later successional stages (Jansen, 1991; Termorshuizen, 1990; cf. also Dighton & Mason, 1985). However, some of these species are common in coniferous forests in mountainous areas and their preferential occurrence in young stands in the Netherlands may represent an abnormal situation, due to eutrophication.

In the "plenter system", i.e. selective logging, trees that can be used are removed from the forest, but the forest ecosystem itself remains intact. In such forests the ectomycorrhizal mycoflora may be favoured by the mechanism of colonization of late-state fungi (Mason et al., 1983) from roots of old trees to roots of young trees (Read, 1991).

So far, no comparisons of this way of forest use with the traditional use regarding the mycoflora have been made.

(2) **Grazing**. In completely natural forest ecosystems large grazing animals play an important role. In western Europe, this situation hardly occurs anymore (see Section 5.). In the Netherlands an increasing forest area is grazed by cattle or sheep as a measure of restoration management. The aim is to suppress an undesired abundant development of grasses (e.g. *Deschampsia flexuosa, Molinia caerulea*) and to create an open forest structure.

In a comparison of a grazed part with an ungrazed part of the Imbosch (prov. Gelderland, the Netherlands), a remarkable increase in ectomycorrhizal and dung-inhabiting and terrestrial saprotrophic fungi was noted in the grazed forest area (Table 2; pers. comm. L. Laarman-Hageraats; cf. also Edwards & Hollis, 1982).

(3) **Fertilizing**. The effects on the mycoflora of fertilizing forests has been reviewed by Kuyper (1989). In experiments N-fertilizer was applied with the purpose to stimulate the growth of trees or to compensate for (harmful) influences of air pollution. Nitrogen fertilization has a strong inhibitory effect on ectomycorrhizal symbiosis, in laboratory as well as in field experiments (Kamminga-van Wijk, 1991, Wästerlund, 1982; Keizer, 1993a). Only a few species increase after fertilizing, e.g. *Paxillus involutus* (Batsch: Fr.) Fr.; *Laccaria bicolor* (Maire) Orton or are indifferent (*Lactarius rufus* (Scop.: Fr.) Fr.; Ohenoja, 1988a, b). Many species decrease or even disappear completely, e.g. many *Cortinarius* and *Suillus* species. It is not yet apparent whether the mycorrhizas themselves are affected as well; the fructification seems to be most sensitive (Menge & Grand, 1978). NPK fertilizer application also has a strong negative effect on most ectomycorrhizal fungi, comparable with N fertilizer alone (Hall, 1978).

In view of these striking results, it is assumed that the strong decline of ectomycorrizal fungi in the Netherlands is caused by atmospheric deposition of nitrogen compounds (Arnolds, 1988; Termorshuizen, 1990).

Fertilizing causes a decline in some of the saprotrophic species and a shift in species composition towards eutraphent species. In a fertilization experiment in a Scots pine forest in the Netherlands, *Mycena sanguinolenta* (Alb. & Schwein.; Fr.) Kummer, *Clitocybe vibecina* (Fr.) Quél., *Entoloma cetratum* (Fr.) Moser decreased, whereas *Mycena galopus* (Pers.: Fr) Kummer, *Clitocybe ditopus* (Fr.: Fr.) Gillet and *C. metachroa* (Fr.) Kummer increased after application of fertilizer (Kuyper, 1989).

Liming of the forest soil is sometimes carried out in order to compensate for soil acidification and decreased tree vitality caused by air pollution. However, liming causes an increase in the decomposition of the organic layer. The nitrogen therein becomes rapidly available in the soil causing a decrease of numbers of sporocarps of ectomycorrhizal fungi. Therefore, the inhibitory effect of liming is most pronounced in soils with a thick organic layer. In the saprotrophic fungi a shift in the species-composition towards the eutraphent species can be observed.

The influence of application of other minerals on macromycetes in forests seems to be complex and will not be treated here.

(4) **Litter collecting and sod removal**. In most forests in W. Europe litter was collected for agricultural purposes (Ellenberg, 1964). This activity has ceased, among other reasons

TABLE 2. Comparison of the mycoflora of a grazed and an ungrazed forest area in the Netherlands.

A species is (arbitrarily) considered to have a preference if it occurs with more than two sporocarps and has produced more than two times the number of sporocarps in one area than in the other. The numbers are the total number of carpophores found during five excursions in 1988, i.e. three years after the start of the experiment. During each excursion the same route was followed. The researched routes in the grazed and the ungrazed forest were of approximately equal length. Figures with * are estimated numbers of sporocarps. The forest is dominated by *Pinus sylvestris* with some scattered broad-leaved trees with undergrowth dominated by *Deschampsia flexuosa*, on dry sandy soil. The grazing density is 1 to 2 cows of the Scottish Highlands breed per 100 hectares of the grazed area; in addition there were some deer and roedeer in both areas. Data from L. Laarman-Hageraats (unpubl.).

Mycorrhizal species	**grazed area**	**area not grazed**	**Coprophilous and terrestrial saprotrophs**	**grazed area**	**area not grazed**
Species with preference for grazed area:					
Suillus bovinus	216	12	Mycena galopoda v. gal.	230	12
Lactarius hepaticus	71*	13	Hygrophoropsis aurantiaca	188	2
Scleroderma citrinum	55	8	Clitocybe vibecina	160	16
Inocybe lacera	58*	2	Marasmius androsaceus	55*	7
Russula fragilis	45	1	Conocybe tenera (s.l.)	20	2
Inocybe mixtilis	30	6	Collybia butyracea	16	1
Lactarius rufus	31		Tubaria conspersa	16	
Suillus luteus	21	2	Cystoderma amianthinum	15	
Xerocomus badius	23		Galerina hypnorum	9	3
Amanita gemmata	17	2	Clitocybe metachroa	9	2
Inocybe dulcamara	17		Conocybe rickeniana	11	
Laccaria bicolor	8		Phallus impudicus	11	
Lactarius theiogalus	7		Mycena leptocephala	8	2
Russula atropurpurea	7		Bovista plumbea	8	
Inocybe rimosa	6		Mycena sanguinolenta	5	
Dermocybe cinnamobadia	4		Panaeolus sphinctrinus	4	
Rhizipogon luteolus	3				
Species with preference for the ungrazed area:					
Laccaria amethystea		47	Galerina pumila	2	12
Lactarius subdulcis		9	Collybia maculata		9
Russula velenovskyi		8	Hygrocybe miniata		6
Russula sardonia		5	Galerina calyptrata		3
Russula vesca		4			
Species without preference:					
Laccaria proxima	102*	62	Stropharia semiglobata	97	85
Russula ochroleuca	78*	50	Mycena galopoda v. nigra	10	5
Paxillus involutus	13	17	Entoloma sericeum	2	2
Amanita muscaria	8	16	Entoloma cetratum	1	2
Amanita rubescens	4	4	Conocybe rickenii	2	
Gomphidius roseus	4	2	Coprinus miser	2	
Amanita fulva	2		Galerina atkinsoniana	2	
Russula emetica	1		Lycoperdon molle	2	
			Clitocybe clavipes	1	
			Clitocybe ditopa	1	
			Galerina laevis	1	
			Lepista nuda	1	
			Micromphale perforans	1	
Number of species	25	19		30	17
Number of carpophores	831	270		890	171

because it was thought that it had a negative, impoverishing influence on the forest. In view of the above-mentioned observations that a rich ectomycorrhizal flora correlates with a thin or absent organic layer, litter collecting probably leads to an increase of ectomycorrhizal fungi. Therefore, in places where no other interests interfere, litter collecting can be applied to enhance ectomycorrhizal fungi.

In coniferous forests with a thick undergrowth of *Deschampsia flexuosa* (which has developed because of air pollution) sod removal is carried out on a small experimental scale in the Netherlands to investigate the effect of this measure on the macromycetes. Preliminary results clearly indicate an increase of sporocarp numbers of ectomycorrhizal fungi in treated plots (Baar, this volume), although this increase is partly due to "ground disturbance-indicators", such as *Laccaria* spp. and *Inocybe* spp. Terrestrial saprotrophs are strongly reduced by this measure. Sod removal followed by litter collecting is a method that probably enhances the ectomycorrhizal fungi, but it is probably too expensive to apply large scale sod cutting in forests as a nature management measure.

(5) **No measures**. In W. Europe only a few small relics of the natural forest are left. In addition, some parts of woodland reserves are appointed to develop as natural forest, i.e. no interference is allowed. So far, no comparative studies have been carried out in natural forests and production forests to evalute the current silvicultural methods regarding the mycoflora.

In natural forests with an open structure large dead trunks are present and a number of wood-inhabiting fungi are restricted to this substrate, e.g. *Aurantioporus croceus* (Pers.: Fr.) Kotl. & Pouzar, *Buglossoporus pulvinus* (Pers.) Donk on *Quercus* (Jahn, 1979), *Amylocystis lapponica* (Rom.) Singer, *Laurilia sulcata* (Burt) Pouzar, *Pycnoporellus fulgens* (Fr.) Donk, *Skeletocutis tschulymica* (Pilát) Keller in boreal Picea forest (Ingelog et al, 1987). This author uses wood-inhabiting saprotrophs as indicator organisms for natural forests.

In reserves where the "no measures" treatment has started recently, an initial decrease of ectomycorrhizal fungi can be expected caused by litter accumulation and decrease of exotic, planted ectomycorrhiza forming tree species. In view of the experience that some more or less natural forest types can have a rich flora of ectomycorrhizal fungi, one may expect that on the long term the number of ectomycorrhizal species will rise again, provided that the large-scales environmental pollution will be drastically reduced.

The effects of several forest management measures on the mycoflora are summarized in Table 3. Fertilizing of forests is unfavourable for most fungi and is not recommended. Grazing and collecting of forest litter probably enhances many ectomycorrhizal fungi and some saprotrophs. Rare saprotrophs of large wood volumina only occur in forests which are lest untouched for a long time.

TABLE 3 Expected effects of various management measures in forests on fungi of four ecological groups

MR = ectomycorrhizal fungi, SA = terrestrial saprotrophic fungi, LI = lignocolous fungi, CO = coprophilous fungi. Effects with regard to the present state in traditionally used forests: + = expected increase, - = expected decrease, ± = unknown or no change expected.

Measure	MR	SA	LI	CO
"Traditional use"	±	±	-	±
Grazing low intensity	+	+	±	-
Fertilizing/liming	-	±/-	±	±
Litter collecting	+	-	±	-
No measures > natural forest	±	+	+	±

Grasslands

In grasslands a rich mycoflora can be present, with many characteristic species, depending on local soil and climatological features and on the vegetation type. Contrary to the ectomycorrhizal fungi, this dependence is generally indirect and not species specific. Many threatened (cf. Table 1) species occur in grasslands with a low availability of nitrogen and phosphorus, with a low-intensive use for hay production or grazing, with a long, unchanged agricultural use (Neuhoff, 1949, Arnolds 1981, Winterhoff, 1987). Such grasslands usually are rich in green plant species, have a short turf and a well-developed bryophyte layer. In many grassland types the number of fungal species exceed the number of green plant species. Characteristic for such grasslands are many species of the genera *Hygrocybe* (Fr.) Kummer, *Entoloma* (Fr.) Kummer, *Clavulinopsis* Overeem, *Geoglossum* Pers. (Arnolds, 1980, 1981; Nitare, 1988b).

Such grassland types were widespread in Europe but have greatly decreased as a consequence of changing agricultural methods. This implies abandonment of the least-profitable areas and strong intensification of agricultural use of other areas. Therefore, in large areas where the low-intensive, small-scaled farming tradition has ceased, the agricultural fields gradually turn into scrub and forest by natural succession processes and loose their species richness in green plants as well as in macromycetes. The Swedish situation is well documented (Lindström, 1980; Jahn & Jahn, 1986; Nitare, 1988a, b), where only 15% of the investigated grassland sites remained unchanged during the last 20 years (Nitare, 1988a). This is representative for other countries as well. Recently, it is estimated that in the EEC area about 10,000,000 ha agricultural land will be left in the near future, when more than one third of the farmers will have to give up their farms (Fig. 1). For many species which belong in the various seminatural grassland communities, this development is disastrous. Many communities will disappear even before anything is known about the inhabiting macromycetes.

In places with intensive, "modern" agricultural methods, the originally high species-richness in green plants and macromycetes of the grasslands drops (has dropped) sharply caused by (excessive) fertilizing. In the Netherlands small remains of the species-rich grasslands can only be found in nature reserves. Here, 78% of the fungi characteristic for dry, nutrient-poor grasslands are considered threatened but only 15% of the fungi occurring in fertilized pastures (Arnolds, 1989a). Mycologically species-rich grasslands are always old, at least 20–30 years but frequently several centuries, with constantly the same use. Therefore, such communities have an extremely low "makability" and deserve protection and careful maintenance.

Much is known about the management of grasslands with regard to green plants (references in Bakker, 1989). Generally, a continuation of the traditional agricultural practices that were applied when the species-rich grasslands were common should be aimed at. An output of nutrients from the system is often necessary, to prevent succession towards more nutrient-rich vegetation types and/or to compensate for imput of nutrients (mainly nitrogen containing compounds) by atmospheric deposition. In poor dry grasslands grazing with a low cattle density (e.g. one cow per 2–3 ha.) or mowing with removal of the hay followed by grazing, without or with slight fertilizing are often applied methods to maintain these communities. With this management the accompanying mycoflora is generally maintained as well. Mowing should preferably be done in late summer or early autumn because an open, short vegetation structure is favourable for many fungi (Arnolds 1991a). Moist grassland vegetation types (e.g. Calthion palustris, Junco Molinion, Caricion davallianae) are usually less species-rich in macromycetes than the dry grasslands. Nevertheless, a large part of these species are rare, restricted to these habitats and very sensitive to environmental changes. Here too, mowing with removal of the hay is preferable.

Although many grassland communities rich in macromycetes are often rich in green plant species as well, this is not always true. This can be the case in road verges, old cemeteries, old lawns, etc. (see section 5). These areas have their own specific mycological value, with their own ecological information, which justifies care and protection. This habitat should be managed in the same way as other nutrient-poor grasslands.

The effects of fertilization on a poor grassland are illustrated by the investigation of Arnolds (1989b) and Keizer (1993a, in prep.; see section 5). Arnolds found 55 species of terrestrial fungi in a poor, weakly fertilized grassland in the Netherlands before treatment with artificial NPK fertilizer and liquid manure and after this treatment 41 species. Species characteristic for poor grasslands disappeared within a few years, e.g. *Hygrocybe ceracea* (Wulfen: Fr.) Karsten, *Entoloma conferendum* (Britz.) Noordel., *Clavulinopsis helveola* (Fr.) Corner, *Geoglossum glutinosum* Pers. whereas a number of ruderal, eutraphent species strongly increased or appeared, e.g. *Marasmius oreades* L (Bolton: Fr.) Fr., *Panaeolina foenisecii* (Pers.: Fr) Maire, *Panaeolus acuminatus* (Schaeffer) Quel. The number of dung-inhabiting fungi increased. The total production of fungal sporocarps did not change significantly. The vegetation changed: almost complete disappearance of the bryophyte layer, decrease of *Festuca rubra*; strong increase of *Lolium perenne* and *Dactylis glomerata*.

Conservation of poor grassland communities means a continuation of the original low-intensive agricultural use. Enrichment with nutrients should be prevented.

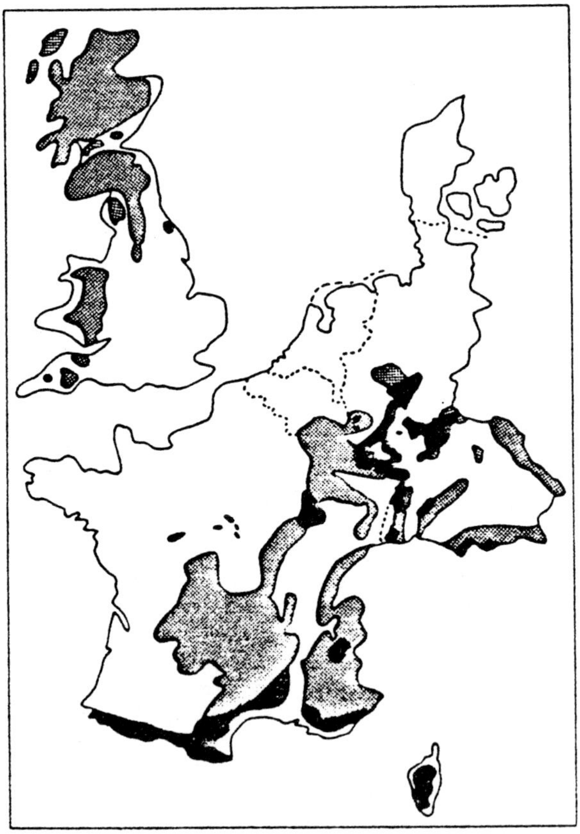

FIG. 1. Location of regions with large areas (black) and of small abandoned agricultural areas, but threatened in the near future (grey) in northwest Europe (EEC Committee, 1979; cited in Bakker, 1989).

Generally, optimal management for green plants is favourable for macromycetes as well. Some grasslands lack rare or "interesting" green plant species but are rich in macromycetes. These need special attention and protection.

Heathlands

Most heathlands are replacement communities of forests due to overexploitation. Without any interference a succession towards woodland would take place.

Traditionally, heathlands are used for grazing, mainly by sheep, and some sod cutting for fertilizing of arable fields or as fuel. Most of these functions no longer exist and large areas of heathland were disused, afforested or reclaimed for agricultural use. In the Netherlands a large proportion of the remaining heathlands is nowadays threatened by undesired succession into grassy or semiruderal vegetation types, mostly dominated by *Molinia caerulea* or *Deschampsia flexuosa*, caused by atmospheric deposition of nutrients. However, heathlands form a valuable habitat for many organisms, including some macromycetes, e.g. *Clavaria argillacea* Pers.: Fr., *Entoloma fernandae* (Romagn.) Noordel., *Cantharellula umbonata* (Gmel.: Fr.) Singer., *Mycena uracea* Pearson.

The most applied restoration and maintenance managements in heathlands are (mechanized) sod removal and grazing. The former treatment creates gaps where some pioneer heathland fungi preferably occur, e.g. *Gymnopilus fulgens* (Favre & Maire) Singer., *Clavaria argillacea*, *Entoloma conferendum* (de Mars, 1987). Grazing causes a strong increase of dung-inhabiting fungi, among which many species that do not occur in fertilized grasslands, e.g. *Anellaria semiovata* Sow: Fr) Pearson & Dennis, *Poronia punctata* (Linn.: Fr) Fr., many *Coprinus* and *Panaeolus* species. Effects of grazing in heathlands on the terrestrial saprotrophs has not yet been investigated, but the expectation is that they will increase after grazing of former heathland which has turned into a *Molinia* monoculture.

Thus, the management developed for the maintenance of green plants in heathlands is also to be preferred above any treatment regarding the macromycetes.

"Tree-meadows", parks and alleys

The mycoflora of the old landscape where cattle grazed in a small-scaled mix of open forest and grassland patches is largely unknown, but must have been extremely rich, both in saprotrophic and ectomycorrhizal fungi. A faint impression of the mycological quality of this habitat can be obtained from Jahn (1982), Jahn & Jahn, (1986), Nitare (1988a, b), Ryman & Holmåsen (1984), who described this "tree-meadow" landscape in south and central Sweden (the "Laubwaldhaine", "Birkenhage" or "björk-" or "ekhagar"). Although in some places some impoverished fragments of this habitat can be found, e.g. Öland (Sweden), Estonia, Ireland, in most of Western Europe nothing is left and no detailed knowledge of the mycoflora in this landscape is available.

Some parks belonging to old estates and roadsides planted with (ectomycorrhiza-forming) trees can be considered as landscape types which are probably mycologically related to the above described old agricultural landscape because it is basically built up of the same elements. Both parks and tree planted road verges consist of old trees with a (usually) unfertilized, grassy undergrowth. Both habitats share a sometimes large number of ectomycorrhizal fungi as well as fungi characteristic for unfertilized grasslands (e.g. Runge, 1989, Keizer & Sullock Enzlin, 1988).

In the northern part of the Netherlands, a mycocoenological study of road verges planted with Beech and Oak trees was carried out (Keizer, 1993b, in prep.). In addition, a field experiment was set up to evaluate five different management treatments (Keizer, 1993a, in prep.) Some results of these studies are presented here because they may be illustrative for effects of nature management in such habitats.

The results of the phytocoenological study of these road verges indicate that they have a moderately species-rich green plant vegetation, but with a very low proportion of rare or threatened species, in accordance with Sykora et al. (1988). This is undoubtedly due to disturbing influences of traffic and neighbouring agricultural fields. However, some road verges have a rich mycoflora, especially in ectomycorrhizal species. The mycoflora of the studied road verges with Oak is partly related to the mycoflora of Oak forests (Jansen, 1984), of road verges with Beech partly related to Beech forests (Opdam, 1991), particularly regarding the ectomycorrhizal symbionts. But on the other hand, many ectomycorrhizal species occur in road verges which are largely absent from forests. These are considered to belong in "tree-meadows" in the Drente area (Table 4). The

TABLE 4. Some examples of ectomycorrhizal fungi that are considered to be characteristic for "tree meadows" and for forests with Oak or Beech on nutrient-poor sandy soil in the Drenthe area.

The former are species differential for road verges, that latter are differential for forests (after Keizer & Arnolds, 1993, in prep.).

Tree-meadows	**Forests**
Amanita spissa	Amanita fulva
Amanita pantherina	Boletus erythropus
Chalciporus piperatus	Cantharellus cibarius
Clitopilus prunulus	Cortinarius bolaris
Cortinarius saniosus	Cortinarius casimiri (incl. decipiens s. Henry)
Cortinarius flexipes	Cortinarius elatior
Cortinarius striaepilus	Cortinarius fusisporus
Inocybe griseolilacina	Cortinarius obtusus
Inocybe petiginosa	Cortinarius paleaceus
Inocybe sindonia	Dermocybe cinnamomea s. lat.
Lactarius blennius	Hebeloma longicaudum
Lactarius vellereus	Hydnellum spongiosipes
Lactarius serifluus	Hydnellum concrescens
Russula odorata	Inocybe napipes
Russula graveolens f. purpurata	Inocybe sambucina
Russula pectinatoides	Laccaria amethystea
Russula ionochlora	Lactarius camphoratus
Russula amoenolens	Lactarius theiogalus
Russula nigricans	Lactarius proxima
Russula graveolens f. cicatricata	Lactarius chrysorrheus
Russula grisea	Russula ochroleuca
Russula graveolens f. graveolens	Russula vesca
Russula velenovskyi	Russula adusta
Russula chamaeleontina	Russula fragilis
Tricholoma ustale	Russula emetica
Xerocomus chrysenteron	Sarcodon scabrosus
Xerocomus porosporus	Thelephora terrestris
Xerocomus rubellus	Tricholoma portentosum
	Tricholoma columbetta
	Tricholoma virgatum
	Tylopilus felleus
	Xerocomus subtomentosus
	Xerocomus badius

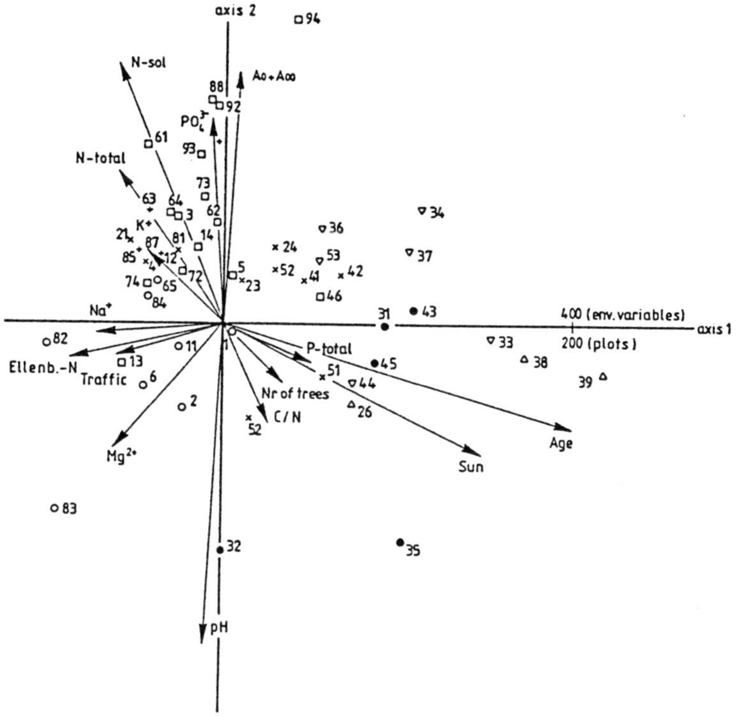

× = *Xerocomus rubellus* type - inops variant
+ = *Xerocomus rubellus* type - typical variant
□ = *Russula ochroleuca* type
○ = *Cortinarius erythrinus* type, subtype of *Laccaria amethystea*
● = *Cortinarius erythrinus* type, subtype of *Boletus edulis*
△ = *Hebeloma mesophaeum* type, inops variant
▽ = *Hebeloma mesophaeum* type, typical variant

FIG. 2a. Ordination diagram of environmental variables and plots of roadside verges planted with Common Oak on the basis of data of the ectomycorrhizal fungi. Plots belonging to the *Xerocomus rubellus* type are poor in species, those of the *Russula ochroleuca* type are mediocre and the plots of the *Cortinarius erythrinus* type are rich in species. A factor represented by a long arrow indicates that the environmental factor is more important than a factor with a short arrow. An arrow in approximately the same direction as an ordination axis correlates strongly with that axis. The numbers in the diagram refer to plot numbers.
Traffic = traffic intensity; Ellenberg-N = Ellenberg-indication value for nitrogen, based on the vegetation (Ellenberg, 1979); A0 + A00 = thickness of organic layer in the soil; Age = year of germination of the tree; Sun = amount of potential direct sunshine in the plot (hours/day in October); pH = pH($CaCl_2$); Mg^+, Na^+, K^+, NO_3, NH_4^+, PO^{43-}, P total = concentration of ions in the soil (mg/kg), N-sol = concentration of soluble nitrogen (mg/kg), N-total = % nitrogen; C/N = carbon - nitrogen ratio; Nr of trees = number of trees per plot (after Keizer, 1992a).

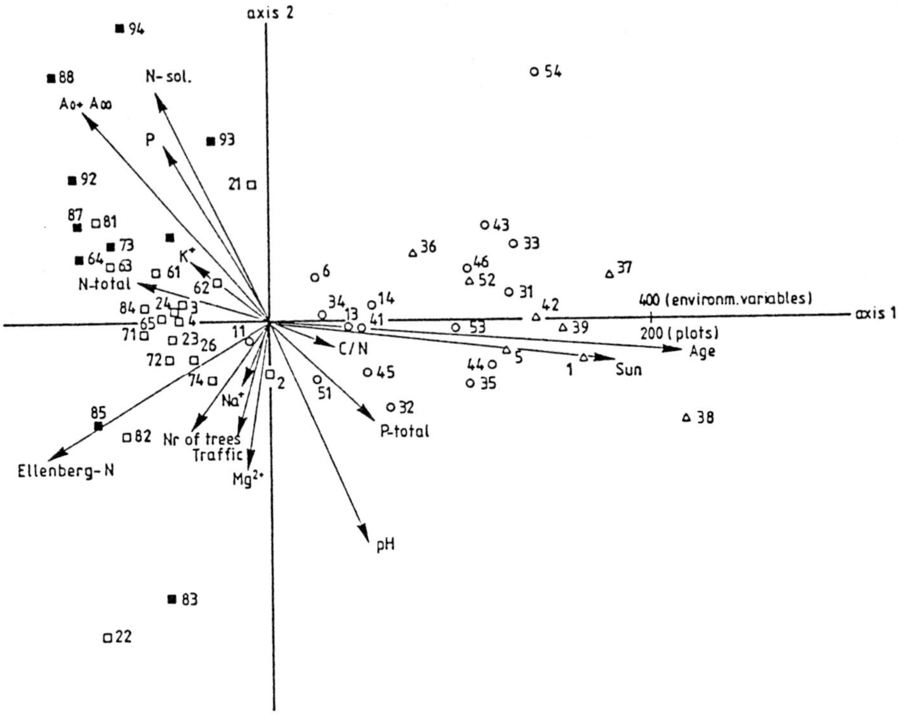

■ = *Psathyrella fulvescens Quercus* type, subtype of *Clitocybe vibecina*
□ = *Psathyrella fulvescens Quercus* type, subtype of *Lycoperdon foetidum*
○ = *Mycena avenacea* type
△ = *Collybia cookei* type

FIG. 2b. Ordination diagram of environmental variables and plots of roadside verges planted with Common Oak on the basis of data of the saprotrophic fungi. Plots of the *Psathyrella fulvescens* type are rather species-rich and predominantly in shady places, those of the *Mycena avenacea* type are also rather species-rich and predominantly in open places, those of the *Collybia cookei* type are species-poor plots. For further explanation see fig. 2a (after Keizer, 1992a).

saprotrophic fungi of the roadsides with a nutrient-poor vegetation was related to the mycoflora of poor grasslands. In Fig. 2a and 2b the most important environmental factors are indicated which explain the variation in ectomycorrhizal species in road verges with Oak. The most species-rich plots have a thin or absent organic layer on the soil, generally a vegetation with species with a low Ellenberg nitrogen availability indicating value (Ellenberg, 1979) and old trees. For the saprotrophic fungi the factors exposition of the plot (shady or not), a mediocre Ellenberg N-value and a thin organic layer correlated strongly with the occurrence of fungi characteristic of poor grasslands.

In the field experiment is it investigated how various management treatments influenced the mycoflora in a road verge along a canal, planted with 100 years old oaks on poor sandy soil in the northern part of the Netherlands (for a more extensive description of the experiment and the results, cf. Keizer, 1993a in prep.). The following treatments were carried out sixfold in 50 × 4.5 M² plots: (a) cutting sods in spring 1987 followed by mowing without removal of the hay; (b) mowing in late summer with subsequent removal of the hay; (c) no treatment; (d) fertilization with Nitrogen (nitrate and ammonium): 150 kg/ha in 1987, 300 kg/ha in 1988 and 1989, no fertilization in 1990 and 1991, all combined with mowing without removal of the hay; (e) mowing without removal of the hay. Treatment e. served as control. The vegetation was recorded in 1987 and 1989. The macromycetes were counted in the autumns 1987–1991 with intervals of 3–4 weeks. Lignicolous species and parasites of arthropods and fungi were not included.

In Table 5 the effects of the treatments on some plant species are shown. In plots where sods were cut the coverage of plants was still lower in 1989 than in the other plots and the layer of pleurocarpous bryophytes had not yet regenerated. In fertilized plots the coverage of plants has increased considerably but the bryophyte layer had practically disappeared. In untreated plots broom scrub had developed under which most other plants had disappeared. There was only little difference between mown plots with and without removal of the hay, probably owing to the poor soil and therefore rather low above-ground biomass production.

The effect of the treatments on the species numbers of macromycetes are shown in Table 6. The large differences in the total number of species and in treatment e. (control) in 1987–1991 reflect climatologically favourable (1987 and 1989) and unfavourable years (1990 and 1991) for the fructification of macromycetes. Therefore only comparisons within years can be made.

Cutting sods initially has a negative effect on the numbers of ectomycorrhizal fungi; however, after three years this effect is already less clear. The species composition is still

TABLE 5. Increase or decrease of some plant species in a road verge planted with Common oaks in the period 1987–1989, based on frequency and average coverage, in plots with different treatments (each treatment sixfold).

In the treatments "mowing with" and "mowing without removal of the hay" no differences were found (after Keizer 1993a in prep.).

Species influenced by sods cutting:

Increase	**Decrease**
Aira praecox	*Anthoxanthum odoratum*
Holcus lanatus	*Leontodon autumnalis*
Ceratodon purpureus	*Cerastium fontanum*

Species influenced by N-fertilization:

Increase	**Decrease**
Agrostis capillaris	*Festuca rubra*
Lolium perenne	*Festuca ovina*
Holcus mollis	*Hieracium pilosella*
Dactylis glomerata	*Hypochaeris radicata*

Species influenced by "no treatment":

Increase	**Decrease**
Cytisus scoparius	*Calluna vulgaris*
Phragmites australis	

TABLE 6. Percentages of the species numbers of macromycetes found in a road verge planted with Common oaks with different treatments (each treatment sixfold).

Treatment "mowing without removal of the hay" served as control (= 100%). In parentheses are the absolute species numbers (after Keizer, 1993a in prep.).

Mycorrhizal fungi	1987	1988	1989	1990	1991
Treatment					
Mowing without removal of hay	100 (42)	100 (36)	100 (43)	100 (30)	100 (23)
Mowing with removal of hay	79	81	93	83	104
No treatment	98	83	91	83	87
N-fertilization	76	36	56	43	65
Sods cutting	81	94	114	93	78
Total	(68)	(55)	(66)	(30)	(39)
Saprotrophic fungi					
Mowing without removal of hay	100 (43)	100 (22)	100 (48)	100 (20)	100 (20)
Mowing with removal of hay	128	91	85	140	120
No treatment	74	82	75	70	100
N-fertilization	70	77	65	50	135
Sods cutting	21	36	48	50	75
Total	(58)	(42)	(73)	(36)	(41)

different from the control. Some species decreased, e.g. *Lactarius chrysorrheus* Fr., *Russula fragilis* (Fr.) Fr. and some other species indicate the ground disturbance caused by the treatment, e.g. *Inocybe* spp., *Lactarius serifluus* (DC: Fr.) Fr. Some species which were not known as pioneers increased after sod removal: *Clitopilus prunulus* (Scop.: Fr.) Kummer, *Cortinarius erythrinus* Fr., *Cortinarius saniosus* (Fr.) Fr. Saprotrophic fungi had largely disappeared after sod cutting, which can be explained by the fact that with the sods most of the organic matter of the upper soil layer is removed. The recovery of saprotrophs was only slow and not yet completed in 1991.

After N-fertilization, carried out in order to simulate the effects of atmospheric N-deposition or N-input from the adjacent arable fields, a strong decline of ectomycorrhizal fungi was recorded, among others *Russula velenovskyi* Melz. & Zvara, *R. nigricans* Fr.,

R. cyanoxantha (Schaeffer) Fr., *Cantharellus cibarius* Fr. Only one species, viz. *Xerocomus chrysenteron* (Bull.) Quel. increased. Among the saprotrophs, there was a decrease in the species number and a shift in species composition. Species of poor soils and those that are associated with bryophytes decreased or disappeared, e.g. *Hygrocybe conica* (Scop.: Fr.) Kummer, *Hygrocybe nivea*, *Clavulinopsis helveola*, *Mycena vitrea* (Fr.) Quel. Some eutraphent or ruderal species appeared: *Calocybe carnea* (Bull.: Fr) Donk, *Clitocybe metachroa* (Fr.) Kummer, *Lepista nuda* (Bull.: Fr.) Cooke.

"No treatment" seems to lead to lower species numbers and production of ectomycorrhizal and saprotrophic fungi, although the differences are not significant. For the saprotrophs this is remarkable because in untreated plots an accumulation of dead organic material can be seen (dead grass and leaves). Apparently this is an unsuitable substrate for many species. It is expected that in untreated plots the vegetation will ruderalize in the long run with a consequent decrease of fungi that are associated with mosses and species confined to an open, short grassland vegetation.

There were only weak, non-significant differences in the fungi of mown plots with and without the removal of the hay. As with the green plant vegetation, the explanation is the poor soil with low above-ground biomass production where the difference between mowing with and without removal of the hay will appear after a longer period than the studied five years.

Conclusions

1. The effects of the studied management treatments are summarized in Table 7. There is a clear negative influence of N-fertilization and therefore of atmospheric or lateral input of nitrogen compounds on the macromycetes in road verges planted with trees. Removal of sods, in many places carried out in road verges by the local authorities, causes a weak decline and a shift in species composition of ectomycorrhizal species. For the saprotrophs this treatment is still unfavourable, even after five years. This agrees with the observations that poor grassland communities rich in saprotrophic fungi occur only in old grasslands. Cutting sods is therefore not recommended as a management measure for road verges. An exception can be the removal of the upper soil layer in recently enriched places, followed by mowing with removal of the hay in order to restore the nutrient-poor situation.

TABLE 7. Effects of different treatments on the mycoflora in a road verge planted with Oaks on nutrient-poor sandy soil.

Treatment: mowing without removal of the hay served as control. Nsp = number of species: 0 = no effect detected; - = significant negative effect; (+) = unsignificant positive effect; (-) = unsignificant negative effect.

	Mycorrhizal fungi		Saprotrophic fungi	
	Nsp	Production	Nsp	Production
Treatment:				
mowing with removal of hay	0	0	0	(+)
no treatment	0	(-)	0	0
N-fertilized	-	-	-	0
cutting sods	0	0	-	-

Although no significant differences were recorded between the control and mown with removal and "no treatment" plots, it is recommended that these road verges are mown, preferably with removal of the hay. This measure best maintains the open structure of the road verge ecosystem: nutrients which have been taken up by the trees are not recycled into the soil because most of the leaves are blown away in the autumn. "No treatment" causes a higher, ruderalized vegetation in which leaves are caught and the diversity in green plants as well as in macromycetes will decrease.

2. This is a fine example of a mycologically rich and valuable road verge. More than 200 species of macromycetes were recorded among which 31 are considered threatened,e.g. *Hydnellum concrescens* (Pers.) Banker, *Phellodon confluens* (Pers.) Pouzar, *Russula albonigra* (Krombh.) Fr., *Cantharellus cibarius*. For some ectomycorrhizal species road verges planted with trees are the only remaining localities in the Netherlands.

3. Some of the saprotrophic fungi are characteristic of nutrient-poor grassland and many ectomycorrhizal species are mainly found in forests. Road verges planted with trees therefore unite elements of both habitats. In addition, some of the recorded ectomycorrhizal species are nowadays hardly found in forests; it is assumed that these are relics of the tree-meadow habitat which has disappeared from this area.

4. Planting of (ectomycorrhiza forming) trees along roads and canals is recommended: it creates growth possibilities for many macromycetes and also adds to the landscape.

Nature Development

In several places in western Europe large areas of former agricultural land will possibly become available for other destinations of which nature development is an important one. The habitats created in nature development communities are usually dominated by few plant species and are also poor in fungal species. (This again stresses the necessity to take care of old, species-rich communities.) However, in a long-term period these grounds offer possibilities to restore the tree-meadows if ectomycorrhiza forming trees are planted in areas with nutrient-poor soil combined with grazing of the undergrowth with a low density of cattle.

Acknowledgements

I am indebted to Dr. T.W. Kuyper and Dr. E. Arnolds, who made important improvements in earlier versions of this paper.

References

Arnolds, E., (1980). De oecologie en sociologie van Wasplaten. *Natura* **77**: 17–44.

Arnolds, E., (1981). Ecology and coenology of macrofungi in grasslands and heathlands in Drenthe, the Netherlands. Part 1. Introduction and synecology. *Biblioth. Mycol.* **83**. 407 pp.

Arnolds, E., (1988). The changing macromycete flora in the Netherlands. *Trans. Brit. Mycol. Soc.* **90**(3): 391–406.

Arnolds, E., (1989a). A preliminary red data list of macrofungi in the Netherlands. *Persoonia* **14**(1): 77–125.

Arnolds, E., (1989b). The influence of increased fertilization on the macrofungi of a sheep meadow in Drenthe, the Netherlands. *Opera Bot.* **100**: 7–21

Arnolds, E., (1991a). Mycologists and Nature conservation. In D.L. Hawksworth (editor). Frontiers in Mycology. Honorary and general lectures from the Fourth International Mycological Congress, Regensburg, 1990. pp. 243–264. International Mycological Institute, Kew.
Arnolds, E., (1991b). Decline of ectomycorrhizal fungi in Europe. *Agric. Ecosyst. Environm.* **35**: 203–244.
Bakker, J.P., (1987). Restoration of species-rich grassland after a period of fertilizer application. In J. van Andel, J.P. Bakker, and R.W. Snaydon (editors). Disturbance in grasslands, pp. 185–200. Junc, Dordrecht.
Bakker, J.P., (1989). Nature management by grazing and cutting. Kluwer Academic Publishers (Thesis). (Geobotany, 14).
Barkman, J.J., (1974). Le Dicrano-Quercetum, nouvelle association des chênaies acidophiles aux Pays-Bas. In J.M. Géhu (editor). Colloques phytosociologiques 3, La végétation des forêts caducifoliées acidophiles, Lille, pp. 251–254. J. Cramer, Vaduz.
Dahlberg, A., (1991). Ectomycorrhiza in coniferous forest: structure and dynamics of populations and communities. Thcsis. Sveriges Landbruksuniversitet, Uppsala.
Darimont, F., (1973). Recherches mycologiques dans les forêts de Haute Belgique. *Mém. Inst. Roy. Sci. Nat. Belgique* **170**: XIV + 220pp.
Derbsch, H. and Schmitt, J.A., (1984). Atlas der Pilze des Saarlandes Teil 1. Verbreitung und Gefährdung. 536 pp. Minister für Umwelt des Saarlandes, Saarbrücken. (Aus Natur und Landschaft im Saarland, 2.).
Dighton, J. and Mason, P.A. (1985). Mycorrhizal dynamics during forest tree development. In D. Moore, L.A. Casselton, D.A. Woods and J.C. Frankland (editors). Developmental Biology of Higher Fungi. pp. 117–139. Cambridge University press, Cambridge.
Edwards, P.J. and Hollis, S., (1982). The distribution of excreta on New Forest grassland used by cattle, ponies and deer. *J. Appl. Ecol.* **19**: 953–964.
Ellenberg, H., (1964). Stickstoff als Standortsfaktor. *Ber. Deutsch. Bot. Ges.* **77**: 82–92.
Ellenberg, H., (1979). Zeigerwirte der Gefässpflanzen Mitteleuropas. 2. Auflage. *Scripta Geobot.* **9**: 1–122.
Floravårdskommittén för svampar, (1991). Kommenterad listta över hotade svampar i Sverige. *Windahlia* **19**: 87–130.
Gabrielson, I.N., (1947). Management of nature reserves on the basis of modern scientific knowledge. Proc. and papers 6th technological Meeting I.U.C.N. pp. 27–35. London.
Haas, H., (1932). Die Bodenbewohnende Grosspilze in den Waldformationen einiger Gebiete von Württemberg. *Beih. Bot. Zentralbl.* **50B**: 35–134.
Hall, D., (1978). Experiments with NPK fertilizers in relation to the growth of toadstools in beechwoods. I. Preliminary study. *Vasculum* **63**: 25–41.
Hallingbäck, T., (1988). A preliminary list of threatened fungi in Sweden. Rept. Dept. Ecology and Environmental Research Univ. Uppsala.
Ingelög, T., Thor, G. and Gustafsson, L. (1987). Floravård i skogsbruket Del 2-Artdel, Skogstyrelsen, 551 83 Jönköping.
Jahn, H., (1979). Pilze die an Holz wachsen. Buranek & Frost. Herford.
Jahn, H., (1982). Über *Lactarius pubescens* und *L. favrei* sp. nov., *Int. J. Mycol. Lichenol.* **1**(1): 75–116.
Jahn, H., (1986). Der "Satanspilzhang" bei Glesse (Ottenstein), Süd-Niedersachsen. Zur Pilzvegetation des Seggen-Handbuchenwaldes (Carici-Fagetum) im Weserbergland und ausserhalb. *Westfäl Pilzbriefe* **10–11**(8b): 289–351.
Jahn, H. and Jahn M.A., (1986). Konstanz und Fluktuation der Pilzvegetation in Norra Warleda (Uppland). Beobachtungen auf einem Schwedischen Bauernhof. *Westfäl. Pilzbriefe* **10–11**(8b): 352–377.
Jansen, A.E., (1984). Vegetation and Macrofungi of acid oakwoods in the north-east of the Netherlands. Wageningen (Agricultural Research reports 923).
Jansen, A.E., (1991). The mycorrhizal status of Douglas Fir in The Netherlands: its relation with stand age, regional factors, atmospheric pollutants and tree vitality. *Agric. EcoSyst. Environm.* **35**: 191–208.
Jansen, A.E. and de Vries, F.W., (1988). Qualitative and quantitative research on the relation between ectomycorrhiza of *Pseudotsuga menziesii*, vitality of the host and acid rain. Report 1985–1987. Bilthoven.

Jansen, E.J. and van Dobben, H.F., (1987). Is decline of *Cantharellus cibarius* in the Netherlands due to air pollution? *Ambio* **16**: 211–213.

Kamminga-van Wijk, J., (1991). Mycorrhizal and nonmycorrhizal Douglas Fir grown in hydroculture. Thesis, Rijksuniversiteit Groningen.

Keizer, P.J. (1993a in prep.). The effects of various management treatments on the macromycetes in a road verge planted with Common Oak (*Quercus robur* L.).

Keizer, P.J. (1993b in press). Mycocoenology of roadsides planted with Common Oak (*Quercus robur* L.) in Drente, the Netherlands. *Nova Hedwigia.*

Keizer, P.J. and Arnolds, E., (1993 in prep.). The macromycete flora in road verges planted with trees in comparison with related forest types.

Keizer, P.J. and Sullock Enzlin, R.A.F., (1988). Het landgoed Vennebroek, een mycologisch juweel. *Coolia* **31**: 101–109.

Kuyper, Th.W., (1989). Auswirkungen der Walddüngung auf die Mykoflora. *Beitr. Kenntn. Pilze Mitteleuropas* **5**: 5–20.

Lange, M., (1982). Fleshy fungi in grass fields. Dependence on fertilization, grass species, and the age of the field. *Nordic J. Bot.* **2**: 131–143.

Leeuwen, C. van, (1966). A relation theoretical approach to pattern and process in vegetation. *Wentia* **15**: 25–46.

Lindström, H., (1980). Hackslått — en förswinneande biotop i mellersta Norrland. *Svensk Bot. Tidskr.* **74**: 281–306.

Lisiewska, M., (1983). Udzial macromycetes w zbiorowiskach roslinnych na wydmach i w borach nadmorskich w slowinskim parku narodowym. *Badan. Fizjogr. Polsk. Zachodn. B* **34**. 23–47.

Mars, de H., (1987). De mycosociologie van de geplagde heide waaronder het Lycopodio-Rhynchosporetum. Resultaten 1985–86. Internal report Biological Station Wijster.

Mason, P.A., Wilson, J., Last, F.T. and Walker, C., (1983). The concept of succession in relation to the spread of sheathing mycorrhizal fungi on inoculated tree seedlings growing in unsterile soils. *Pl. & Soil* **71**: 247–156.

Menge, J.A. and Grand, L.F., (1978). Effect of fertilization on production of epigeous basidiocarps by mycorrhizal fungi in loblolly pine plantations. *Canad. J. Bot.* **56**: 2357–2362.

Natuurbeschermingsraad, (1991). Wie het kleine niet eert... Ongewervelde dieren en het terreinbeheer. Report Natuurbeschermingsraad, Utrecht.

Neuhoff, W., (1949). Die Pilzflora holsteinischer Viehweiden in den Jahren 1946–1948. *Z. Pilzk.* **1949**(4): 1–6.

Nitare, J., (1988a). Förändringer i svampfloran-forskning och artbevarande. *Svensk Bot. Tidskr.* **82**: 485–489.

Nitare, J. (1988b). Jordtungor, en svampgrupp på tillbakagång i naturliga fodermarker. *Svensk Bot. Tidskr.* **82**: 341–368.

Ohenoja, E., (1988a). Behaviour of mycorrhizal fungi in fertilized forests. *Karstenia* **28**: 27–30.

Ohenoja, E., (1988b). Effect of forest management procedures on fungal fruitbody production in Finland. *Acta Bot. Fenn.* **136**: 81–84.

Opdam, A., (1991). Paddestoelen-gemeenschappen van beukenbossen. Unpublished Report Agricultural Univ. Wageningen.

Read, D.J., (1991). Mycorrhizas in Ecosystems — Nature's Response to the "Law of the Minimum". In D.L. Hawksworth (editor). Frontiers in Mycology. Honorary and general lectures from the Fourth International Mycological Congress, Regensburg, 1990, pp. 101–130. International Mycological Institute, Kew.

Runge, A., (1989). Wegränder-Ersatzstandorte für gefährdete Pilzarten. L.Ö.L.F.-Mitteilungen, Heft. 1.

Ryman, S. and Holmåsen, I., (1984). Svampar, en Fälthandbok. Interpublishing, Stockholm.

Siepel, H., van de Bund, C.F., van Wingerden, W.K.R.E., Bink, F.A. Bongers, W., Mabelis, A.A., Roelofsen, G.J., Meijer, J., and den Boer, M.H., (1987). Beheer van graslanden in relatie tot de ongewervelde fauna: ontwikkeling van een monitorsysteem. Research institute for Nature management, Leersum. (Report 87/29).

Sykora, K.V., de Nijs, L. and Pelsma T., (1988). Plantengemeenschappen in de Nederlandse wegbermen en de zeldzaamheidswaarde van de bermflora. *Levende Natuur* **89**(1): 14–20.

Termorshuizen, A.J., (1990). Decline of carpophores of mycorrhizal fungi in stands of *Pinus sylvestris*. Thesis. 127 pp.
Wästerlund, I., (1982). Försvinner tallens mykorrhizasvampar vid gödsling? *Svensk Bot. Tidskr.* **76**: 411–417.
Westhoff, V., (1945). Biologische problemen der natuurbescherming. In M.J. Goris et al (1991). Bibliografie Victor Westhoff. pp. 110–121.
Westhoff, V., (1949). Schaakspel met de natuur. *Nat. Landschap* **3**: 54–62.
Westhoff, V. and van Dijk, J., (1952). Experimenteel successieonderzoek in natuurreservaten, in het bijzonder in het Korenburgerveen bij Winterswijk. *Levende Natuur* **55**: 5–16.
Westhoff, V., (1983). Man's attitude towards vegetation. In W. Holzner., M.J.A. Werger., and I. Ikusama (editors). Man's impact on vegetation, pp. 7–24. Junk, Den Haag.
Wilkins, W.H., Ellis, E.M. and Harley, J.L., (1937). The ecology of larger fungi: I. Constancy and frequency of fungal species in relation to certain vegetation communities, particularly oak and beech. *Ann. Appl. Biol.* **24**: 703–732.
Wilkins, W.H., Harley, J.L. and Kent, G.C., (1938). The ecology of the larger fungi: II. The distribution of the larger fungi in part of Charlton forest, Sussex. *Ann. Appl. Biol.* **25**: 472–489.
Winterhoff, W., (1984). Vorläufige rote Liste der Grosspilze (Macromyzeten). In J. Blab, E. Nowak, W. Trautmann and H. Sulopp (editors). Rote Liste der gefährdete Tiere und Pflanzen in der Bundesrepublik Deutschland. 4. Aufl. pp. 162–184. Kilda, Greven.
Winterhoff, W., (1987). Die Grosspilzflora der Schafweidenim Eselsburger Tal bei Herbrechtingen (Schwäbische Alb). *Beitr. Kenntn. Pilze Mitteleurópas* **3**: 343–354.

G. Pacioni (1993). Philosophy of Mushroom and Truffle Conservation in Italy. In D.N. Pegler, L. Boddy, B. Ing & P.M. Kirk (Editors). Fungi of Europe: Investigation, Recording and Conservation, pp. 271–273. Royal Botanic Gardens, Kew.

21. PHILOSOPHY OF MUSHROOM AND TRUFFLE CONSERVATION IN ITALY

G. PACIONI

Dipartimento Scienze Ambientali, Universita', L'Aquila, Italy

Abstract

Protection of macrofungi is carried out by the control of gathering. No Red List exists and the regulation for mushrooms and truffles have two long and distinct histories. In 1992 the Italian Parliament was on the point of passing a national law specifically for mushroom collection. The basic components of this law stem from local laws issued since 1972. They are: 1) mushrooms are an important component of eco-systems; 2) amounts and methods of harvesting must be restricted to protect mycelium; 3) only some edible species can be gathered. In some local law pathogenic mushrooms, like, *Armillaria mellea*, are specifically not protected. Truffle hunting is a very complex situation. The first national law (1970) changed in 1985 and a third law is being discussed at present because of the considerable conflicting interests putting pressure on legislators. In addition to the general regulations controlling methods, tools and hunting system, the most valuable species of truffles have a calendar for hunting. Truffle preserves, farming and new reafforestations with mycorrhizal trees are supported.

The regulations concerning mushrooms and truffles have two, long and distinct, histories but both have their fountain-head in Latin law. Since macrofungi are a soil product, a preliminary point for any regulation to make clear is the one concerning the property of these spontaneous 'fruits'. According to Roman law, the basis of modern laws, they are '*res nullius*' or rather '*res communes omnium*', that is 'everybody's things', but only if they grow in public lands '*ager publicus*'. When the producion occurs in private lands, macrofungi are property of the owner ('*dominus*'). Later, small communities sometimes need to own lands in common, with property undivided ('*universitas*' that is the modern 'town territory'), from where local people got various economic benefits including mushroom and truffle gathering. This privilege of land use survives also in today's Italian law ('*Usi Civici*') and it is in some way similar to a private property. When the ownership of a land is not clear or the hunting of macrofungi is not specifically forbidden, they are considered '*res derelictae*', even in private lands, on a par with the ears of wheat after harvesting that can be picked up by everyone, in agreement with the 'gleaning right'. According to these legal principles of ownership, a protection of macrofungi can affect only their condition of '*res nullius*' or '*res derelictae*'. When they are '*dominus*', the owner is exempted from the observance of the law. The idea is that of the '*pater familias*' the good father: the owner has respect for his property and guarantees the best use of it.

In Italy, mushrooms and truffles have two different systems of protection, both implemented through the control of the material collected. No Red List exists for macrofungi like those for plants and animals. Early this year (1992) a national law regulating mushroom gathering and marketing was passed by the Chamber of Deputies. It is a frame law and the Italian regions must adapt their laws to it. The basic principles of this law stem from local laws issued since 1972. They are: (1) mushrooms are an important component of ecosystems; (2) quantity and methods of harvesting must be restricted to protect mycelia; (3) usually only some edible species can be gathered.

The first regulation of mushroom gathering dates back to 1962 and it was a law for the protection of alpine flora, issued by the Autonomous Region "Trentino-Alto Adige" (L.R. 28 June 1962 n.10). The province of Bolzano (Bozen), one of the two provinces of Trentino-Alto Adige, issued a law targeting mushroom gathering (L.P. 28 June 1972 n.12) in the lands classified as 'disturbed'. The article n.1 of this law as well as some fundamental principles have been included in all the following laws. According to the first article, the regulation was issued "In order to preserve the benefits for plant ecosystems resulting from the presence of edible and uneatable mushrooms". All regional and local laws successively issued have accepted this basic aim together with the main principles of this regulation. In particular, a year later, the province of Trent extended this regulation to all its territory. After the provinces of Bolzano and Trent, laws for mushroom protection were passed in all regions of north and central Italy (Veneto, Lombardy, Liguria, Piedmont, Friuli-Venezia Giulia, Molise, the Abruzzi, Tuscany, Umbria, Emilia-Romagna, the Marches, Val d'Aosta), and they are being issued in Calabria and Sardinia. Usually they protect only mushrooms but sometimes mushrooms are included in a wider natural context: flora, lower fauna (beetles, snails, or frogs) or undergrowth brush fruits (straw-, rasp-, blue-, black- and juniper berries) object of widespread hunting.

In the province of Bolzano, as well as elsewhere in Italy, mushroom hunting is very active for commercial or gastronomic purposes. As a consequence, the most effective system of control has been to restrict the allowed daily amount per person. In the laws of Bolzano and Trent every species of mushroom collected was included in the allowed amount, however in the subsequent law of Veneto only the collection of edible species is permitted and an additional rule also prohibits the destruction, crushing and damage of any fruiting bodies.

Some regions give a list of fungi which are allowed to be collected and basically the Italian frame law allows people to collect *Boletus* (*B. edulis* Bull.: Fr, *B. pinicola* (Vitt.) Vent., *B. aereus* Bull.: Fr., *B. reticulatus* Schaeffer, *Cantharellus* (*C. cibarius* Fr., *C. lutescens* (Pers.) Fr., *Amanita caesarea* (Scop.: Fr.) Pers., *Morchella esculenta* Fr. and allied species, leaving to the single regions the possibility to extend the list of species according to customary practices. These lists rule out the collection of all other species of mushrooms '*a priori*'. Among the previous laws, the one of Piedmont is noteworthy for the lack of restraint on the collection of *Armillaria mellea* (Vahl: Fr.) Kummer, this is in order to damage such a highly pathogenic mushroom.

Mushroom hunting may be absolutely forbidden in areas where some drastic environmental change has occurred, while it is forbidden for the first five years in reafforestations.

The law also exercises vigilance on mycelia, in fact all laws include specific restrictions about mushroom harvesting methods. The wording is basically the following: "Rakes, hooks or other tools that can damage the soil humus layer are interdicted". As a matter of fact some people, especially those looking for young fruiting bodies, are used to resorting to such tools to remove the leaf or needle covering in the forest, with a consequent damage on mycelia and roots. Some regions go as far as to issue special regulations, meant to protect mushroom life, which safeguard spore dispersion. For example in Tuscany it is forbidden to pick up fruiting bodies too small, with a cap diameter less than 4 cm for *Boletus* and *Amanita* and 2 cm in the case of *Hygrophorus marzuolus* (Fr.) Bres. and in Piedmont it is suggested to detach fruiting bodies with a hand torsion while in Umbria, the harvesting must be carried out in pierced baskets to allow spore dispersion (sic!). However the most effectual measures are those of didactic or scientific nature, but these, unfortunately, are planned only by Umbria and Piedmont respectively. At any rate, the new national frame law, should have an article specifically aimed at this cultural promotion.

In the past mushrooms, also because somewhat risky, were generally regarded as low food, good for villagers or poor people at large (the so-called 'poors' flesh'). By contrast, truffles have always been highly prized and most of the time set apart for the rich. What with their high value and the rather scanty information about their biology, truffles have been considered different from mushrooms for a long time, this in spite of the fact that Micheli had fully understood their fungal nature as early as seventeen-twenty-nine.

The local history of the Italian regions is studded with edicts, orders and contracts about truffles. The first national law regulating only truffle hunting was issued in 1927 (L. 16 June 1927, n.1766) and reaffirmed by Royal Decree n.332 a year later (26 February 1928). However, an organic law by Italian Republic appeared only in 1970 (L. 17 July 1970 n.568); it listed seven taxa of Tuberales, six *Tuber* species and *Terfezia leonis* Tul. which were allowed to be marketed. Among the different articles of this law, number 6 is the most important for truffle conservation. This article specifies: (a) truffle hunters must have a licence; (b) the hunting must be carried out with dogs or pigs; (c) after the unearthing of a truffle, the dug hole must be filled with soil. In this way, both the amount and the names of the truffle hunters should be under control; the use of trained animals avoids the casual and harmful digging of truffle habitats and the covering of holes prevents further damage to mycelia and roots. After this law, some regions issued their own laws until the new national frame law was passed in 1985 (L. 16 December 1985, n.752). This was a more complete and articulate law introducing several changes to the previous one and pointedly asking the regions to issue regional laws dealing with local situations. All regions in northern and central Italy were ready to comply with such a request, but in the end it resulted in a great confusion and imbalance because of the high number of inconsistencies and contradictions. Understandably, this was mainly due to conflicting interests and the consequent action of pressure groups lobbying on legislators. The number of truffle hunters is very high, to the point that in some regions they are even able to influence political elections. All this compounded by the fact that the economy of wide areas is heavily dependant on truffle marketing. At present a third national law should be in the pipeline. Notwithstanding this murky situation, the main objectives of the law remain the preservation, as best as possible, of truffle habitats and, where possible, their extension. The cornerstones of the law are the following: (a) the possibility of collecting nine taxa of *Tuber* (*T. magnatum* Pico, *T. melanosporum* Vitt., *T. brumale* Vitt., and var. *moschatum* Ferry, *T. aestivum* Vitt., and subsp. *uncinatum* (Chatin) Fischer, *T. borchii* Vitt., *T. macrosporum* Vitt., *T. mesentericum* Vitt.); (b) improvement of truffle habitats and new reafforestations with mycorrhizal trees, generally 'truffle farming', is strongly supported; (c) it is mandatory for truffle hunters to have a licence which is issued after an examination testing their knowledge about truffle biology and harvesting regulations. Every infringement is punished with heavy fines; (d) to locate truffles, hunters need dogs and must dig with an appropriate tool only in the point scented by them; (e) every species of truffle may be hunted during a specific period; (f) the casual hoeing of truffle habitats, the gathering of immature truffles, the unobservance of the hole filling rule, the hunting during a forbidden period and the failure of abiding by the daily amount of collectable truffle (one kilogram) are all serious infringements. Truffle hunters argue against the annual licence fee and local hunting calendars both differ from region to region and, of course, the limited daily amount of truffles. The guidelines of today's policy appear to be, on the one hand, the maintenance of this limit to the hunting on public land and, on the other, the expansion of areas of truffle farming where no limit is imposed. A considerable amount of public money is invested every year to this purpose and truffle habitats are expanding as well as the wooded surface.

The same policy is now beginning to be adopted in forest mushroom farming such as *Boletus* and *Amanita*. This is in keeping with the basic concept that mushrooms are a component of ecosystems and any conservation must, of necessity, be a conservation of the ecosystems they live in.

J. Baar & Th.W. Kuyper (1993). Litter Removal in Forests and Effect on Mycorrhizal Fungi. In D.N. Pegler, L. Boddy, B. Ing & P.M. Kirk (Editors). Fungi of Europe: Investigation, Recording and Conservation, pp. 275–286. Royal Botanic Gardens, Kew.

22. LITTER REMOVAL IN FORESTS AND EFFECT ON MYCORRHIZAL FUNGI

J. BAAR & TH. W. KUYPER

Biological Station of the Agricultural University, Kampsweg 27, 9418 PD Wijster, the Netherlands.

Abstract

It has been observed that removal of litter and humus layers by management practices or by wind had a positive effect on the ectomycorrhizal flora. Field experiments have been set up in stands of *Pinus sylvestris* to study the effects of removing the ectorganic layer in a more detailed way. Sod-cutting had a positive effect on the number of ectomycorrhizal species and on the number of carpophores. Adding sods had a negative effect on the number of ectomycorrhizal species and on the number of carpophores. Laboratory experiments were carried out to gain causal understanding of these processes. Extracts of Swedish Scots pine needles, Dutch Scots pine needles originating from a fallen tree and Dutch Scots pine litter reduced the growth of the ectomycorrhizal fungus *Laccaria proxima* (Boud.) Pat. completely, but had a stimulatory effect on *L. bicolor* (Maire) Orton. The growth of *Paxillus involutus* (Batsch: Fr.) Fr. was inhibited by high concentrations of the extracts and the growth of *Xerocomus badius* (Fr.) Gilb. was significantly reduced by all concentrations. The sensitivity of different species only partly accorded with observations made in the field. The desirability of litter removal as a management practice to restore or maintain ectomycorrhizal species diversity is discussed.

Introduction

During the last decades the diversity and abundance of ectomycorrhizal fungi have shown a dramatic decline in large parts of Europe (Arnolds, 1991). This decline is especially evident in older stands and differs clearly from the slight decline in ectomycorrhizal diversity during forest succession.

Ecological research is needed both to elucidate the causes and mechanisms of this decline and to counteract this impoverishment. Several types of ecological experiments, viz. natural experiments, field experiments, and laboratory experiments, are useful tools in this respect.

In natural experiments the researcher has selected sites where a disturbance (or a desired measure) has taken place, either naturally or by humans other than the ecologist. Many field observations in selected sites actually belong to this category. In field experiments only one or a limited number of variables are manipulated in a predetermined way. In laboratory experiments all relevant variables can be manipulated independently and easily replicated. However, laboratory experiments often suffer from a lack of realism and generalism. Temporal and spatial scales in such experiments are often limited and the number and kinds of species that are studied are often unacceptably small, implying that they are of not much use in community ecology (Cody, 1986; Hairston, 1989).

In this paper we show how field observations and natural experiments led to hypotheses about the role of the ectorganic (humus and litter) layer in the decline of mycorrhizal fungi. These hypotheses were tested in field experiments that indicated the role of additional factors (age of the stand, nitrogen content of organic material, phenolic composition, etc.). The role of several additional factors was investigated in the laboratory and the results were compared with field data.

This research strategy is meant to form the basis for measures to restore or maintain the ectomycorrhizal species diversity in various Dutch forest types (Klap & Schmidt, 1992).

Natural Experiments

Field observations suggested that accumulation of organic material in the forest floor might cause a decline in mycorrhizal diversity. Vries et al. (1985) found a strong negative correlation between ectomycorrhizal species richness and the thickness of organic layers. Markkola & Ohtonen (1988) also noted a negative effect of a thick humus layer on ectomycorrhizas.

In deciduous forests in south Sweden Tyler (1991) observed sporocarp abundance of some ectomycorrhizal genera to be larger on sites with below-average litter accumulation.

The high ectomycorrhizal diversity along some roadsides is probably due to the favourable effects of wind, blowing away leaf litter (Keizer, this volume). A striking dominance of mycorrhizal species was also observed in beech forests exposed to wind that prevented accumulation of litter (Jahn, 1986).

In the beginning of this century Kallenbach noted that management practices like removal of litter positively affected ectomycorrhizal fungi, especially hydnaceous fungi (cited in Grosse-Brauckmann & Grosse-Brauchmann, 1978). Termorshuizen (1991) studied ectomycorrhizal fungi in non-disturbed Scots pine stands in the Netherlands. During his study a positive effect on the abundance of carpophores was found when on a small scale the humus and litter layer including the vegetation was removed illegally by unknowns. On this spot 94 times more carpophores of ectomycorrhizal fungi were counted than on a comparable area with humus and litter present.

Field Experiments

Based on these natural experiments field experiments have been set up. In a small-scale field study (plots 5m^2, 10 replicates) Tyler (1991) investigated the effects of removing and adding of freshly fallen beech litter on the mycoflora. The removal of beech litter stimulated the fructification of species of *Russula*, but decreased the number of carpophores of *Lactarius subdulcis* (Bull.: Fr.) S.F. Gray. Adding litter affected *Russula* and *Lactarius* negatively.

We carried out two field experiments that will be discussed in more detail.

A small-scale experiment was set up in three old (40–80 years) secondary stands of Scots pine (*Pinus sylvestris* L.) in the northeastern part of the Netherlands. The litter and humus layer (including the vegetation) was removed manually in two plots of 100m^2 and two other plots were used as controls. The stands were investigated for ectomycorrhizal carpophores from 1985–1989.

Results were tested for significance with the Kruskal-Wallis and Mann-Whitney U-tests (Siegel, 1956) or with Student's t-test (Sokal & Rohlf, 1981). Correlations were calculated and tested with the Spearman rank correlation test (Siegel, 1956).

Sod-cutting had a significantly positive effect on both species richness and fruitbody abundance (Fig. 1). The first two years after the treatment no significant effects have been established; three years after treatment sod-cutting positively affected the number of ectomycorrhizal species and carpophores.

As shown in Table 1 where the species are ranked on the basis of preference for sod-cut versus control sites most species were clearly stimulated by sod-cutting. Comparison of the data in Table 1 with those in Termorshuizen (1991), where ectomycorrhizal species are ranked on the basis of a preference for young versus old stands, indicated a a significant agreement (Spearman's $R = 0.75$; $n = 13$; $p < 0.05$).

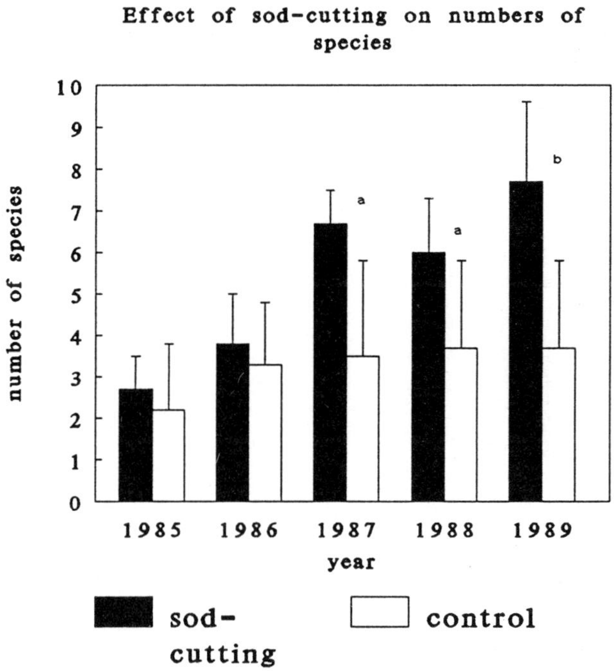

Fig. 1a

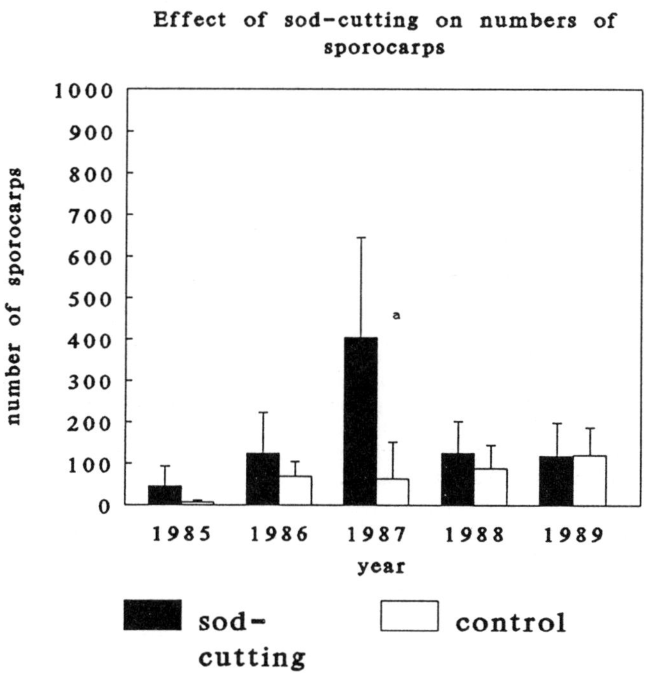

Fig. 1b

FIG. 1. Effects of sod-cutting on numbers of species and fruitbodies of ectomycorrhizal fungi in old Scots pine stands. a: significant at $p < 0.05$; b: significant at $p < 0.01$.

TABLE 1. Relative abundance (%) of ectomycorrhizal species in control and sod-cut plots in old (40–80 years) Scots pine stands in Netherlands from 1985 until 1989. The species have been arranged in a sequence of decreasing affinity for plots poor in organic material. C = control, S = sod-cut; n = total number of carpophores.

	S	C	n
Inocybe lacera (Fr.: Fr.) Kumm.	100	0	235
Inocybe brevispora Huijsman	100	0	68
Cantharellus cibarius Fr.: Fr.	100	0	30
Rhizopogon luteolus Fr.	100	0	13
Suillus bovinus (L.: Fr.) O. Kuntze	100	0	13
Russula adusta (Pers.: Fr.) Fr.	100	0	12
Amanita pantherina (DC.: Fr.) Krombh.	100	0	4
Amanita rubescens Pers.: Fr.	98	2	62
Cortinarius fusisporus Kühner	98	2	177
Laccaria bicolor (R. Maire) P.D. Orton	90	10	3130
Laccaria proxima (Boud.) Pat.	89	11	415
Xerocomus badius (Fr.: Fr.) E.J. Gilb.	71	29	7
Lactarius rufus (Scop.: Fr.) Fr.	61	39	105
Thelephora terrestris Willd.: Fr.	46	54	101
Lactarius hepaticus Plowr.	41	59	2309
Russula emetica Pers.: Fr.	19	81	117
Paxillus involutus (Batsch: Fr.) Fr.	11	89	58
Russula paludosa Britz.	0	100	2
Inocybe ovatocystis Bours. & Kühner	0	100	52

This correlation implies that sod-cutting for a large part reverts the ectomycorrhizal succession to earlier stages, suggesting that age of the humus profile is a major determinant of ectomycorrhizal species composition. However, the presence of late-stage species like *Russula adusta* (Pers.: Fr.) Fr. and *Amanita pantherina* (DC: Fr.) Krombh. indicate that tree age is important as well. For that reason tree age and age of ectorganic layer had to be included as relevant parameters in a subsequent experiment.

In June 1990 a subsequent large-scale experiment has been started in secondary Scots pine stands of different age in the northeastern part of the Netherlands. In two young (3 and 10 years at the start of the experiment) and two middle-aged (16 and 27 years) stands the humus and litter layer (including the vegetation) was removed (plots 225 m², 4 replicates). Areas of 400 m² were sod-cut in two old (50 and 66 years) stands. Control treatments were included. In the young and middle-aged stands the organic layer was covered by sods to simulate the thick layer present in old stands. By this experimental set-up we created old stands with young humus profiles (sod-cutting in stands >50 years old) and young stands with old humus profiles (stands <30 years after addition of organic material). We were therefore able to treat tree age and age of ectorganic layer as independent variables. In the autumn of 1990 and 1991 carpophores of ectomycorrhizal fungi were counted. The countings will be continued in 1992 and 1993. The results shown are therefore interim results.

The number of carpophores have statistically been analysed by the Mann-Whitney U-test (Siegel, 1956). Comparisons between numbers of carpophores are only possible within years and not between years due to the yearly variation.

Sod-cutting significantly negatively affected the number of carpophores in the middle-aged and old stands in 1990, but had a significantly positive effect in the middle-aged stands in 1991 (Table 2). In 1990 adding sods significantly negatively affected the number of carpophores in the young and middle-aged stands in 1991 (Table 2). In 1990 adding sods significantly negatively affected the number of carpophores in the young and middle-aged stands, but did not have a significant negative effect in 1991.

TABLE 2A. Average number of carpophores of ectomycorrhizal fungi/1000 m^2 with standard deviation. Pairwise comparisons (Mann-Whitney U-test) between treatments followed by different letters are significantly ($p < 0.05$) different; S: Sod-cut, C: control, and A: sods added

	1990			1991		
	S	C	A	S	C	A
Stand age						
young	571a	380a	53b	2621a	1799ab	90b
	713	338	80	2469	2121	20
middle-aged	111b	2068a	398b	1029b	300a	203a
	103	1864	511	522	380	259
old	64b	293a	—	91a	29a	—
	126	186	—	27	27	—

TABLE 2B. Average number of ectomycorrhizal species per plot with standard deviation.

	1990			1991		
	S	C	A	S	C	A
Stand age						
young	1.8a	2.3a	.8b	3.3a	2.1ab	0.6b
	.7	.7	1.0	1.6	1.5	0.5
middle-aged	2.1ab	4.1a	2.0b	3.1a	2.0ab	1.0b
	1.5	2.2	.7	.9	1.6	.9
old	1.3b	3.3a	—	1.5	1.9	—
	1.1	1.3	—	.6	.9	—

In 1990 the number of species were reduced by sod-cutting in the old stands, and adding sods significantly reduced the number of species in the young and middle-aged stands. In 1991 both treatments did not significantly affect the number of species.

In all stands sod-cutting caused a shift in the ectomycorrhizal species composition compared to the control plots (Tables 3–4). In the young stands sod-cutting had a

TABLE 3. Relative abundance (%) of ectomycorrhizal species in control and sod-cut plots and in plots where sods had been added in two young (5 and 12 years) Scots pine stands in the Netherlands in 1990 and 1991. C: control, S: sod-cut, A: sods added; n: total number of carpophores.

	S	C	A	n
Inocybe lacera	100	0	0	48
Laccaria bicolor	97	3	0	148
Lactarius hepaticus	89	5	6	345
Thelephora terrestris	63	37	0	1333
Laccaria proxima	54	43	3	2258
Scleroderma citrinum Pers.: Pers.	25	75	0	4
Paxillus involutus	17	62	21	24
Lactarius rufus	0	100	0	14
Suillus bovinus	0	100	0	8
Xerocomus subtomentosus (L.: Fr.) Quél.	0	15	85	13
Xerocomus badius	0	7	93	14

TABLE 4. Relative abundance (%) of ectomycorrhizal species in control and sod-cut plots and in plots where sods had been added in two middle-aged (18 and 29 years) Scots pine stands in the Netherlands in 1990 and 1991. C: control, S: sod-cut, A: sods added; n: total number of carpophores.

	S	C	A	n
Inocybe lacera	100	0	0	115
Coltricia perennis (L.: Fr.) Murr.	100	0	0	7
Laccaria bicolor	90	10	0	721
Russula ochroleuca Pers.	50	50	0	6
Paxillus involutus	25	25	50	4
Lactarius rufus	18	58	24	156
Laccaria proxima	10	89	17	21
Lactarius hepaticus	5	77	18	1953
Russula emetica	0	100	0	13
Amanita rubescens	0	100	0	6
Cortinarius fusisporus	0	100	0	4
Xerocomus badius	0	84	16	19
Inocybe ovatocystis	0	2	98	62

positive effect on the abundance of *Inocybe lacera* (Fr.: Fr.) Kummer, *Laccaria bicolor*, and *Lactarius hepaticus* Plowr. Adding sods positively affected the abundance of *Xerocomus badius* and *X. subtomentosus* (L.: Fr.) Quél. (Table 3). In the middle-aged stands *Inocybe lacera, Coltricia perennis* (Linn.) Murr., and *Laccaria bicolor* were positively influenced by sod-cutting. Adding sods stimulated *Inocybe ovatocystis* Bours. & Kühner (Table 4). In old stands sod-cutting had mainly a positive effect on the abundance of *Laccaria bicolor* and a negative effect on most species occurring in the controls (Table 5).

TABLE 5. Relative abundance (%) of ectomycorrhizal species in control and sod-cut plots (52 and 68 years) Scots pine stands in the Netherlands in 1990 en 1991. C: control, S: sod-cut; n: total number of carpophores.

	S	C	n
Amanita muscaria (L.: Fr.) Pers.	100	0	1
Amanita rubescens	100	0	1
Laccaria bicolor	94	6	525
Laccaria proxima	42	58	109
Paxillus involutus	11	89	55
Lactarius hepaticus	7	93	1253
Xerocomus badius	3	97	32
Lactarius rufus	0	100	1
Russula paludosa	0	100	1
Russula ochroleuca	0	100	2
Inocybe ovatocystis	0	100	3
Russula emetica	0	100	15
Laccaria laccata (Scop.: Fr.) Cooke	0	100	25

Laboratory Experiment

To better understand the results of the field experiments laboratory experiments were set up. Experiments to study the sensitivity of ectomycorrhizal species for needle, Scots pine litter and humus (H-layer) extracts were carried out as well as experiments to investigate effects of nitrogen on the growth of some ectomycorrhizal species. Only one of these laboratory experiments will be discussed in detail.

In the autumn of 1990 *Laccaria proxima* and *L. bicolor* were collected in a young Scots pine stand (3 years), *Xerocomus badius* in a middle-aged stand (27 years), and *Paxillus involutus* in an old stand (50 years), all secondary forests. Needles were collected in the autumn of 1990. The needles originated from the litter layer of a Dutch 50-year-old Scots pine forest or from a 50-year-old fallen Scots pine tree in the same stand. In the autumn of 1989 needles were collected from Scots pine trees in south Sweden. After sampling the needles were dried (50°C, 48 hr). The dry needles were ground roughly and 90 g ground needles were added to 675 g distilled water. Extracts were made by shaking (22°C, 24 hr.). After autoclaving for 20 min (0.5 atm, 110°C) the extracts were added to modified Melin Norkrans medium in two concentrations: 7.5 ml extract/100 ml MMN and 37.5 ml extract/100 ml MMN. For inoculation mycelial plugs were laid on the media with the extracts or control media (no extracts added). The agar plates were incubated at 22°C. Radial growth of a fungus was determined by measuring the colony diameter every three days as long as the growth of the fungi in the controls was linear. The colony diameter of *L. bicolor* was measured for 31 days, of *L. proxima* for 32 days and of *X. badius* for 48 days. Measuring the colony diameter of *P. involutus* was finished when the fungus reached the edge of the Petri dish (after 56 days).

The date were analysed with the Kruskal-Wallis and the Mann-Whitney U-test (Siegel, 1988).

The growth of *L. bicolor* was enhanced by the needle extracts, especially by those of Dutch origin ($p < 0.05$) (Fig. 2a). The growth rate of *L. proxima* was significantly negatively affected by all needle extracts ($p < 0.05$) (Fig. 2b). The high concentrations of the needle extracts reduced the growth of *P. involutus*, but the low concentrations had a stimulatory effect ($p < 0.05$) (Fig. 2c). The extracts of the Swedish needles and the treatments with high concentrations reduced the growth of *X. badius* significantly ($p < 0.05$) (Fig. 2d). The low concentration of the Dutch needle extracts did not have an effect.

Discussion

A. Laboratory experiment

In this laboratory experiment the ectomycorrhizal fungi were affected by the needle extracts which was probably either due to substances like phenolics and/or to nitrogenous compounds. Earlier investigations confirmed the sensitivity of mycorrhizal fungi to phenolics (Melin, 1946; Rose et al., 1983; Coté & Thibault, 1988).

The reduced growth of the early-stage fungus *L. proxima* is in accordance with suggestions that early-stage fungi are more sensitive to substances in forest floor material than late-stage fungi (Perry & Choquette, 1987). *Paxillus involutus* and *X. badius* were sensitive to all high concentrations; the growth of *X. badius* was reduced by the extracts of the Swedish needles. These findings are in contrast with the suggestion that late-stage fungi are able to derive nutrients from recalcitrant litter (Last et al., 1987).

Nitrogen might have played a role as well. The stimulated growth of *L. bicolor* by the Dutch needle extracts was probably due to nitrogen. Jongbloed & Borst-Pauwels (1990) showed that high concentrations of ammonium did not inhibit the growth of *L. bicolor*.

B. Translation of the laboratory experiment to the field

These data give only limited support to the theory that ectomycorrhizal succession can be explained by effects of accumulated nitrogen-rich litters.

Late-stage fungi were more sensitive than Last et al. (1987) suggested. *Laccaria bicolor* was stimulated by the needle extracts which was probably due to nutrients of the needles, especially nitrogen, as growth of *L. bicolor* was stimulated by high nitrogen (unpubl. obs.).

Discrepancy between field and laboratory experiments might be caused by the experimental set-up of the laboratory experiment. In these experiments only the fungus was studied and not the fungus in symbiosis with the tree. Subsequent laboratory experiments will include the fungus in symbiosis with the tree.

Furthermore, the perfomance of the fungi was assessed on the basis of radial growth. We preferred radial growth, which indicates the absorbing area of trees in the soil, as a parameter above mycelium weight, which probably reflects carbon drain on the plant.

C. Relation between natural, field and laboratory experiments

The positive effects of the removal of the organic layers on the ectomycorrhizal flora in the field experiments is in agreement with the results of natural experiments. The results of the small-scale experiment show a clear positive effect of removing the humus and litter layer on the ectomycorrhizal flora. And although the large-scale experiment has recently been set up the interim results show that sod-cutting positively affected the number of ectomycorrhizal carpophores and adding sods negatively.

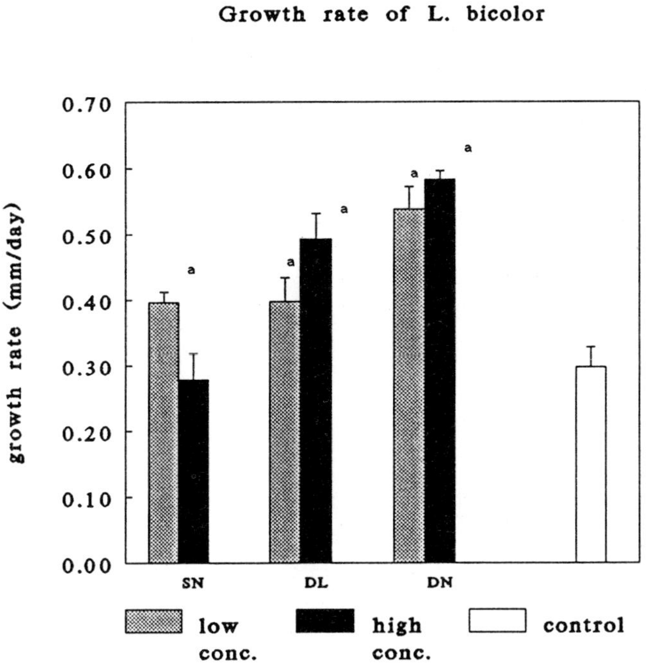

Fig. 2a

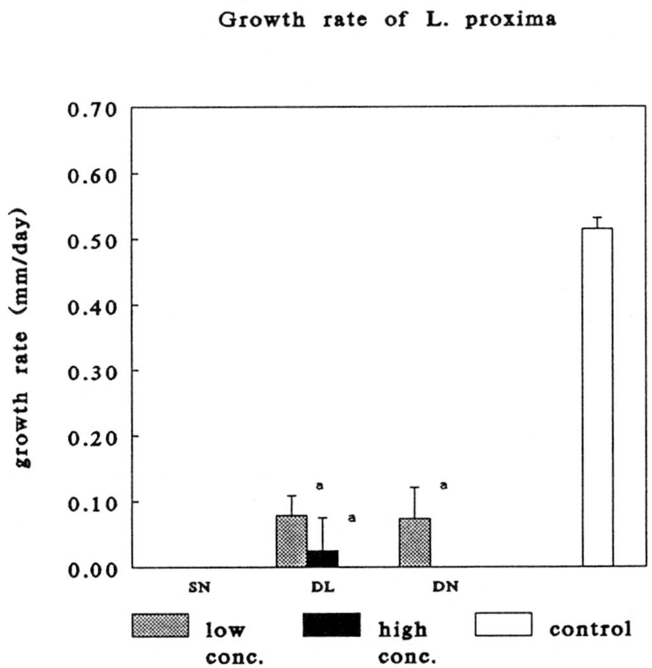

Fig. 2b

FIG. 2. Average growth of *Laccaria proxima* (a), *L. bicolor* (b), *Paxillus involutus* (c), and *Xerocomus badius* (d) on MMN agar with extracts of needles and litter of *Pinus sylvestris* in two concentrations. SN: Swedish needles; DL: Dutch litter; DN: Dutch needles. Significant differences ($p < 0.05$) with the control treatment (Mann-Whitney U-test) are indicated by an (a).

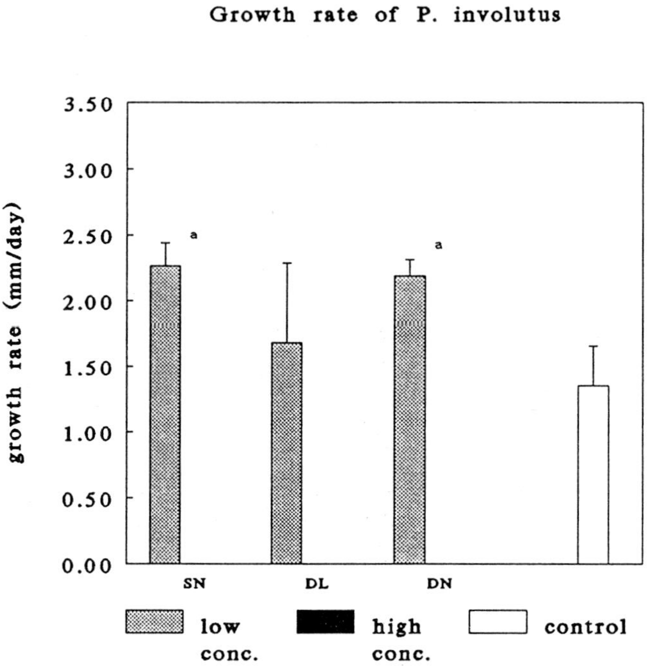

Fig. 2c

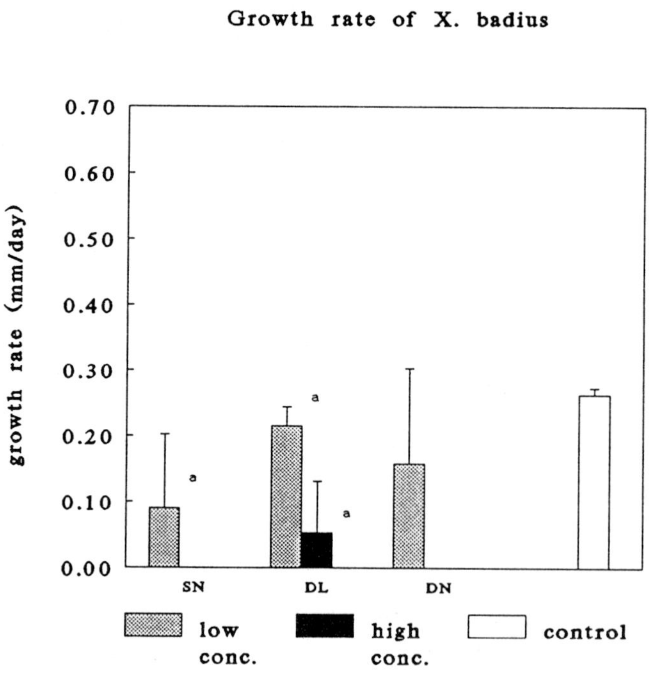

Fig. 2d

FIG. 2. Average growth of *Laccaria proxima* (a), *L. bicolor* (b), *Paxillus involutus* (c), and *Xerocomus badius* (d) on MMN agar with extracts of needles and litter of *Pinus sylvestris* in two concentrations. SN: Swedish needles; DL: Dutch litter; DN: Dutch needles. Significant differences ($p < 0.05$) with the control treatment (Mann-Whitney U-test) are indicated by an (a).

The decision to use *L. bicolor*, *L. proxima*, *P. involutus*, and *X. badius* in the laboratory experiment discussed before was partly based on the results of the field experiments. These species reacted in a different way to the treatments. *Laccaria bicolor*, *L. proxima*, *P.involutus*, and *X. badius* have also been chosen for practical reasons. These fungi can easily be cultured and grow relatively fast (about 0.3–1 mm/day). It should have been very interesting to use other species like *Inocybe lacera*, *Coltricia perennis*, *Sarcodon imbricatus* (L.: Fr.) P. Karst., and species of *Tricholoma* in the laboratory experiment as well, but it was not possible to culture these species or the growth rate of the fungi is very low (about 0.5–1 mm/month).

D. Implications of removal of litter

Removing the organic layer in Scots pine stands caused removal of nutrients and a fairly large amount of roots of Scots pine trees. Especially the high amounts of nitrogen due to air pollution in the litter and humus layer had been removed by sod-cutting. The removal of nutrients and roots might negatively affect the growth of the Scots pine trees. The removal of much of the nitrogen might positively affect the ectomycorrhizal fungi, and hence the vitality of the trees as well. Although removal of litter and humus layer affected the trees somewhat negatively just after the treatment, it had a positive effect on the development of ectomycorrhizal fungi. The trees recovered shortly after the treatment (unpubl. obs.).

Sod-cutting had a positive effect on the number of ectomycorrhizal species and carpophores, but rare ectomycorrhizal species like *Hygrophorus hypotheius* (Fr.: Fr.) Fr., *Sarcodon imbricatus*, *Tricholoma albobrunneum* (Pers.: Fr.) Kumm., *T. flavovirens* (Pers.: Fr.) Lundell, *T. portentosum* (Fr.: Fr.) Quél. found in primary Scots pine forests (pers. obs.) have not been found in sod-cut plots in secondary Scots pine forests until now. This might be due to the high content of organic material in the soils of the secondary stands, as the soils of a secondary young and middle-aged stand in the large scale experiment are podzolic. This organic material has not been removed by sod-cutting and might still affect the ectomycorrhizal flora.

Further manipulations of the organic matter content of the soil (e.g. by removal of only fresh little litter material, or the whole ectorganic horizon) will therefore be necessary to improve management success. However, the results of this experiment are already sufficiently positive to suggest that this might be executed in other forest types as well. For that reason the organic layer in some oak forests on former drift sands will be removed in the future with the aim to remove nitrogen and to recover the ectomycorrhizal flora (Klap & Schmidt, 1992).

Acknowledgements

This is communication 478 of the Biological Station Wijster. The investigations were supported by the Foundation for Biological Research (BION), which is subsidized by the Netherlands Organization for Scientific Research (NWO). We thank Dr. E. Arnolds and Dr. L. Brussaard for critical reading of the manuscript.

References

Arnolds, E. (1991). Decline of ectomycorrhizal fungi in Europe. *Agric. Ecosyst. Environm.* **35**: 209–244.

Cody, M.J. (1986). Overview: Laboratory experiments, field experiments and natural experiments. In J. Diamond and T.J. Case (editors). Community Ecology, pp. 1–22. Harper and Row, New York.

Coté, J.-F. and Thibault, J.-R. (1988). Allelopathic potential of raspberry foliar leachates on growth of ectomycorrhizal fungi associated with black spruce. *Amer. J. Bot.* **75**: 966–970.

Gross-Brauckman, H. and Grosse-Brauckmann, G. (1978). Zur Pilzflora der Umgebung von Darmstadt vor 50 Jahren and heute (ein Vergleich der floristischen Befunde Franz Kallenbachs aus der Zeit von 1918 bis 1942 mit dem gegenwärtigen Vorkommen der Arten). *Z. Mykol.* **44**: 257–269.

Hairston, N.G. (1989). Ecological experiments. Purpose, design and execution. Cambridge University Press, Cambridge.

Jahn, H. (1986). Der "Satanspilzhang" bei Gless (Ottenstein), Süd-Niedersachsen. Zur Pilzvegetation des Seggen-Hangbuchenwaldes (Carici-Fagetum) in Weserbergland and ausserhalb. *Westfäl. Pilzbriefe* **10–11**: 289–351.

Jongbloed, R.H. and Borst-Pauwels, G.W.F.H. (1990). Effects of ammonium and pH on growth of some ectomycorrhizal fungi in vitro. *Acta Bot. Neerl.* **39**: 349–358.

Keizer, P.J. (1993). The influence of nature management of the macromycete flora. In D.N. Pegler, L. Boddy, P. Kirk and B. Ing (editors). Fungi of Europe: investigation, recording and mapping. pp. 251–269. Royal Botanic Gardens, Kew.

Klap, J.M. and Schmidt, P. (1992). Maatregelen om effekten van eutrofiëring en verzuring in bossen tegen te gaan. *Hinkeloord Reports* **3**: 1–140.

Last, F.T., Dighton, J. and Mason P.A. (1987). Succession of sheathing mycorrhizal fungi. *Trends Ecol. Evol.* **2**: 157–161.

Markkola, A.M. and Ohtonen, R. (1988). The effect of acid deposition on fungi in forest humus. In A.E. Jansen, J. Dighton & A.H.M. Bresser (editors), Ectomycorrhizae on acid rain, pp. 122–126. Commission of the European Communities, Brussels.

Melin, E. (1946). Der Einfluss von Waldstreuextrakten auf das Wachstum von Bodenpilzen, mit besonderer Berücksichtigung der Wurzelpilze von Bäumen. *Symb. Bot. Upsal.* **8** (3): 1–116.

Perry, A.P. and Choquette, C. (1987). Allelopathic effects on mycorrhizae. Influence on structure and dynamics of forest ecosystems. In G.R. Waller (editor), Allelochemicals: role in agriculture and forestry, pp. 185–195. American Chemical Society, Washington.

Rose, S.L., Perry, D.A., Pilz, D. and Schoeneberger, M.M. (1983). Allelopathic effects of litter on the growth and colonization of mycorrhizal fungi. *J. Chem. Ecol.* **9**: 1153–1162.

Siegel, S. (1956). Nonparametric statistics for the behavioural sciences. 312 pp. International student edition, New York.

Sokal, R.R. and Rohlf, F.J. (1981). Biometry. Second edition. 859 pp. Freeman and Company. San Francisco.

Termorshuizen, A.J. (1991). Succession of mycorrhizal fungi in stands of *Pinus sylvestris* in the Netherlands. *J. Veg. Sci.* **2**: 555–564.

Tyler, G. (1991). Effects of litter treatments on the sporophore production of beech forest macrofungi. *Mycol. Res.* **95**: 1137–1139.

Vries, B.W.L. de, Jansen, A.E. and Barkman, J.J. (1985). Verschuivingen in het soortenbestand van fungi in naaldbossen van Drenthe, 1958–1983. In E. Arnolds (editor), Veranderingen in de paddestoelenflora (mycoflora). *Wetensch. Meded. Kon. Ned. Natuurhist. Ver.* **167**: 74–83.

Helfer, S. (1993). Rust fungi — A conservationist's dilemma. In D.N. Pegler, L. Boddy, B. Ing & P.M. Kirk (Editors). Fungi of Europe: Investigation, Recording and Conservation. pp. 287–294, Royal Botanic Gardens, Kew.

23. RUST FUNGI — A CONSERVATIONIST'S DILEMMA

S. HELFER

Royal Botanic Garden Edinburgh, Inverleith Row, Edinburgh EH3 5LR, U.K.

Abstract

There are at least 102 species, depending on interpretation and taxonomy, of rare and endangered vascular plants in the British Isles. Fifty-two of these are hosts to one or more rust fungi (Uredinales). In many cases rust fungi are host specific. Whilst the rusts, therefore, are endangered themselves, they contribute to the environmental threat and ecological performance of their hosts. On the other hand there are 58 native rust species which are reported as rare in the British Isles. Some of these occur on common hosts or are capable of infecting a number of host species, others depend on specific host combinations, and others again have reached epidemic proportions in plantations of monocultures elsewhere. Should parasitic fungi be discouraged on endangered and rare vascular plants or should they themselves be conserved? Should the occurrence of rare rusts be monitored in similar ways as that of phanerogams, and should they enjoy similar protection? Examples of rare rusts and their importance in the conservation of their endangered host plants are discussed.

Introduction

The distribution and relative abundance of rust fungi and other parasitic organisms of non-cultivated plants is poorly documented in the literature. Only a few of the standard reference works on the Uredinales (e.g. Grove, 1913; Wilson & Henderson , 1966) state the 'conservation status' of rust species consistently, grouping them into the categories: abundant, frequent, common, scarce, rare and very rare. These groupings have been made mainly on the basis of incidental collections, and no systematic research into the abundance of wild plant parasites in British Isles has been conducted so far.

On a European scale the conservation of fungi and other cryptogams has been publicised for a number of years (e.g. Jansen & Lawrynowicz, 1991) but parasitic fungi have so far not been included in conservation efforts.

The present literature and herbarium survey attempts to present the case for a thorough and ongoing monitoring programme for wild plant parasites in relation to their conservation status, and to point out the possible role these parasites play in the vulnerability of the endangered native plants.

Material and Methods

The 'conservation status' of the native rusts was extracted from Grove (1913) and Wilson & Henderson (1966). This was confirmed and extended by a herbarium survey in the herbaria of the Royal Botanic Gardens at Kew (K) and Edinburgh (E), and by limited own field experience. Rare rusts on cultivated or alien plants were not included. Infected host plants were demonstrated from herbarium sheets and confirmed from the literature (Wilson & Henderson, 1966). Their abundance was verified in the literature (Clapham et al., 1981; Stace, 1991).

The conservation status of the flora of the British Isles was extracted from Stace (1991) with amendments and additions provided by P. Lusby (Scottish Rare Plants Project, RBG Edinburgh, pers. comm.). Actual and possible rust parasites were extracted from Gäumann (1959), Brandenburger (1963) and Wilson & Henderson (1966). Rusts were only included where their occurrence on a given host species is notably mentioned in the literature or where herbarium specimens exist.

Results and Discussion

Table 1 shows the rare rust parasites of native plants of the British Isles. Of the 58 species, 38 (65.5%) occurred on ± common host plants (common, frequent, locally common, locally frequent & without conservation status sensu Stace, 1991; group 1 in Table 1). These rusts are not uncommon because of the scarcity of their host plants, and other limiting factors, genetic or environmental are prevailing to restrict their abundance. However, as no systematic sampling procedure was employed to interpret their rarity, factors of selective sampling are most likely contributing to this phenomenon. Some of the rusts in this group appear to be more abundant in other countries, a fact reflected by the number of foreign herbar ium vouchers in K and E, and *Melampsora ribesii-viminalis* Kleb. has been found to cause widespread damage in epidemic levels in trial energy plantations in Sweden (unpublished results). *Pucciniastrum guttatum* (Schroet.) Hyl., Jørst. Nannf. and *Puccinia umbilici* Duby have probably been widely overlooked and there are several new collections in the herbaria: their conservation status need to be confirmed. Similarly *Uromyces chenopodii* (Duby) Schroet. and *U. salicorniae* de Bary seem to have been collected preferentially in the past, but no recent specimens are deposited in K or E.

The remaining 20 rust species (34.5%) occur on ± rare host plants (group 2 in Table 1). Some of these fungi (e.g. *Uredo goodyerae* on *Goodyera repens*) are found relatively regularly where the host plant occurs. Most rusts in this group are presumably rare because of the rarity of their hosts.

In the herbaria at K and E there are few recent records of rare British rusts. Great care has to be taken not to damage populations by over-collecting, especially where rare host plants are involved. Nevertheless, a systematic recording of the parasites to monitor population dynamics is essential, taking adequate specimens (e.g. on microscope slides accompanied by a photographic record).

Table 2 shows the 102 native British plant species currently considered to be very rare and endangered or vulnerable (Stace, 1991, Lusby, pers. comm.), and their actual or potential rust parasites. Fifteen species of these (14.7%) are infected by relatively common rusts (common or frequent) and are therefore conceivably at a greater risk from rust attack than the remainder. Many of the rusts in this group are polyphagous and may not be specifically aggressive to these hosts. For the three *Hieracium* species only anecdotal evidence exists of the presence of rust (Tennant pers. comm.), and no subspecies of *P. hieracii* Martin has been named. For 20 rare host plants (19.6%) only rare (scarce, rare or very rare) rust parasites are known. This group represents more specific host parasite combinations, and the scarcity of the rust is linked to that of the host plant. The rust of *Cystopteris dickieana*, *Woodsia ilvensis* and *W. alpina* (*Hyalopsora polypodii* (Diet.) Magn.) has not been confirmed from the British Isles but its occurrence is highly likely, as this rust is very polyphagous on many ferns. For the 67 remaining species (65.7%) either no rust species is reported in the literature or the abundance of the rust is unknown in Britain (but is presumably very rare: no British herbarium material available) or the rust may only have been confirmed outside the British Isles.

Conclusions

There is a lack of recent information on the rare rusts of the British Isles, and of the rusts of rare vascular plants. It is therefore desirable that a systematic programme is established to generate reliable distribution data for these orgin sms as well as to monitor their population dynamics. For this purpose the author seeks the cooperation with other mycologists and vascular plant botanists.

Acknowledgement

I wish to thank Mr Phil Lusby for his expert advice on British rare plants.

References

Brandenburger, W. (1963). Vademecum zum Sammeln parasitischer Pilze. Stuttgart.
Clapham, A.R., Tutin, T.G. and Warburg, E.F. (1981). Excursion Flora of the British Isles. 3rd edition. Cambridge University Press.
Gäumann, E. (1959). Die Rostpilze Mitteleuropas. *Beitr. Kryptogamenfl. Schweiz* **12**; 1407pp.
Grove, W.B. (1913). British Rust Fungi. Cambridge.
Jansen, A.E. and Lawrynowicz, M. (editors) (1991). Conservation of Fungi and other Cryptogams in Europe. *Lodzkie Towarzystwo Naukowe, Szlakami Nauki (Science Tracks)* **18**; 120pp.
Stace, C.A. (1991). New Flora of the British Isles. Cambridge.
Wilson, M. and Henderson, D.M.H. (1966). British Rust Fungi. Cambridge University Press.

TABLE 1. Rare rusts on native plants, their hosts and potential spread

Group 1: rusts of +/- frequent plants (c, f, lc, lf & - *)

Rusts	rust status	Hosts (host status *)
Ochropsora ariae	rare	[*Anemone nemorosa* (f)]; *Sorbus aucuparia* (-)
Uredinopsis filicina	very rare	[*Abies mayriana*]; *Phegopteris connectilis* (c)
Milesina murariae	rare	*Asplenium ruta-muraria* (c)
Milesina carpatorum	rare	*Dryopteris filix-mas* (c)
Milesina whitei	rare	*Polystichum setiferum* (c)
Hyalopsora aspidiotus	rare	[*Abies alba*]; *Gymnocarpium dryopteris åal* (f)
Pucciniastrum guttatum	rare	*Galium odoratum* (c), *G. palustre* (c), *G. saxatile* (c), *G. uliginosum* (c), *G. verum* (c), *Sherardia arvensis* (c)
Melampsora vernalis	rare	*Saxifraga granulata* (lc)
Melampsora ribesii-viminalis	rare 1)	[*Ribes uva-crispa* (-)]; *Salix viminalis* (c)
Melampsora ribesii-purpureae	rare	[*Ribes alpinum* (r), *R. uva-crispa* (-)]; *Salix purpurea* (c)
Phragmidium acuminatum	very rare	*Rubus saxatilis* (-)
Frommea obtusa	rare	*Potentilla erecta* (-), *P. reptans* (c)
Triphragmidum filipendulae	rare	*Filipendula vulgaris* (lf)
Puccinia eutremae	rare	*Cochlearia officinalis* (-), *C. danica* (-)
Puccinia argenata	very rare	*Adoxa moschatellina* (c)
Puccinia umbilici	rare	*Sedum roseum* (c), *Umbilicus rupestris* (c)
Puccinia hydrocotyles	rare	*Hydrocotyle vulgaris* (lc)
Puccinia oxyriae	rare	*Oxyria digyna* (-)
Puccinia convolvuli	very rare	*Calystegia sepia* (c)
Puccinia clintonii	rare	*Pedicularis palustris* (c), *P. sylvatica* (c)
Puccinia betonicae	rare	*Stachys officinalis* (c)
Puccinia thymi	rare	*Thymus polytrichus* (c), *T. pulegioides* (lf), *Origanum vulgare* (lc)
Puccinia commutata	rare	*Valeriana officinalis* (f)
Puccinia eriophori	very rare	*Solidago virgaurea* (c); *Trichophorum caespitosum* ssp. *germanicum* (c), *Eriophorum gracile* (rrr)
Puccinia ribis-nigri-lasiocarpae	rare	*Ribes* spp.; *Carex lasiocarpa* (lc)
Puccinia microsora	rare	*Carex vesicaria* (f)
Puccinia longissima	rare	[*Sedum acre* (c)]; *Koeleria macrantha* (lc)
Puccinia pratensis	very rare	*Helicotrichon pratense* (-)
Puccinia borealis	rare	*Thalictrum alpinum* (l); *A. canina* (f)
Puccinia schismi	rare	[*Allium* spp. (c)]; *Vulpia bromoides* (lc)
Uromyces chenopodii	rare	*Suaeda vera (r)*, *S. maritima* (c)
Uromyces salicorniae	rare	*Salicornia europaea* (c), *S. ramosissima* (c)
Uromyces minor	rare	*Trifolium dubium* (c), *T. incarnatum* ssp. *molinerii* (rr)
Uromyces punctatus	rare	*Euphorbia cyparissias* (lc); *Astragalus danicus* (l)
Uromyces pisi sens. stricto	rare	*Euphorbia cyparissias (lc)*; *Lathyrus pratensis* (c)
Uromyces tuberculatus	rare	*Euphorbia exigua* (c)
Uromyces eugentianae	rare	*Gentianella amarella agg.* (c)
Uredo quercus	rare	*Quercus robur* (c)

*) host status: (c) common; (f) frequent; (lc) locally common; (lf) locally frequent; (s) scattered; (r) rare; (rr) very rare; (rrr) vulnerable; (-) no specific information available (but not rare); 1) this rust has reached epidemic proportions in energy plantations in Sweden

TABLE 1. Continued

Group 2: rusts of +/- rare plants (s, r & rr)

Rusts	rust status	Hosts (host status *)
Hyalopsora adianti-capilli-veneris	very rare	*Adiantum capillus-veneris* (r)
Pucciniastrum pyrolae	rare	*Pyrola media* (r), *P. minor* (s), *P. rotundifolia* (r), *Orthilia secunda* (r)
Uredo goodyerae	rare	*Goodyera repens* (r)
Chrysomyxa pirolata	rare	*Pyrola minor* (s), *P. rotundifolia* (r)
Nyssopsora echinata	rare	*Meum athamanticum* (r)
Puccinia ribis	rare	*Ribes spicatum* (r)
Puccinia thesii	rare	*Thesium humifusum* (r)
Puccinia bulbocastani	rare	*Bunium bulbocastanum* (rr)
Puccinia bupleuri	rare	*Bupleurum tenuissimum* (r), *Bupleurum falcatum* (rrr)
Puccinia cicutae	rare	*Cicuta virosa* (r)
Puccinia libanotidis	very rare	*Seseli libanotis* (rr)
Puccinia physospermi	rare	*Physospermum cornubiense* (rr)
Puccinia rugulosa	very rare	*Peucedanum officinale* (rr)
Puccinia polemonii	rare	*Polemonium caeruleum* (rr)
Puccinia albulensis	rare	*Veronica alpina* (r)
Puccinia asparagi	rare	*Asparagus officinalis* ssp. *prostratus* (rr)
Puccinia gladioli	very rare	*Gladiolus illyricus* (rrr)
Puccinia cancellata	rare	*Juncus acutus* (r)
Puccinia silvatica	rare	*Taraxacum officinale* (c); *Carex capillaris* (r)
Uromyces gageae	rare	*Gagea lutea* (r)

*) host status: (c) common; (f) frequent; (lc) locally common; (lf) locally frequent; (s) scattered; (r) rare; (rr) very rare; (rrr) vulnerable; (-) no specific information available (but not rare);

TABLE 2. Rare and endangered plants of the British Isles and their potential rust parasites.

Group 1: Host plants of common rusts (c & f *)

Host plant	possible or actual rust (& status *)
Arenaria ciliata	*Puccinia arenariae* (c), *Uredo arenariae-ciliatae* (-)
Minuartia stricta	*Puccinia arenariae* (c)
Dianthus gratianopolitanus	*Uromyces dianthi* (f)
Polygonum maritimum	*Uromyces polygoni-avicularae* (c)
Rumex rupestris	*Uromyces rumicis* (c)
Althaea hirsuta	*Puccinia malvacearum* (c)
Euphorbia peplis	*Melampsora euphorbiae* (c)
Mentha pulegium	*Puccinia menthae* (c)
Veronica spicata ssp. spicata	*Puccinia veronicae* (f)
Melampyrum arvense	*Coleosporium melampyri* (f), *Coleosporium tussilaginis* (c)
Rhinanthus angustifolius	*Coleosporium euphrasiae* (c), *Coleosporium melampyri* ? (f)
Phyteuma spicatum	*Coleosporium campanulae* (f), *Uromyces caricis-sempervirentis* (-), *Uromyces phyteumatum* (-)
Hieracium attenuatifolium	*Puccinia hieracii* ? (f)
Hieracium northoense	*Puccinia hieracii* ? (f)
Hieracium zetlandicus	*Puccinia hieracii* ? (f)

Group 2: Host plants of +/- rare rusts (s, r & rr *)

Host plant	possible or actual rust (& status *)
Cystopteris dickieana	*Hyalopsora polypodii* (s)?
Woodsia ilvensis	*Hyalopsora polypodii* (s)?
Woodsia alpina	*Hyalopsora polypodii* (s)?
Saxifraga cernua	*Caeoma cernuae* (-), *Puccinia saxifragae* (s)
Saxifraga rosacea ssp. *hartii*	*Puccinia saxifragae* ? (s)
Saxifraga caespitosa	*Melampsora alpina* (-), *Puccinia saxifraga* (s)
Cotoneaster itegerrinus	*Gymnosporangium confusum* (s)
Bupleurum falcatum	*Puccinia bupleuri-falcati* (r)
Apium repens	*Uromyces lineolatus* (s)
Gentianella ciliata	*Puccinia gentianae* (r)
Gentiana verna	*Puccinia gentianae* (r), (*Cronartium flaccidum*? [s])
Gentiana nivalis	*Puccinia gentianae* (r)
Pulicaria vulgaris	*Uromyces junci* ? (s)
Artemisia campestris	*Puccinia absinthi* (s), *Puccinia artemisiicola* (-)
Eriophorum gracile	*Puccinia eriophori* ? (rr)
Polygonatum verticillatum	*Puccinia digraphidis* (s), *Puccinia smilacearum-festucae* (-)
Allium sphaerocephalon	*Melampsora allii-populina* (-), *Melampsora allii-fragilis* (s?), *Melampsora alii-salicis-albae* (f?), *Puccinia allii* (s), *Puccinia fragosoi* (r), *Puccinia lagori-chamaemoly* (-), *Uromyces ambiguus* (s)
Gladiolus illyricus	*Puccinia gladioli* (rr)
Orchis militaris	*Melampsora repentis* (s), *Puccinia orchidearum-phalaridis* (s)
Ophrys fuciflora	*Puccinia orchidearum-phalaridis* (s)

TABLE 2. Continued

Group 3: Host plants of rusts of unknown abundance.

Host plant	possible or actual rust (& status *)
Equisetum ramosissimum	no rust known in British Isles
Ophioglossum lusitanicum	no rust known in British Isles
Trichomanes speciosum	no rust known in British Isles
Ranunculus ophioglossifolius	no rust known in British Isles
Fumaria reuteri	no rust known in British Isles
Chenopodium vulvaria	no rust known in British Isles
Arenaria norvegica	no rust known in British Isles
Minuartia recurva	no rust known in British Isles
Scleranthus perennis	*Uromyces scleranthi* (-)
Corrigiola litoralis	*Puccinia corrigiolae* (-)
Petrorhagia nantenillii	no rust known in British Isles
Limonium recurvum	no rust known in British Isles
Viola persicifolia	no rust known in British Isles
Arabis alpina	no rust known in British Isles
Arabis scabra	no rust known in British Isles
Thlaspi perfoliatum	*Puccinia thlaspeos* (-)
Cakile wrightii	no rust known in British Isles
Phyllodoce caerulea	no rust known in British Isles
Diapensia lapponica	no rust known in British Isles
Crassula aquatica	no rust known in British Isles
Pyrus cordata	no rust known in British Isles
Ononis reclinata	no rust known in British Isles
Lythrum hyssopifolia	no rust known in British Isles
Eryngium campestre	*Puccinia eryngii* (-)
Bupleurum baldense	no rust known in British Isles
Selinum carvifolia	no rust known in British Isles
Centaurium tenuiflorum	no rust known in British Isles
Gentianella anglica	no rust known in British Isles
Gentianella uliginosa	no rust known in British Isles
Cynoglossum germanicum	*Puccinia cerinthes-agropyrina* ? (-)
Teucrium scordium	no rust known in British Isles
Teucrium botrys	*Puccinia chamaedryos* (-)
Clinopodium menthifolium	no rust known in British Isles
Salvia pratensis	*Puccinia stipina* (-)
Limosella australis	no rust known in British Isles
Veronica triphyllos	no rust known in British Isles
Orobanche caryophyllacea	no rust known in British Isles
Orobanche reticulata	no rust known in British Isles
Orobanche artemisiae-campestris	no rust known in British Isles
Scorzonera humilis	*Puccinia scorzonerae* (-)
Lactuca saligna	*Puccinia opizii* (-)
Cicerbita alpina	*Puccinia mulgedii* (-)
Crepis foetida	*Puccinia barkhausiae-rhoedifolia* (-)
Filago lutescens	no rust known in British Isles
Gnaphalium luteoalbum	no rust known in British Isles
Inula salicina	*Coleosporium inulae* (-)
Erigeron borealis	*Puccinia dorrensis* (-)

TABLE 2. Continued

Group 3: Host plants of rusts of unknown abundance.

Host plant	possible or actual rust (& status *)
Oranthus maritimus	no rust known in British Isles
Senecio paludosus	*Puccinia senecionis-acutiformis* (-)
Homogyne alpina	*Puccinia conglomerata* (-), *Uromyces veratri* (-)
Alisma gramineum	no rust known in British Isles
Damasonicum alisma	no rust known in British Isles
Elodea verticillata	no rust known in British Isles
Najas marina	no rust known in British Isles
Schoenoplectus triqueter	no rust known in British Isles
Cyperus fuscus	*Puccinia romagnoliana* (-)
Carex depauperata	no rust known in British Isles
Lloydia serotina	*Puccinia bessei* (-)
Romulea columnae	no rust known in British Isles
Cypripedium calceolus	no rust known in British Isles
Cephalanthera rubra	no rust known in British Isles
Epipactis youngiana	no rust known in British Isles
Epipogium aphyllum	no rust known in British Isles
Liparis loeselii	no rust known in British Isles
Orchis simia	no rust known in British Isles
Himantoglossum hircinum	no rust known in British Isles
Ophrys sphegodes	no rust known in British Isles

*) Rust status: c common, f frequent, s scarce, r rare, rr very rare, - no information available for British Isles.

CONGRESS OF EUROPEAN MYCOLOGISTS
REVISED STATUTES, 1992

Article 1. Status: The Congress of European Mycologists is recognized by the International Mycological Association (International Union of Biological Sciences, Section for General Mycology). The Standing Committee of CEM serves as the Committee for Fungi in Europe within IMA.

Article 2. Objectives: The objectives are to (i) promote all forms of mycological investigation within Europe; (ii) enhance contact between mycologists within Europe; (iii) to exchange results, information and opinions by means of paper-readings, poster sessions, field meetings, workshops and committees; (iv) to prepare a European input to the IMA, other international bodies and initiatives.

Article 3. Location and timing of Congress: The Congress meets approximately every three years, each within a different European country, preferably alternating with the International Mycological Congresses. Invitations must be distributed to the Standing Committee three months before the previous Congress, voted on that Committee and their recommendation made to the Congress. Where possible congresses shall alternate between eastern and western Europe.

Article 4. Organizing Committee: The organisation of each Congress is undertaken by an Organizing Committee appointed by the host country. The composition of that Committee shall be made clear in the proposal to the Standing Committee. The host country accepts full responsibility for all matters related to the organization, finance, and programme of each Congress.

Article 5. Announcement of Congress: The First Circular, announcing the place and date of the next Congress, shall be issued at least 18 months ahead of the Congress. The Second and Final Circular shall be issued 12 months ahead of the Congress.

Article 6. Official Language: The official language for Congresses shall be English. In case of difficulties, papers may be presented in French, German, or Russian, provided than an English abstract or translation is made available.

Article 7. Standing Committee: The Standing Committee comprises one representative for each country, elected by the members present from each participating country either at or before the start of the Congress. The results are to be announced by the President at the Opening Session. The Standing Committee must have a Chairman and Secretary, and serves until the subsequent Congress. Members cannot be re-elected on more than two successive occasions. The President and the Secretary-General of the Congress can be elected to the Standing Committee.

Article 8. Congress Officials: The Organizing Committee nominates a President, two Vice-Presidents, a Secretary-General, and, if required, two Assistant Secretaries to serve during the Congress. Honorary Presidents may also be selected. All Congress officials are formally elected at the Opening Session.

Article 9. Functions of the Standing Committee: The Standing Committee deals with all matters concerning organisation and policy. Resolutions for the final plenary session must first be submitted to the Standing Committee, which shall make recommendations to the Congress. The decisions of the Standing Committee are announced by the President at the Closing Plenary Session. The country and date for the next Congress is recommended by the Standing Committee through the Secretary-General and announced prior to the Closing Ceremony (see Art. 3).

Article 10. Organisation of Next Congress: The Congress officials terminate their duties at the closure of the Congress. The Secretary-General is to contact the Chairman of the next Organizing Committee and transfer all matters relating specifically to the regulations and organization.

Article 11. Termination of Congresses: In the event of a proposal to terminate the series of CEMs, a two-thirds majority vote of the current Congress shall be required. Any outstanding assets will be forwarded to the International Mycological Association.

ALPHABETICAL LIST OF PARTICIPANTS TO XI CONGRESS OF EUROPEAN MYCOLOGISTS

AINSWORTH, Dr. A., I.M.I., Bakeham Lane, Egham, Surrey TW20 9TY, UK.
AITCHISON, Ms L., I.M.I., Bakeham Lane, Eegham, Surrey TW20 9TY, UK.
ARNOLD, Dr. G. Friederich Schiller Univ. Jena, Biologische Facultät, D6- 53300 Weimar, Germany.
ARNOLDS, Dr. E., (Biological Station, Centre for Soil Ecology, Kampsweg 27, GU18 PD Wijster, The Netherlands).
ASTIER, M. P., 65 Bd de la Liberation, 13001 Marseille, France.
ASTIER, M. J., 65 Bd de la Liberation, 13001 Marseille, France.
BAAR, Ms J., Biological Station, Kampsweg 27, 9410 PD Wijster, The Netherlands.
BAPTISTA-FERREIRA, Prof J., Centro de Micologia/Faculadade de Ciencias, P-1700 Lisboa Codes, Edif C2-Campo Grande, Portugal.
BAPTISTA-FERREIRA, Mrs I., Portugal.
BARATA, Mrs M., Largo da Graca, 82-2QE, 1100 Lisboa, Portugal.
BARLUZZI, Mr M., Dipartimento di Biologia Ambientale, Via P.A. Mattioli 4, 53100 Siena, Italy.
BARLUZZI MARTINI, Mrs C., Dipartmenlo di Biologia Ambientale, Via P. A. Mattioli, 53100 Siena, Italy.
BELLU, Dr F., Via C. Battisti, 34-Bolzano, 39100, Italy.
BENDIKSEN, Mr E., Norwegian Institute for Nature Research (NINA), PO Box 1037 Blindern, N-315 Oslo 3, Norway.
BENKERT, Dr D., Siemensstrasse 9, D-1590 Potsdam, Berlin, Germany.
BERNICCHIA, Ms A., Istituto di Patologia Vegetale, Universita di Bologna, Via F. Reg 8, 40126 Bologna, Italy.
BLOOR, Mr K., Stoke-on-Trent City Museum, Hanley, Stoke-on-Trent, Staffordshire ST1 3DN, UK.
BODDY, Dr L., School of Pure and Applied Biology, University of Wales College of Cardiff, CF1 3TZ, UK.
BONDARCEVA, Dr M, Botanical Institute, Academy of Science, 2 Popov Street, St Petersburg 197 022, RUSSLA.
BOTTALICO, Prof A., Istituto Tossine e Micotossine de C.N.R., viale Luigi Einaudi, 51-70125 Bari, Italy.
BUJAKIEWICZ, Dr A. M., Dept. of Plant Ecoiogy, Adam Mickiewicz Univ., Al. Niedpodleglosci 14, P61-713 Poznan, Poland.
BUSCOT, F. F., Spez. Botanik Mycologie Inst. für Bodenbiologie 1, Auf der Morgenstelle 1, D 7400 Tubingen, Germany.
BUTTERFILL, Ms G., The Herbarium, R.B.G., Kew, Surrey, TW9 3AE, UK.
BUYCK, Prof B. B., Faculty of Science, Univ. Burundi, BP 2700 Bujumbura, Burundi.
CALONGE, Prof F. D., Real Jardin Botanica, Plaza de Murillo 2, 28014 Madrid, Spain.
CANDUSSO, Dr M., via Grandi 28, Saronno 21047, Italy.
COOK, Mrs R., 1, Summerville Gardens, Stockton Heath, nr Warrington, Cheshire, UK.
COURTECUISSE, Dr R., Botany Dept., Fac. of Biological & Pharmaceutical Sciences, Univ. Lille, Rue du Prof Laguesse, BP 83, 59006 Lille cedex, France.
DAHLBERG, Dr A., Dept. of Forest Mycology & Pathology, Swedish Univ. Agricultural Science, PO Box 7026, 75007 Uppsala, Sweden.
DAOV, Dr S., ZPUC 'Dimitar Vlahov', 92 400 Strumica, Macedonia, 'Yugoslavia'.
DECLERCQ, Dr B., Axelwoordeken 28, B-9185 Wachtebeke, Belgium.
DEMOULIN, Dr V., Dpt. Botanique, B22, Univ. Liège, Liège, B-4000, Belgium.
DIAMANDIS, Dr S., National Agricultural Research Foundation, Forest Research Institute, 570 06 Vassilka, Thessaloniki, Greece.

DICKERSON, Mrs S., The Herbarium, R.B.G., Kew, Surrey, TW9 3AE, UK.
DICKSON, Mr G., Flagstones, 72 Catisfield Lane, Fareham, Hants PO15 SNS, UK.
D'VRIGNY, Dr C. J., 17 Chemin Desvallieres, Ville d'Avray, F-92410, France.
ECKBLAD, Prof F.-E., Div. Botany, Univ. Oslo, PO 1045, Blindern, N-0316 Oslo, Norway.
EMMETT, Mr E. E, Alasdair, Farnham Lane, Haslemere, Surrey, GU27 IHD, UK.
EMPAIN, Dr A., Jardin Botanique National de Belgique, Domaine de Bouchout, B-1860 Meise, Belgium.
FELLNER, Dr R., Forestry & Game Management Research Institute, CS-156-04, Prague-Zbrazlav, 'Czechoslovakia'.
FISHER, Dr P. J., Hatherly Laboratory, Biological Sciences, Univ. Exeter, Devon, EX4 4PS, UK
FOITZIK, Dr O., Herbarium Haussknecht, Friedrich-Schiller-Univ. Jena, Schlossgasse, Universitätsgebäude, D-6900 Jena, Germany.
FOURRE, M G., 152 Rue Jean Jaurs, Niort, F-7900, France.
FOURRE, Mme P., c/o G Fourre, France.
FOX, Dr F., Copse End, Copse Lane, Jordans, Bucks, HP9 2TA, UK.
FRAITURE, Ing. A., Jardin Botanique National de Belgique, Domaine de Bouchout, B-1860 Meise, Belgium.
FRANKLAND, Dr J. C., Institute of Terrestrial Ecology, Merlewood Research Station, Grange-over-Sands, Cumbria, CA17 4NI, UK.
FULLER, Mr H., Botany Dept., Univ College Dublin, Belfast, Dublin 4, Ireland.
GAHNE, Ms B., Rasbo, Grän, S-75596 Uppsala, Sweden.
GAMS, Dr W., CBS, PO Box 273, 3740 AG Baarn, The Netherlands.
GAY, Mlle S., 4 bd St Denis, 92400 Courbevoie, France.
GESSNER, Dr E., Am Bagno 5, D-W4405 Nottuln, Germany.
GIBBS, Dr J., Forest Research Station, Alice Holt Lodge, Farnham, Surrey GUIO 4LH, UK.
GINNS, Mrs A., c/o Dr J Ginns.
GINNS, Dr J., Saunders Bldg., Experimental Farm, Canada Agriculture, Ottawa, Ontario, Canada, K1A OC6.
GREGORY, Mr M. J., Duke's Orchard, Spring Elms lane, Little Baddow, Chelmsford, Essex, CM3 4SG, UK.
GRIFFITHS, Dr A., IMC5 Secretariat, c/o Venue West, 645-375 Water St, Vancouver, BC V6B 5C6, Canada.
GRGURINOVIC, Ms C., Australian Biological Resources Study, GPO Box 636, Canberra, ACT 2601, Australia.
GROSSBARD, Dr E., 29 Perrott Close, North Leigh, Oxfordshire OX8 6RU, UK.
GULDEN, Dr G., Botanical Garden and Museum, Trondheimsun 23B, N 0562 Oslo, Norway.
HARKONEN, Dr M. E., Univ. Helsinki, Dept. of Botany, Unioninkatu 44 SF-00170 Helsinki, Finland.
HALLENBERG, Dr N., Dept. Systematic Botany, Carl Skottsburg Gata 22, S-413 19 Goteburg Sweden.
HAMMON, Dr K., Dept. Zoology, Arizona State Univ., Tempe, AZ 85281-1501, USA.
HARRINGTON, Mr T. J., Chemistry & Life Sciences Dept., Univ. Limerick, Limerick, Ireland.
HAWKSWORTH, Prof D. L., I.M.I., Bakeham Lane, Egham, Surrey TW20 9TY, UK
HELFER, Dr S., R.B.G., Edinburgh, EH3 SLR, UK.
HENRICI, Mr A., 57, Clarendon Road, London W11 4JD, UK.
HILBER, Dr O. R. A., Research Institute for Mushroom Cultivation, Huettenallee 235, D-4150 Krefeld, Germany.
HOILAND, Mr K., Norwegian Institute for Nature Research, PO Box 1037, Blindern, N-0315 Oslo 3, Norway.
HOLUBOVA-JECHOVA, Dr V., (c/o V. Hobbs, The Royal Society, 6 Carlton House Terrace, London SWIY 5AG) Czechoslovakia

HUHTINEN, Dr S., Dept. Biology, Univ. Turku, SF-20500 Turku, Finland.
ING, Dr B., Chester College of Higher Education, Chester CHI 4BJ, UK.
ISILOGLU, Dr M., Inonu Univ., Science & Art Faculty, Dept. Biology, Kampus Nalatya, Turkey.
IVANCEVIC, Mr B., Natural History Museum, Njegoseva 51, Belgrade 11OOO, 'Yugoslavia'.
JONES, Prof E. B. Gareth, School Biological Sciences, Univ. Portsmouth, King Henry I St., Portsmouth, Hampshire, UK.
KARADELEV, Mr M., St Marsal Tito 79 ll/9, 91 480 Gevgelija, Macedonia, 'Yugoslavia'.
KEIZER, Dr P. J., Biologisch Station, Kamspweg 27, NL 9418, PD Wijster, The Netherlands.
KINSEY, Dr G., Xenova Ltd., 545 Ipswich Road, Slough, Berkshire SL1 4EQ, UK.
KIRK, Dr P. M., I.M.I., Bakeham Lane, Egham, Surrey TW20 9TY, UK.
KNUDSEN, Dr H., Botanical Museum, Gothersgade 130, Copenhagen 1123K, Denmark.
KORTEKANGAS, Mrs E. I., Palovartijantie 6 B 92, 00750 Helsinki, Finland.
KOVALENKO, Prof A., Komarov Botanical Institute of Russian Academic Science, 2 Popov Street, St Petersburg 197 376, Russia.
KREISEL, Prof H., Hans-Beimler-str 10, Greifswald, D-2200, Germany.
KUYPER, Dr T., Biological Station, Kampsweg, Wijster, NL 9410 PD, The Netherlands.
LADO, Dr C., Real Jardin Botanico, Pza. Murillo 2, Madrid 28014, Spain.
LAESSOE, Mr T., Herbarium, R.B.G., Kew, Richmond, Surrey, TW9 3AE, UK.
LANGE, Mr C., Europaplads 8, 1.TH., DK-8000 Aarhus, Denmark.
LAWRYNOWICZ, Prof M., Institute of Environmental Biology, Univ. Lodz, PL-90 237 Lodz, ul. Banacha 1216, Poland.
LENNIE, Mr J., 861 Keeler Avenue, Berkeley, California, 94708, USA.
LEONARD, Mrs A. C, 12 Argyle Road, Christchurch, Dorset BH23 3NW, UK.
LEUCHTMANN, Dr A., Geobotanisches Inst. ETH-Zurich, Zollikerstrasse 107, CH-8008, Zurich, Switzerland.
LUCAS, Dr M. T., Estaao Agronomica Nacional, Oeiras, P-2780, Portugal.
MAGGI, Dr O., Dipartimento di Biologiavegetale, Univ. 'La Sapienza', Largo Cristina di Svezia, 24 00165 Roma, Italy.
MARCINKOWSKA, Dr J., Warsaw Agricultural Univ., UL Nowoursynowsku 166, Warsaw 02-766, Poland.
MATAVULJ, Dr M., Botanical Inst., Univ. Regensburg, Postfach 1010 42, D-8400 Regensburg, 6FR, Germany.
MELO, Dr I., Museu, Laboratorio e Jardim Botanico, Rua da Escola Politecnica 58, Lisboa P 1294, Portugal.
MOLITORIS, Prof P. H., Bot. Inst. Univ., Postf. 8400 Regensburg, Germany.
MONTANINI, Ms N., Lepetit Research Center, Via Roberto Lepetit 34, 21040 Gerenzano (Varese), Italy.
MONTEMARTINI CORTE, Prof A., Istituto Botanico 'Hambury', Corso Dogali 1 C, I 16136 Genova, Italy.
MUKHIN, Mr V., Institute Plant & Animal Ecology, Russian Academy Sciences, 8 Marta St., 202 Ekaterinburg, 620219, Russia.
MUROMTSEV, Prof G. S., Institute of Agricultural Biotechnology, Bldg 4, 12 Pskovskaya St., Moscow 127253, Russia.
NAUTA, Dr M., Institute for Forestry and Nature Research, PO Box 23, 6700 AA Wageningen, The Netherlands.
NIEMELA, Prof T., Dept. Botany, Univ. Helsinki, Unioninkatu 44, SF-00170 Helsinki, Finland.
NODDER, Mr T., Friends of Kew, Cambridge Cottage, R.B.G., Kew, TW9 3AB, UK.
NOORDELOOS, Dr M. M., Solingenstr. 12, 2804 XT Gouda, The Netherlands
OHENOJA, Dr E., Botanical Museum, Univ. Oulu, SF-90570 Oulu, Finland.
PACIONI, Prof G., Dipart. Scienze Ambientali, Univ. L'Aquila 67100, Italy.
PARMASTO, Prof E., Insitute of Zoology and Botany, 21 Vanemuise Str., 202400 Tartu, Estonia.

PEGLER, Dr D. N., The Herbarium, R.B.G., Kew, Richmond, Surrey, TW9 3AE, UK.
PELIZZARI, Dr F., Laboratorio Farmaceutico Lofarma, Viale Cassala 40, 20143 Milano, Italy.
PERINI, Dr C., Dipartimento di Biologia Ambientale, Via P. A. Mattioli 4, 53100 Siena, Italy.
PERSIANI, Dr A. M., Dipartimento di Biologia Vegetale, Univ. Roma 'La Sapienza', Largo Cristina di Svezia, 24-0165 Roma, Italy.
PETERS, Mr M., TU Berlin, Sekr OE5/1 Franklin stv 29, 1000 Berlin 10, Germany.
PETKOVSEK, Mrs S., Erico Velenje, Zavod ZA Ekoloske Raziskave, Koroska 64, 63320 Velenje, Slovenija.
PIERART, Prof P., 74 Avenue de Fléron 1180, Brussels, Belgium.
PILTAVER, Mr A., Zofke Kvedrove 24, Ljubljana 6100, Slovenija.
QUANTEN, Ms E. J., L. U. C. Univ. Campus, Diepenbeek, B-3590, Belgium.
RENVALL, Mr P., Bolanical Museum, Univ. Helsinki, Unioninkatu 44, SF-00170 Helsinki, Finland.
REXER, Dr K.-H., Univ. Tübingen, Botanisches Institut Spezielle, Botanik/Mykologie, Auf der Morgenstelle 1, D-7400 Tübingen, Germany.
ROTHEROE, Mrs P., Fern Cottage, Falcondale, Lampeter, Dyfed SA48 7RX, UK.
ROTHEROE, Mr M., Fern Cottage, Falcondale, Lampeter, Dyfed SA48 7RX, UK.
RUNGE, Dr F., Diesterwegstrasse 63, D-4400 Munster, Germany.
RUNGE, Frau A., Diesterwegstrasse 63, D-4400 Munster, Cermany.
RYVARDEN, Prof L., Botany Dept., Univ. Oslo, PO Box 1045 Blindern, N-0316 Oslo, Norway.
SAARIMAKI, Miss T. M., Univ. Helsinki, Dept. Botany, Unioninkatu 44, SF-00170 Helsinki, Finland.
SALCEDO LARRALDE, Dr I., Lab. Botanica, Dplo. Biologia Vegetal y Ecologia, Fac. Ciencias, Univ. PaIs Vasco, Apdo. 644, 48080 Bilbao, Spain.
SALO, Dr V., Dept. Botany, Univ. Helsinki, Unioninkatu 44, SF 00170 Helsinki, Finland.
SCHMITT, Miss S., Fr Biochemie, Univ. des Saarlandes, D-6600 Saarbrücken, Germany.
SCHMITT, Dr J. A., Fr Biochemie, Univ. des Saarlandes, D-6600 Saarbrücken, Germany.
SENN-IRLET, Dr B., Systematisch-Geobotanisches Institut der Univ. Bern, Altenbergrain 21, CH-3013 Bern, Switzerland.
SKIRGIELLO, Prof A., Aleje Ujazdowskie 4, Warsaw, PL-00-478, Poland.
SPOONER, Dr B. M., The Herbarium, R.B.G., Kew, Richmond, Surrey, TW9 3AE, UK.
STALPERS, Dr J., C.B.S., PO Box 273, 3740 AC Baarn, The Netherlands.
STENLID, Dr J., Forest Mycology & Pathology, Swedish Agricultural Univ., Box 7026, S-75007 Uppsala, Sweden.
STRUNNIKOVA, Dr O. K., Inst. Agric. Microbiology, Podbelsky, 3, St Petersburg-Pushkin, 189620, Russia.
SVENSSON, Mr S., Botanical Museum, Ostra Vallgaten 18, S-223 61, Lund, Sweden.
SVRCEK, Dr M., Narodni Museum, Vaclavske Nam 68, 115-79 Praha 1, 'Czechoslovakia'.
TELLERIA, Dr M. T., Real Jardin Botanico, Plaza de Murillo 2, 28014 Madrid, Spain.
THOMAS, Dr M. A. J., Rua Professor Gomes Teixeira 5 - 6 Dto, 1300 Lisboa, Portugal.
TORKELSEN, Ms A.-E., Botanical Garden and Museum, Univ. Oslo, Trondheimsvn. 23B, N-0562 Oslo, Norway.
VAN BRUMMELEN, Dr J., Rijksherbarium, PO 9514, NL-2300 RA, Leiden, The Netherlands.
VAN DER GUCHT, Mrs K., Labor. voor Morf. Syst. & Ecologie v.d. Planten, Rijksuniv. Gent K.L. Ledeganckstr. 35, Ghent 9000, Belgium.
VANEV, Dr S. G., Institute of Botany, Acad. G. Bonchev Str., BL. 23, Sofia 1113, Bulgaria.
VEGHELYI, Dr K., Enterprise for Research and Development in Fruit Growing & Ornamentals, 1223 Budapest Park u. 2., Hungary.
VELLINGA, Dr E. C., Research Institute for Forestry and Nature Management, PO Box 9201, 6800 HB Arnhcm, The Netherlands.

VINCENZO, Dr D., Dipartimento di Biologia Ambientale, Via P.A. Mattioli 4, 53100 Siena, Italy.
WASSER, Prof S., N G Kholodny Inst. Botany, Acad. Sci. Ukraine, 252601 Kiev-GSP-1, Repin Street 2, Ukraine.
WATKINSON, Dr S. C., Dept. Plant Sciences, Univ. Oxford, South Parks Road, Oxford OX1 3RA, UK.
WATLING, Dr R., R. B. G., Edinburgh EH3 5LR, Scotland, UK.
WEBSTER, Mrs M. E., c/o Prof J Webster, UK.
WEBSTER, Prof J., Hatherly Laboratories, Prince of Wales Road, Exeter, Devon, EX4 4PS, UK.
WEIR, Mr A., 32 Hartburn Lane, Stockton-on-Tees, Cleveland TS17 3QH, UK.
WHALLEY, Prof A. J. S., Dept. Microbiology, School of Biomolecular Sciences, The Liverpool John Moores University, Liverpool L3 3AF, UK
WOLLWEBER, Dr H., In den Birken 73, D-56 Wuppertal-A, Germany.
WOLLWEBER, Frau E., In den Birken 73, D-56 Wupperlal-A, Germany.

INDEX